FUNDAMENTALS
OF SOIL SCIENCE

FUNDAMENTALS OF SOIL SCIENCE

SIXTH EDITION

HENRY D. FOTH

Professor of Soil Science
Michigan State University

John Wiley & Sons New York · Santa Barbara· · London · Sydney· · Toronto

Library of Congress Catalog Number: 77-86509

Henry D. Foth.
Fundamentals of Soil Science

Printed in the United States of America

10 9 8 7 6 5 4 3 2 1

To my college advisors

*To Dr. Robert J. Muckenhirn at the University of Wisconsin,
who served as my advisor for the B.S. and M.S. degrees.*

*To Dr. Frank Riecken of Iowa State University,
who served as my major professor for the PhD.*

710 081913 -2

PREFACE

This textbook will serve two primary goals. First, it is a textbook for college students that will maximize their learning of soil science. Topics have been developed by drawing on our common interests. Many new illustrations have been developed just for this edition. Since learning is increased by interest in subject, more examples have been included that relate to the diverse backgrounds of today's students.

Second, this textbook is updated; it contains the most useful knowledge about soil science. The readers of this textbook should become better evaluators of important national and international issues such as soil resource conservation, land use, environmental quality, and food production as it relates to population and international relations.

Many persons have contributed to this book. My colleagues at Michigan State University have given generously of their time to clarify subject matter. Drs. Terence Cooper and Bruce Dickson of California Polytechnic University at San Luis Obispo and Drs. Bernie Knezek and James Tiedje of Michigan State University critiqued chapters. I have used *Soil Taxonomy* freely as a source and am indebted to the Soil Survey Staff of the Soil Conservation Service. I thank Dr. W. H. Allaway for permission to use material from "Effects of Soils and Fertilizers on Human and Animal Health." My daughter, Mary, developed drawings, and Nate Rufe took and printed many of the new photographs. Dr. Renata Snyder of Michigan State University provided soil-animal specimens for photographing, and Dr. Max Mortland of Michigan State University provided counsel on clay mineralogy and photographs. Photographs were provided by Dr. J. B. Fehrenbacher of the University of Illinois, Dr. H. Lyon of Cornell University, Dr. Henry Smith, University of Washington, Drs. David Marx, B. P. Dickerson, and Eugene Shoulders of the United States Forest Service, and Dr. Robert Campbell of the Weyerhaeuser Company. I acknowledge the extensive use made of materials of the United States Department of Agriculture.

Henry D. Foth

CONTENTS

1

CONCEPTS OF SOIL

SOIL. Can you think of a substance that has had more meaning for humanity? The close bond that ancient civilizations had with the soil was expressed by the writer of Genesis in these words: "the Lord God formed Man from the dust of the earth—and Man became a living being." There has been and is a reverence for soil or the earth. You and I may take a more scientific view of soil; however, most of the world's people are tillers of the soil and, in a very real sense, the soil is their source of life.

The word soil is derived through Old French from the Latin *solum*, which means floor or ground. A more concise definition is made difficult by the great diversity of soils in the world. The goal of this book is not to arrive at a carefully developed definition of the word "soil" or to provide definite answers to specific questions, but to enable the reader to develop the basis required for intelligent thinking about soils. This chapter is devoted to the development of important concepts that will be expanded throughout the remaining chapters.

General Definitions of Soil

In general, soil refers to the loose surface of the earth and moon as distinguished from solid rock. Many people, when they think of the word soil, have in mind the material that nourishes and supports growing plants. This meaning is even more general, since it includes not only soil in the common sense, but also rocks, water, snow, and even air—all of which are capable of supporting plant life. The farmer, of course, has a more practical conception of soil, and regards it as the medium in which crops grow. The civil engineer, on the other hand, looks upon soil as the material that supports buildings and roads. In short, the word soil has many meanings and will be used in various ways throughout this textbook.

Soil as the Earth Underfoot

The earliest humans must have been conscious of the soil as the ground that supported their movement and habitation. Soils influenced the location of trails and campsites. Early mankind undoubtedly recognized the varying ability of different areas to produce plants and animals, but it is unlikely that these differences were associated with soil differences. The recognition of soil as a medium for plant growth occurred much later.

Soil as a Medium For Plant Growth

With the beginning of agriculture about 10,000 years ago mankind came to view the soil as a medium for plant growth. By 3000 B.C. farming villages existed as far north as Scandinavia in Europe. These early agriculturists recognized soil differences and had a preference for silty soils (soils developed from loess, a windblown silt).

The oldest classification of soils is believed to have been developed in China 4000 to 5000 years ago. The classification was based on the ability of soil to produce crops and was used to determine taxes. The Greeks and Romans were also aware of soil differences for agriculture, and they developed many soil management practices that are still used today. Of all the concepts of soil, the most important to most people is the concept of soil as a medium for plant growth. This concept becomes more important with each passing day as the world's population continues to increase.

Soil as a Mantle of Weathered Rock

Geologists became interested in soil as a product of weathering. The early scientific work on soils in the United States was dominated by geologists, and soils were classified as alluvial, residual, limestone, siliceous, sandy, clayey, and so forth. The addition of organic matter to the upper part of the weathered mantle was recognized and soil was viewed as synomonous with the weathered mantle or *regolith*. Since the regolith is subject to erosion and is eventually transported to the oceans, soil can be considered as "rock enroute to the sea." Enroute, however, the soil serves as a medium for plant growth.

Soil as a Mixture of Materials

About 400 B.C. earth or soil was considered one of the four basic components of all matter along with fire, water, and air. As knowledge of soils increased, people became more aware of the components making up the soil. The concept of the soil as a mixture of materials is useful in discussing soil as an engineering material, soil as a three-phase system, and soil as a manufactured product.

Fig. 1-1 Landslide in Great Smoky Mountains National Park.

Engineering Concept of Soil

Soil as an engineering material is any unconsolidated material composed of discrete solid particles with liquids and gases occupying the spaces between the particles. Geologists and engineers view soils similarly. Engineers are interested in the ability of soils to resist compression and to remain in place. Land- and rockslides along highways are a common engineering problem in the mountains (Fig. 1-1). Muck soils are subject to settling and are unsuited as foundation material for roads and buildings.

Soil as a Three-Phase System

Soil can be defined as a three-phase system consisting of solids, liquids, and gases. In most soils the solid phase consists of mineral particles that form a sketal framework onto which humus or organic particles are adsorbed. Pore spaces exist between the particles of the solid phase. The pore spaces are jointly filled with the liquid and gas phases. The liquid phase is mainly water from precipitation, which exists as films surrounding particles of the solid phase and occupies the smaller pore spaces. The larger pore spaces are filled with gases unless the soil is saturated with water. There is an exchange of gases between the soil and the atmosphere. Biological activities, such as root respiration and decomposition of organic matter, consume oxygen and produce carbon dioxide. As a consequence, there is a continuous

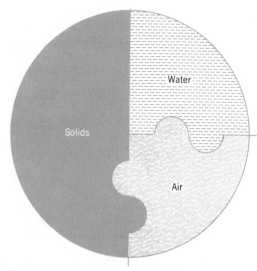

Fig. 1-2 The soil volume occupied by solids, water, and air is variable.

diffusion of oxygen from the atmosphere into the soil and of carbon dioxide from the soil into the atmosphere.

The volume of the soil occupied by the various phases varies from time to time and place to place (Fig. 1-2). The volume of air and water bear a direct reciprocal relationship with each other. Entrance of water into the soil excludes air. As water is removed, by drainage, evaporation, or plant growth, pore space that was occupied by water becomes filled with air once more. Subsoils are generally characterized by considerably less organic matter than surface soils. An organic soil, like a muck or peat, has a greater volume occupied by organic matter than by mineral matter.

Fig. 1-3 Mixing equal amounts of topsoil, peat, and perlite (material similar to sand) together to produce soil for growing greenhouse plants.

The volume of the components in a "desirable" lawn, garden, or field soil will be approximately as follows: 45 percent mineral matter, 5 percent organic matter, 25 percent water, and 25 percent air. About half of the soil volume is pore space (see Fig. 1-2).

Manufactured Soils

Soils used in greenhouses are manufactured in the sense that topsoil, sand, and organic matter are mixed together to provide a desirable proportion of the four components (Fig. 1-3).

Soil as ordinarily found in the field is not suited for golf greens. When the soil surface is wet, traffic causes soil compaction, and the maintenance of favorable air and water relationships is difficult. In the construction of a golf course, the sites where the greens will be located are excavated and refilled with a base layer high in sand and gravel. This is overlain by a layer that is commonly 12 inches (30 centimeters) thick and composed of a mixture of sand, topsoil, and organic matter (peat).

Soils for bedding plants and hanging planters are usually made with a large proportion of organic matter so the soil is light and is easy to handle and transport.

There are many instances of soil manufacturing or "man-made" soils in the world. Canal mud is returned to cultivated fields in China. Eroded soil is hauled from the lower to the upper part of the field in many countries. South exposures are highly prized for grapes along the Rhine River. Huge retaining walls are built and backfilled with soil. This operation also creates flat, smooth slopes that are easier to manage. The Incas built a mountain top fortress with stone terraces for growing food (Fig. 1-4). The soil on the terraces was carried up from the valley floor.

Soils as Organized Bodies

The rapid accumulation of knowledge about soils during the nineteenth century created a need for a concept of soil that would accommodate the new facts. A revolutionary way of looking at soil was

Fig. 1-4 The Incas carried soil up from the valley to create these terraces used for growing food at the mountain top fortress of Machu Picchu in Peru.

developed about 1870 in Russia by Dokuchaev. As he traveled about, he observed many different kinds of soils and observed that a given soil was found repeatedly in a given situation. Dokuchaev saw that each kind of soil had a unique morphology resulting from a unique combination of climate, living matter (plants and animals), earthy parent material, topography, and age of the land. The soil was the product of evolution and changed over time.

Evolutionary Nature of Soil

The soils on the earth's surface are undergoing continual change, which escapes a casual study of the soil. Each soil has a life cycle in terms of geologic time. This dynamic and evolutionary nature is embodied in a definition of soil as

> *unconsolidated mineral matter on the surface of the earth that has been subjected to and influenced by genetic and environmental factors of*: parent material, climate (*including moisture and temperature effects*), macro- *and* microorganisms, *and* topography, *all acting over a period of* time *and producing a product-soil-that differs from the material from which it is derived in many physical, chemical, and biological properties, and characteristics.*[1]

Weathering of bedrock produces unconsolidated debris that serves as the *parent material* for the evolution of soils that eventually reflect the integrated effect of climate, living matter, relief, and time. Exposure of parent material to the weather, under favorable conditions, will result in the establishment of photosynthesizing

[1] From the "Glossary of Soil Science Terms," published by the Soil Science Society of America, February 1975, Madison, Wis.

plants. Their growth results in the accumulation of some organic residues. Animals, bacteria, and fungi join the biological community and feed on these organic remains. Breakdown of organic residues sets free the nutrients contained therein for another plant growth cycle. The microorganisms and animals feeding on the organic debris become a part of the total organic matter complex. When the surface layer of the soil attains a reasonable thickness and assumes a darkened color because of the accumulation of organic matter, an *A horizon* comes into existence. A soil with only two horizons is shown in Fig. 1-5. A soil horizon is a layer approximately parallel to the earths' surface that is the product of evolution and has properties differing from adjacent horizons.

The soil in Fig. 1-5 may eventually become many feet thick if the rate of soil removed by erosion is less than the rate of rock weathering. Erosion, however, is a fundamental process on the earth's surface, and most soils form in sediments (a product of erosion) instead of from the direct weathering of bedrock. Where soil evolution occurs in sediments, horizon evolution may proceed rapidly by comparison to evolution directly from hard bedrock. Pore spaces in sediments permit deep rooting by plants and facilitate removal of soluble compounds by percolating water. Suspended colloidal-sized particles (defined in glossary) are translocated by percolating water; however, the suspended colloidal particles tend to move only a few feet before they become lodged or precipitated. The result is the formation of a zone under the A horizon where colloidal particles accumulate. This zone is designated a *B horizon*. The most common colloids accumulating in B horizons are clay, organic matter, and oxides of iron and

A Horizon

R Horizon (Bedrock)

Fig. 1-5 An A/R soil. The A horizon is about 1 foot thick (30 centimeters) and formed by the direct weathering of sandstone in the foothills of Rocky Mountains in Colorado. Soils like this may form in as little as 100 years or in as long as 100,000 years or more, depending on the hardness of the rock and environmental conditions.

aluminum. A soil with a B horizon is shown in Fig. 1-6. The vertical exposure of the soil horizons is the soil *profile*.

Soils that develop under grass typically have thick, dark-colored A horizons resulting from the profuse growth of roots throughout the A horizon. In the forest the addition of organic matter occurs largely from leaves and wood. The addition of leaves and wood on the top of the soil promotes the development of a thin, dark-colored layer enriched with organic matter. The leaching of material out of the A horizons of forest soils in humid regions causes the lower part of the A horizon to become "bleached" or light colored. In these cases

the A horizon is subdivided into an A1 and A2, as illustrated in Fig. 1–7. The relationship of the soil body to the soil profile is also shown in Fig. 1-7.

The concept that modern soils are evolving into soils that will have different properties is well documented and is one of the most useful concepts in soil science today. Some soils appear to have reached an end point in their development, and any additional significant change is hard to imagine. Theoretically, soil development proceeds to the point where the soil becomes incapable of supporting vegetation if left undisturbed for a sufficiently long period of time in a humid environment. The release of

A Horizon-zone of
organic matter
accumulation

B Horizon-zone
of colloid
accumulation

C Horizon-parent
material-a zone of
minimal weathering

Fig. 1-6 A grassland soil profile with the three major horizons, A, B, and C. The soil profile has developed in a sediment consisting of material that weathered from bedrock at some other site and deposited here.

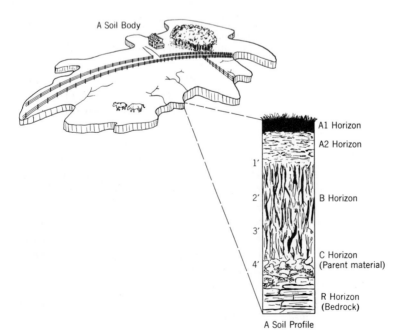

A Soil Body

A1 Horizon

A2 Horizon

1′

2′   B Horizon

3′

C Horizon
(Parent material)

4′

R Horizon
(Bedrock)

A Soil Profile

Fig. 1-7 The A1, A2, B, and C horizon sequence is typical for soils developed under forest vegetation. Also shown is the relationship of soil horizons and profile to the soil body. (Courtesy Dr. F. D. Hole, Soil Survey Div., Wisc. Geol. Nat. Hist. Survey, University of Wisconsin.)

nutrients from minerals by weathering and their removal by water may be nearly completed. The soil in Fig. 1-8, Nipe clay, is representative of soils that are essentially "weathered out." The original rock that served as parent material contained 6 percent magnesium oxide, whereas the soil contains only a trace. The original rock contained only a small amount of iron; however, the soil is about 60 percent iron oxide. The iron in the original rock was converted to very insoluble iron oxide compounds that remained as the more soluble compounds were removed. Extensive areas of Nipe-like soils on the earth's surface are not common because erosion, deposition, volcanism, and other geologic forces are so active.

The Pedon and Polypedon

Soil bodies are large, and there is a need for a smaller unit of soil that can be the object of scientific study. The *pedon* is this unit. A soil pedon is the smallest volume that can be called a soil and is roughly polygonal in shape. The lower limit is the somewhat vague boundary between soil and nonsoil or the approximate depth of root penetration. Lateral dimensions are large enough to represent any horizon. The area of a pedon ranges from 1 to 10 square meters, depending on the variability of the soil. The pedon is to a soil body what an oak tree is to an oak forest. A soil body is composed of many pedons (Fig. 1-9); therefore, a soil body is called a *polypedon*.

The Polypedon as a Landscape Unit

The landscape as a whole can be viewed as being composed of many different soil bodies or polypedons, each contributing to the whole as a piece of a jigsaw puzzle contributes to the overall pattern of the

Fig. 1-8 A very old soil, the Nipe clay of Puerto Rico. The soil is largely an accumulation of iron oxide.

puzzle. Soil properties gradually change over distance in accordance with changes in slope (topography), parent material, vegetation, climate, or age of land surface. Each soil is a continuum with properties that gradually change in all directions. Discrete individual soils comparable to individual plants and animals do not exist. Lateral changes from one soil to another soil are commonly associated with changes in slope. Typically a transition zone exists between adjacent soils in which properties of both soils can be observed. A sharper than usual boundary exists between the two soils shown in Fig. 1-10 because the change in slope is abrupt. Polypedons are large, 1 to over 10 acres in size. We will use the words soil and polypedon interchangeably.

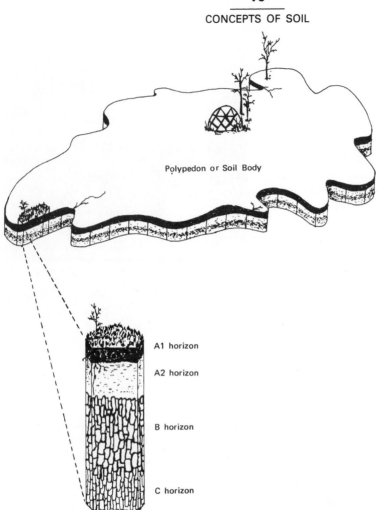

Polypedon or Soil Body

A1 horizon

A2 horizon

B horizon

C horizon

Pedon

Fig. 1-9 The pedon and its relation to the polypedon.

Naming Polypedons

Each polypedon is given a *series* name; there are about 10,500 series recognized in the United States. The series name is an abstract name usually taken from the name of a town or landscape feature near the place where the series was first recognized and described. All soils of the same series have the same horizon sequence and nearly identical properties of the horizons.

Examples of series include Walla Walla in Washington, Molokai in Hawaii, and Okeechobee in Florida. Walla Walla soils are deep brown silty soils formed from wind-deposited silt (loess) and some volcanic ash. Molokai soils are red in color and intensely weathered, as are the Nipe soils of Puerto Rico. Okeechobee soils are mainly composed of organic matter and are found in the Everglades.

The soil *type* is a subdivision of the series and recognizes differences in texture (discussed in Chapter 3) of the A horizon. The Miami series includes the Miami loam, Miami sandy loam, and Miami silt loam in

Fig. 1-10 A sharp boundary exists between two polypedons in this southern Michigan field. An organic soil exists in the foreground and a mineral soil exists on the slope.

Illinois and Indiana. The Miami soils are named after the Miami River in Ohio.

Soil as a Natural Resource

Changes in the kinds of use and intensity of soil utilization are a part of the story of civilization. A brief review of some of these changes will help to clarify the present and future emphasis that is likely to arise regarding the use of soil in our society.

Land Use in a Food-Gathering Society

In a food-gathering society the land supports plants and animals that are gathered for food, clothing, and shelter. This frequently results in a nomadic life, because people follow animal migrations or migrate with the seasons to obtain sufficient food. The land is used as it is, usually having low productivity and capable of supporting a small human population. It has been estimated that 518 hectares or 1280 acres of fertile land in the natural state is required to support one person in a food-gathering economy. On this basis the earth could sup-

port 10 million people. There are over 400 times as many people on the earth today.

Land Use in an Agricultural Society

Man domesticated plants and animals and ushered in the agricultural revolution in the hills surrounding the fertile crescent in south-western Asia about 10,000 years ago. The change from food gathering to food production at first gave more certainty to the source of food than to the abundance of supply. It was not until about 4000 B.C. when irrigation had developed sufficiently on the floodplains of southern Mesopotamia that productivity of the soil was great enough to give rise to cities.

The agricultural revolution spread north and westward into Europe. Most soils of Europe developed under forests in a humid environment and were leached and low in fertility compared to alluvial soils of the Middle East. Farmers discovered that better crops were produced where trees had been cut and burned. Now we know that this was because of plant nutrients in the ashes and the release of nutrients from decomposition of organic matter. In several years, however, the nutrients were depleted, crop yields declined, and the land was abandoned. Another area of forest was cut and burned, and cropping "shifted" to this new land. This form of agriculture is termed *shifting cultivation.*

Land use under shifting cultivation is more intense than with food gathering but less intensive than most agriculture today. Twenty hectares of land are required per person and population density is low. Presently about 200 million people are supported by shifting cultivation in the humid tropics.

Agriculture in China has a long history, as shown in the book *Farmers of Forty Centuries.* Much of China is mountainous, and most Chinese live on alluvial soils. These soils are naturally quite fertile, irrigation is widespread, more than one crop is usually produced each year, soil fertility is maintained by careful use of animal wastes, including human wastes, and canal mud is returned to the land. The result is very intensive use of the land and a high population density. When F. H. King visited several farms shortly after 1900, he found land supporting 1783 people, 212 cattle or donkeys, and 399 pigs per square mile (640 acres or 259 hectares). It is not uncommon today in China and other alluvial areas of southeast Asia to find land supporting over 2000 persons per square mile or 3 persons per acre (7.4 per hectare). Stated in another way, an area equal to many residential city lots (about 100×150 feet) supports one person. How adequate would you feel if you had to produce your food and perhaps some fiber for clothing on such a small parcel of land?

We have seen great differences in intensity of land use in agricultural societies and that land use was importantly related to soils. Let us next consider the extent to which land use in a technological society is related to soils.

Land Use in a Technological Society

Certainly the earliest of farmers recognized that soils had different capacities or abilities to produce a particular crop. With limited knowledge and limited ability to alter the soil, however, they had to use the soil in large part as it was. Today, in an industrial society, many additional alternatives exist. Crops may be grown in a given location because the climate is ideal, markets are close, or water is available for irrigation instead of because "good soil" exists. For example, in Florida, naturally infertile sand soils are used for citrus, and organic soils for vegetable production. The year round growing season and the location near centers of population make it feasible to greatly alter the existing soil so these crops can be grown successfully. Many acres of land have been encased with glass (greenhouses) in the Netherlands to intensify land use (see Fig. 1-11). Thus, some of the most productive soils in society owe their usefulness to human ability to alter their characteristics instead of to their original fertility or nature.

No close relationship exists between natural soil and the productivity of the soil in our society. The emphasis of modern soil science is on changing soil to better serve human needs. The question we commonly ask is, "How much input must be used on a given soil to produce a given quantity of product?" This is comparable to the question, "How much will it cost to produce a million square feet of office space on an acre?"

When the questions like the one just raised are pursued deeply in a technologi-

Fig. 1-11 The apparently uninterrupted glass roof illustrates one of the most intensive uses of land for market gardening in the world. One tenth of the world's market-gardening under glass occurs in this district located between the Hague and Rotterdam in the Netherlands. (Photo courtesy Ministry of Agriculture and Fisheries, Netherlands.)

cal society, one finds fantastic production possibilities. In fact, there is evidence that with current technology, a person's food can be produced on an area smaller than that needed to "live" in. However, many persons are becoming quite concerned about the impact of technology on the quality of the environment.

Soil as the Interface Where We Live

Another concept relevant to soil utilization is that of the soil as the interface between the atmosphere and lithosphere. We live on the soil, this is our home. Besides being a basic resource for food production, the soil collects and purifies water and disposes of wastes. Soil itself can be a pollutant as dust in the air and as sediment in waters. For the future a new concept of soil utilization will need to be forged—a concept that considers the impact of soil utilization on all aspects of life, including the quality of the environment. We may find new meanings in the words of one of America's greatest conservationists, Aldo Leopold.

A thing is right only when it tends to preserve the integrity, stability, and beauty of the community, and the community includes the soil, water, fauna, and flora, as well as the people.

The chapters ahead contain information and ideas that will enable you to formulate some productive concepts about the nature of soil and its importance in your life.

References

Braidwood, R. J., "The Agricultural Revolution," *Sci. Am.*, September 1960.

Brown, Harrison, *Challenge of Man's Future*, Viking, New York, 1954.

Cline, Marlin G., "The Changing Model of Soil," *Soil Sci. Soc. Am. Proc.*, 25:441–451, 1961.

deWit, C. T., "Photosynthesis in Relation to Overpopulation," in *Harvesting the Sun*, Academic, New York, 1967, pp. 315–320.

Kellogg, C. E., "Modern Soil Science," *Am. Scientist*, 36:517–536, 1948.

King, F. H., *Farmers of Forty Centuries*, Mrs. F. H. King, Madison, Wis., 1911.

Russell, M. B., "Physical Properties," in *Soil*, USDA Yearbook, Washington, D. C., 1957, pp. 31–38.

Simonson, Roy W., "Concept of Soil," in *Advances in Agronomy*, Vol. 20, Academic, New York, 1968, pp. 1–47.

Soil Survey Staff, *Soil Taxonomy*, Agr. Handbook 436, USDA, Washington, D.C., 1975.

Taylor, N. H., "Man and the Soil," *New Zealand Soc. Soil Sci.: Proceedings*, 5:2–8, 1962.

Tuan, Yi Fu, *China*, Aldine, Chicago, 1969.

Waterbolk, H. T., "Food Production in Prehistoric Europe," *Science*, 162:1093–1102, 1968.

2

SOIL AS A MEDIUM FOR PLANT GROWTH

Soil influences our livelihood in many ways. It is important for lawns, as a foundation material for engineering structures, for sewage disposal, and for recreation. Some highly weathered soils of the humid tropics are rich in iron or aluminum and are mined as ore. More important, however, the soil is the interface between the living and the dead—where plants combine solar energy and carbon dioxide of the atmosphere with nutrients and water from the soil into living tissue. Even though a significant amount of the photosynthesis on the earth occurs in the oceans, 99 percent of our food is produced from the land. The role of the soil as a medium for plant growth is so important that it warrants this separate chapter.

Factors of Plant Growth

Basically, plants growing on the land depend on the soil for water and nutrient elements. Beyond this, the soil must provide an environment in which roots can function. This requires pore spaces for root extension. Oxygen must be available for root respiration and the carbon dioxide produced must diffuse out of the soil instead of accumulating. An absence of inhibitory factors, such as a toxic concentration of soluble salts or extreme temperature, and pathogens is essential. Roots anchored in the soil also hold the plant erect. A summary of plant growth factors is given in Fig. 2-1, and a discussion of these factors follows.

Support

One of the most obvious functions of the soil is support for the plant. Roots anchored in soil enable growing plants to remain upright. Plants grown by hydroponics are commonly supported by a wire network. There are soils in which the impermeability of the subsoil or B horizon or the presence of a water table close to the surface of the soil often induces shallow rooting. Shallow rooted trees are easily blown over by the wind in a phenomenon called *windthrow*.

Fig. 2-1 A summary of plant growth factors. (The background plants are oats.)

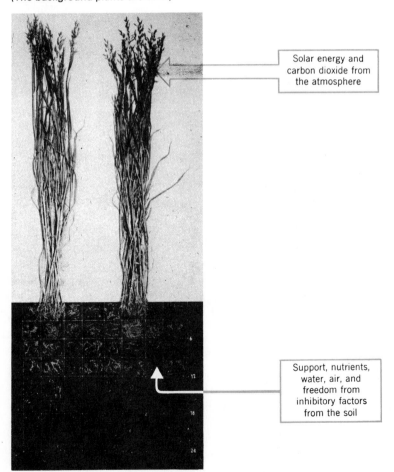

Solar energy and carbon dioxide from the atmosphere

Support, nutrients, water, air, and freedom from inhibitory factors from the soil

Essential Nutrient Elements

At least 16 elements are currently considered necessary for growth of vascular plants. The carbon, hydrogen, and oxygen, combined in the photosynthetic reactions, are obtained from air and water. They comprise 90 percent or more of the dry matter. The remaining 13 elements are obtained largely from the soil. Nitrogen, phosphorus, potassium, calcium, magnesium, and sulfur are required in large quantities and are referred to as the *macro or major* elements. Nutrients required in considerably smaller quantities are called the *micro or trace* elements and include manganese, iron, boron, zinc, copper, molybdenum, and chlorine.

More than 40 additional elements have been found in plants. Some plants accumulate elements that are not essential but have a beneficial effect. The absorption of sodium by celery is an example and results, in this case, in an improvement in flavor.

Most of the nutrients exist in mineral and organic matter and as such are insoluble and *unavailable* to plants. Nutrients become *available* through mineral weathering and organic matter decomposition. It is a rare soil, indeed, that is capable of supplying all of the essential elements for long periods of time in quantities needed to produce high yields.

The nutrients are absorbed from the soil solution or from colloid surfaces as cations and anions. Cations are positively charged; anions are negatively charged. Thirteen essential elements, their chemical symbols, and the forms in which they are commonly absorbed by plant roots are given in Table 2-1.

Table 2-1

Chemical Symbols and Common Ionic Forms of the Essential Elements Absorbed by Plant Roots from Soils

Nutrient	Chemical Symbol	Ionic Forms Commonly Absorbed by Plants
Macronutrients		
Nitrogen	N	NO_3^-, NH_4^+
Phosphorus	P	$H_2PO_4^-$, HPO_4^{2-}
Potassium	K	K^+
Calcium	Ca	Ca^{2+}
Magnesium	Mg	Mg^{2+}
Sulfur	S	SO_4^{2-}
Micronutrients		
Manganese	Mn	Mn^{2+}
Iron	Fe	Fe^{2+}
Boron	B	BO_3^{3-}
Zinc	Zn	Zn^{2+}
Copper	Cu	Cu^{2+}
Molybdenum	Mo	MoO_4^{2-}
Chlorine	Cl	Cl^-

Water Needs of Plants

It requires about 500 grams of water to produce 1 gram of dry plant material. About 5 grams or 1 percent of this water becomes an integral part of the plant. The remainder is lost through the stomates of the leaves during absorption of carbon dioxide. Atmospheric conditions such as relative humidity and temperature play a major role in determining how quickly water is lost and the amount of water plants require.

Since the growth of virtually all economic crop plants will be curtailed when a shortage of water occurs, even though it may be temporary and the plants are in no danger of dying, the ability of the soil to hold water against the force of gravity becomes very important unless rainfall or irrigation is frequent. The need for removal of excess water from soils is related to the need for oxygen and is discussed in the following paragraphs.

Oxygen Requirements

Roots have openings called *lenticels* that permit gas exchange. Oxygen diffuses into the roots cells and is used for respiration, whereas the carbon dioxide diffuses into the soil. Respiration releases energy that the plant needs for synthesis and translocation of organic compounds and the active accumulation of nutrient ions against a concentration gradient.

Some plants (rice, for example) can grow in standing water because they have morphological structures that permit the internal diffusion of atmospheric oxygen down into the root tissues. Successful production of most plants in water culture requires aeration of the solution. Great differences exist between plants in their ability to tolerate low oxygen levels. Sensitive plants may

Fig. 2-2 The soil in which these tomato plants were growing was saturated with water. The stopper at the bottom of the left crock was immediately removed, and excess water quickly drained away. The plant on the right became severely wilted within 24 hours by the saturation treatment.

be wilted or killed by saturating the soil with water for a day, as shown in Fig. 2-2. The wilting is believed to result from a decrease in the permeability of the root cells to water, a result of a disturbance of metabolic processes due to an oxygen deficiency.

Aerobic microorganisms, bacteria, actinomycetes, and fungi utilize oxygen from the soil atmosphere and are primarily responsible for the conversion of nutrients in organic matter into soluble forms that plants can reuse.

Freedom from Inhibitory Factors

A soil should provide an environment free of inhibiting factors such as extreme acidity or alkalinity, disease organisms, toxic substances, excess salts, or impenetrable layers (see Fig. 2-3).

Fig. 2-3 The marked change in the direction of the alfalfa taproot at a depth of 23 inches resulted from the impermeability of the underlying horizon.

Utilization of the Soil by Plants

The density and distribution of roots affect the plants' efficiency in utilizing a soil. Perennials like oak or alfalfa do not reestablish a completely new root system each year; this gives them a distinct advantage over annuals like corn or cotton. This partly explains the difference in the water and nutrient needs of plants. Vegetable and cereal crops often respond to added phosphorus. Established apple trees retain, from year to year, most of the phosphorus necessary for their life processes and rarely respond to phosphorus fertilizer. Let us look at the extensiveness of root systems and the extent to which the soil is in direct contact with root surfaces.

Extensiveness of Root Systems

It is only reasonable to assume that there is as much difference in root systems as in the tops of plants. Root growth is influenced by environment; therefore root distribution and density are a function of both the kind of plant and nature of the root environment.

Root extension occurs by cell division and cell elongation directly behind the root cap. Actively growing roots have been known to increase several inches in length in 24 hours. It is not surprising that the lateral and vertical extension of the roots of some plants is great. By the time corn is "knee high," roots may have ramified soil midway between rows spaced 100 centimeters apart. By late summer they may extend more than 2 meters deep in well-drained, permeable soil. Alfalfa taproots commonly penetrate 2 to 3 meters and have been known to reach a depth of 7 meters. Cereal crops like oats have a moderately deep root system and effectively utilize the soil to a depth of 24 to 30 inches. Tree roots commonly extend 15 meters from the base of the tree at shallow depths (see Fig. 2-4).

The soybean root system shown in Fig. 2-5 was removed from a Conover loam on the University Farm at East Lansing, Michigan. Metal frames were used to collect the soil samples and running water was used to separate roots from soil. The roots appear to have been effective to a depth of about 1 meter. The roots extended 35

SOIL AS A MEDIUM FOR PLANT GROWTH

Fig. 2-4 Abundant root growth occurs in the upper layer of many forest soils because of an abundant supply of water, air, and nutrients. The tree is a 10-year old lobolly pine. (Photo courtesy Ed Kerr of USDA Forest Service.)

centimeters laterally to each side of the row, which indicates that the soil between the rows was effectively used.

Considerable uniformity in lateral distribution of roots through much of the soil (see Fig. 2-5) is explained on the basis of two factors. First, there is a random lateral distribution of large pore spaces, resulting from cracks or channels formed by roots of previous crops or earthworm activity. Second, as roots permeate soil, they remove water and nutrients, which makes such soil a less favorable location for additional root growth. Roots then preferentially grow in areas of the soil where roots have not yet grown. In this way the soil is permeated uniformly by soybean roots and by the roots

of many other plants. (See Figs. 2-1 and 3-9 for the root distribution of oat plants.)

Extent of Root and Soil Contact

A rye plant was grown in a cubic foot of soil for 4 months at the University of Iowa by Dittmer. The root system was carefully removed from the soil by using a stream of running water, and the roots were counted and measured for size and length. It was determined that the plant had 385 miles of roots that had 2550 square feet of surface. Root growth averaged over 3 miles per day. The root hairs had a length of 6600 miles and 4320 square feet of surface. Because a loamy soil was used, at most only 1 percent of the total soil surface was estimated to be in direct contact with root surfaces. The average distance between roots, however, would be in the order of thousands of an inch (a cubic foot of soil would be cut almost 7000 times if root surfaces were spread in a plane. This would give $\frac{12}{7000} \approx 0.002$ inch between root surfaces). The movement of water and the nutrients dissolved in the water, even over very small distances, enables plants to utilize the soil water and nutrients to a remarkable extent.

Pattern of Soil Utilization by Plants

A seed is a dormant plant. Placed in moist soil where the temperature is favorable, it may absorb water by osmosis and swell. Enzymes then become activated and food reserves (carbohydrate, etc.) in the endosperm move to the embryo and are utilized for growth. With the development of green leaves and the initiation of photosynthesis, the plant becomes independent of the seed for food. With extension of roots through the soil, the plant becomes independent of the seed for nutrient elements. The plant

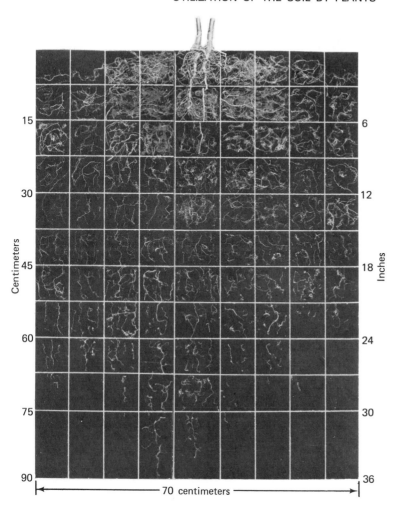

Fig. 2-5 Root distribution of mature soybean plants grown in rows spaced 28 inches (70 centimeters) apart. Roots penetrated at least 3 feet (1 meter) and easily grew to the centers of adjacent rows.

becomes totally dependent on the atmosphere and soil for its sustenance.

In a sense this is a critical period in the life of the plant because the root system is small. It may not be able to contact a sufficiently large amount of soil surfaces or volume of water to supply the phosphorus needs because phosphorus is immobile in soils. Plants frequently show phosphorus deficiency symptoms at this time, although the symptoms gradually disappear as root extension through the soil enables the plant to utilize an ever-increasing supply of phosphorus. The placement of soluble phosphorus fertilizer in the path of young roots has become a common practice in the production of many annual crops. It becomes apparent that the more mobile a nutrient is in the soil water, the more readily it will move to the root and be taken up by the plant. The limited distance that water and dissolved nutrients can readily move requires continual ramification of the soil by the roots in order to sustain the growth of an annual plant.

As the plant continues to grow, extension of roots into the subsoil will probably occur. The subsoil environment will be different in terms of the supply of water, nutrients, and oxygen and in other growth factors. This causes roots in different horizons to perform different functions or the same functions to a varying degree. For example, most of the nitrogen will probably be absorbed by roots from the A horizon because most of the organic matter is concentrated there and nitrogen is made available by organic matter decomposition. In contrast, deeply penetrating roots might penetrate less weathered and leached horizons where little available nitrogen is found but where there is an abundance of calcium. Roots here will tend to absorb a disproportionately large amount of calcium.

The plow-layer frequently becomes depleted of moisture in dry periods while an abundance of moisture still exists in underlying horizons. This results in relatively greater dependence on nutrient and water absorption from the subsoil. Subsequent rains that remoisten the upper layer of soil cause a shift to greater dependence again on the surface soil. This is probably due, in part, to better aeration nearer the surface of the soil. Thus we see that the manner in which an annual plant utilizes the soil is complex and changes continually through the season. In this regard the plant may be defined as "an integrator of a complex and ever-changing set of environmental conditions."

The Concept of Soil Productivity

From the foregoing section it is apparent that the utilization of the soil by plants is complex. Add to this the fact that plant requirements are diverse; it is readily seen that it is impossible for a given soil to be productive for the growth of all plant species. A brief discussion of some of the differences in plant requirements will aid our understanding of the soil productivity concept.

Fig. 2-6 Through the center of the photograph (in the foreground and background) are small white cedar trees. Many years ago a limestone road existed in this area. Red pine was originally planted in rows to serve as a windbreak, but was unable to survive on the alkaline roadway that was later planted to the white cedar. (Photograph taken at the Conservation Nursery, Boscobel, Wisc.)

Diversity of Plant Requirements

The requirement of many economic plants will be satisfactorily met if the soil is well aerated, near neutral in reaction, without layers that inhibit root penetration, without excess salt, and has sufficient water and an ample nutrient supply. Corn and sugar cane are profitably grown under a very wide range of soil conditions, but blueberries and azaleas are capable of survival only under a narrow range of conditions that include very acid soil. Alfalfa, red clover, and table beets are only slightly tolerant of acidity and require nearly neutral soils for highest yields.

Trees like willow, black spruce, white cedar, and tamarack can tolerate wet soil conditions for long periods of time whereas maple, red pine, and fruit trees require well-drained and well-aerated soil environments. In addition, red pine cannot tolerate alkaline soil, whereas white cedar thrives very well (Fig. 2-6).

Soil Productivity Defined

Soil *productivity* is defined as "the capability of a soil to produce a specified plant (or sequence of plants) under a specified system of management." For example, the productivity of a soil for cotton is commonly expressed as kilos of cotton per hectare when using a particular management system in which are specified things such as planting date, fertilization, irrigation schedule, tillage, and pest control. Soil scientists determine soil productivity ratings of soils for various crops by measuring yields (including tree growth or timber production) over a period of time in a "reasonable" number of management systems that are currently relevant. (For an example see Table 3-2.) Included in the measurement

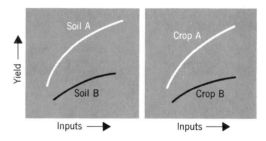

Fig. 2-7 On the left, as inputs are increased, the yield from soil A increases more rapidly than from soil B. Soil A is a more productive soil than soil B. On the right, the soil represented is more productive for crop A than crop B; crop A has a greater profit potential than crop B.

of productivity are the influence of climate and the nature and aspect of slope. Thus, soil productivity is an expression of all the factors, soil and nonsoil, that influence crop yields.

Soil productivity is basically an economic concept and not a soil property. Three things are involved: (1) inputs (a specified management system), (2) outputs (yields of particular crops), and (3) soil type. By assigning costs and prices, net profit can be calculated and used as a basis for determining land value, which is important in loan appraising and tax assessment. For planning management programs, two important aspects of soil productivity are presented in Fig. 2-7. First, different soils have different capacities to absorb inputs for profit maximization. Second, different crops have different capacities to absorb management inputs for profit maximization on a given soil type.

Soil Fertility Versus Soil Productivity

Soil *fertility* is defined as "*the quality* that enables a soil to provide the proper compounds, in the proper amounts, and in the proper balance, for the growth of specified plants when temperature and other factors are favorable." Soil *productivity*, on the other hand, is defined as "*the capability* of a soil for producing a specified plant or sequence of plants under a specified system of management." For a soil to be productive it must be fertile. It does not follow, however, that a fertile soil is productive. Many fertile soils exist in arid regions but, under systems of management that do not include irrigation, they cannot be productive for corn or rice.

Soil Productivity and Land Use

Sometimes the cost involved in making a soil capable of returning a reasonable profit for a certain crop is prohibitive. In these cases more attention is given to the selection of those crops or uses that allow successful utilization of the soil with a minimum of investment. Many soils in semiarid regions are used to produce forage for grazing animals or dryland grains. Sandy soils are frequently used for recreation or forestry.

For timber production emphasis has been placed on selecting a species that will profitably grow on the soil as it is. It is apparent that a thorough knowledge of soil properties is a requisite to intelligent soil utilization. The role of physical properties of the soil will be discussed in Chapter 3.

References

Brown, Lester R., and G. W. Finsterbusch, *Man and His Environment: Food*, Harper & Row, New York, 1972.

Dittmer, H. J., "A Quantitative Study of the Roots and Root Hairs of a Winter Rye Plant," *Am. Jour. Bot., 24*:417–420, 1937.

Foth, Henry D., "Root and Top Growth of Corn," *Agron. Jour., 54*:49–52, 1962.

Harvard University, "The Harvard Forest, 1968–69," *Harvard Black Rock Forest Annual Report*, 1968–1969.

Soil Science Society of America, *Glossary of Soil Science Terms*, Madison, Wis., 1975.

Wadleigh, C. H., "Growth of Plants," in *Soil*, USDA Yearbook, Washington, D.C., 1957, pp. 38–49.

3
PHYSICAL PROPERTIES OF SOILS

The physical properties of a soil have much to do with its suitability for the many uses to which it is put. The rigidity and supporting power, drainage and moisture-storage capacity, plasticity, ease of penetration by roots, aeration, and retention of plant nutrients are all intimately connected with the physical condition of the soil. It is pertinent, therefore, that persons dealing with soils know to what extent and by what means those properties can be altered. This is true whether the soil is to be used as a medium for plant growth or as a structural material in the making of highways, dams, and foundations for buildings, in building golf courses and athletic fields, or for waste disposal systems. Texture is perhaps the most permanent and important characteristic of the soil and will be discussed first.

Soil Texture

The relative size of the soil particles is expressed by the term *texture*, which refers to the fineness or coarseness of the soil. More specifically, texture is *the relative*

25

proportions of sand, silt, and clay. The rate and extent of many physical and chemical reactions important in plant growth are governed by texture because it determines the amount of surface on which the reactions can occur.

Soil Separates—The Particle Size Categories

The mineral soil particles have been divided into groups entirely on the basis of size, that is, without regard to their chemical composition, color, weight, or other properties. The groups of particles are termed soil *separates*.

In Table 3-1 the names of the separates are given, together with their diameters, the number of particles in 1 gram, and the surface area in square centimeters exposed by the particles in 1 gram of the separate. The importance and magnitude of the surface area are presented in Fig. 3-1.

Fig. 3-1 This handful of loam-textured soil has an estimated 5 to 10 acres (2 to 4 hectares) of surface area. Surface area is important in its relation to plant growth, since most of the available nutrients and water are adsorbed on the surfaces of soil particles.

Sand particles are of comparatively large size and hence expose little surface compared to that exposed by an equal weight of silt or clay particles. Because of the small

Table 3-1
Some Characteristics of Soil Separates

Separate	Diameter, mm[a]	Diameter, mm[b]	Number of Particles per Gram	Surface Area in 1 Gram, cm²
Very coarse sand	2.00–1.00	—	90	11
Coarse sand	1.00–0.50	2.00–0.20	720	23
Medium sand	0.50–0.25	—	5700	45
Fine sand	0.25–0.10	0.20–0.02	46,000	91
Very fine sand	0.10–0.05	—	722,000	227
Silt	0.05–0.002	0.02–0.002	5,776,000	454
Clay	Below 0.002	Below 0.002	90,260,853,000	8,000,000[c]

[a] United States Department of Agriculture System.
[b] International Soil Science Society System.
[c] The surface area of platy shaped montmorillonite clay particles determined by the glycol retention method by Sor and Kemper. (See SSSAP, Vol. 23, p. 106, 1959.) Other separates assumed to be spheres and calculation is based on largest size permissible.

surface of the sand separates, the part they play in the chemical and physical activities of a soil is small. Unless present in too small proportion, the sands increase the size of spaces between particles, thus facilitating movement of air and drainage water.

Since sand and silt consist mainly of particles resulting from the physical breakdown of rocks and minerals, they differ primarily in size in a given soil. As a consequence, silt has more surface area per gram (Table 3-1) and a faster weathering rate and release of soluble nutrients for plant growth than sand. Silt particles feel smooth like a powder and have little tendency to stick together or adhere to other particles. Soils with the greatest capacity to retain water against the pull of gravity are characterized as high in clay. Silty soils have great capacity to hold *available* water for plant growth but, at the same time, silt in a roadbed may present a frost heaving hazard for highway construction because of the expansion of water on freezing.

The great increase in surface area per gram of clay, as compared to silt and very fine sand, suggests that the difference between silt and sand and the clay cannot be accounted for on the basis of size alone. Chemical weathering on the surfaces of rocks, sand, and silt particles results in ions that recombine to form fine-sized particles of clay size. The clay fraction in most soils is composed of minerals that differ greatly in composition and properties from the minerals in the sand and silt. The data for the clay separate in Table 3-1 are for an unusually fine-sized, high surface area clay. Since a large part of the water is held as a film on the surface of the clay particles, the amount of clay in the soil has a great influence on its *total* water-holding capacity. In addition, certain available nutrients are held on the surface of clay particles. Therefore, clay

acts as a storage reservoir for both water and nutrients. Clay may have thousands of times more surface area per gram than silt and nearly a million times more surface area than very coarse sand (Table 3-1).

Particle Size Analysis by the Hydrometer Method

Bouyoucos devised the hydrometer method for determining the content of sand, silt, and clay without separating them. The soil sample is first soaked overnight in a sodium pyrophosphate solution in order to facilitate dispersion. Then it is placed in a metal cup with baffles on the inside and is dispersed for several minutes by the soil mixer running at a speed of 16,000 revolutions per minute. The soil mixture is poured into the cylinder and distilled water is added to bring the contents to the 1130 milliliter mark. With the help of a stirrer, the soil suspension is thoroughly resuspended and the time is immediately noted. The rate of fall of suspended particles is related to size—sand settling faster than silt and silt settling faster than clay.

Two hydrometer readings are taken of the soil suspension using a special soil hydrometer. A reading taken at 40 seconds determines the grams of silt and clay remaining in suspension since the sand (2.0 to 0.05 millimeters) has settled to the bottom. Subtraction of the 40-second reading from sample weight gives the grams of sand. The 2-hour reading determines the grams of clay (below 0.002 mm) in the sample (see Fig. 3-2). The silt (0.05 to 0.002 millimeters) is calculated by difference— add the percent sand and the percent clay and subtract from 100.

After the hydrometer readings have been obtained, the soil suspension can be poured over a screen to recover the entire

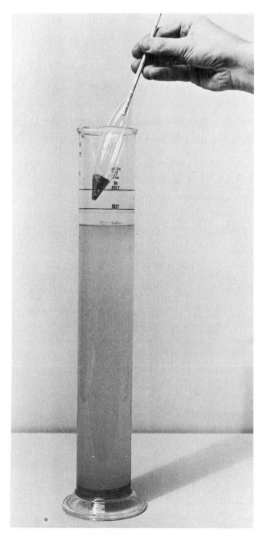

Fig. 3-2 Inserting hydrometer for the 2-hour reading. The sand and silt have settled. The 2-hour reading is grams of clay in suspension and is used to calculate percentage of clay.

sand fraction. After drying, the sands are sieved to obtain the various sand separates listed in Table 3-1.

Soil Classes Used to Designate Texture

Suppose the results of a mechanical analysis showed that a soil contained 15 percent clay, 65 percent sand, and 20 percent silt. The logical question is, "What is the textural class of the soil?"

The proportions of the separates in *classes* commonly used in describing soils are given in the textural triangle shown in Fig. 3-3. The sum of the percentages of sand, silt, and clay at any point in the triangle is 100. Point A represents 15 percent clay, 65 percent sand, and 20 percent silt; the textural class name for this sample is *sandy loam*. A soil containing equal amounts of the three separates is a *clay loam* (point *B* in Fig. 3-3). The various soil classes are separated from one another by definite lines of division in Fig. 3-3. Their properties do not change abruptly at these boundary lines, however, but one class grades into the adjoining classes of coarser or finer texture.

Determining Soil Class by the Feel Method

Estimates or determinations of soil texture are often necessary when examining soils in the field. When soil scientists map soils, they use the feel method to determine the texture of the various horizons of polypedons in order to identify soils by series and type and to distinguish between different soils on the landscape.

A small quantity of soil is moistened with water and kneaded to the consistency of putty to determine how well the soil forms a ribbon. The kind of ribbon formed is related to the clay content and is used to categorize soils as loams, clay loams, and clays (Fig. 3-4). A loam that feels very gritty or sandy is a sandy loam. Smooth-feeling loams are high in silt content and are silt loams. The same applies to clay loams and clays. Sands are loose and incoherent and do not form ribbons.

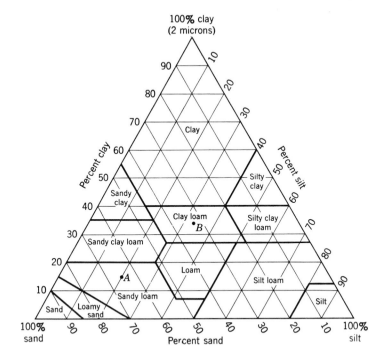

Fig. 3-3 The textural triangle shows the limits of sand, silt, and clay contents of the various texture classes.

Influence of Coarse Fragments on the Class Name

Some soils contain significant amounts of gravel, stones, or other coarse fragments that are larger than the size of sand grains. An appropriate adjective is added to the class name in these cases. For example, a sandy loam with 20 to 50 percent of the volume made up of gravel is a *gravelly* sandy loam. If 50 to 90 percent of the volume was gravel, it would be a *very gravelly* sandy loam. *Rockiness* is used to express the amount of the land surface composed of exposed bedrock.

Organic Soils and Texture Designation

Coarse-textured soils with 20 percent or more organic matter and fine-textured soils with 30 percent or more organic matter by weight have properties dominated by the organic fraction instead of by the mineral fraction. Where such soils are over 1 foot thick, they are called organic soils.

Muck, peat, mucky peat, and peaty muck are used in place of the textural class names for organic soils. *Peat* soils contain organic

Fig. 3-4 Field determination of texture by feel method. Loam soil on left forms a good cast when moist. Clay loam in center forms ribbon that breaks easily. Clay on right forms a long flexible ribbon.

matter that still retains some of the original structure of the plant tissues that can be identified. In *muck* soils the organic matter is in an advanced state of decomposition and in a fine state of subdivision. Peat soils, therefore, can become muck soils upon further decomposition, and this happens when they are drained and used for crop production. Mucky peat and peaty muck are intergrade classifications.

Texture and Use of Soils for Agriculture

Strictly speaking, the class name describes only the particle-size distribution. Plasticity, rigidity, permeability, ease of tillage, droughtiness, fertility, and productivity may be closely related to the textural classes in a given geographical region but, because of the great variation that may exist in the mineralogical nature of the separates, no broad generalizations of the world's soils can be made. Thus, a soil with 25 percent clay, which is montmorillonitic, will be much more plastic than a soil with over 70 percent clay that is composed mainly of

hydrous oxides of iron and aluminum. The sand and silt in some soils are composed mainly of minerals rich in essential nutrients, whereas in others they are dominated by quartz (SiO_2). The infertility of sandy soils is commonly related to their high content of quartz. The discussion of minerals in Chapter 8 will aid in clarifying this point.

Many useful crop-yield and soil-texture relationships have been established for certain geographic areas. The dominant texture in the upper 3 feet of the profile and the yield of four crops in Michigan are given in Table 3-2. Corn and oak yields were highest on loam and yields of potatoes and pine were highest on sandy loam. Data such as these are useful in appraising land for loan or taxation purposes and in planning cropping systems or reforestation programs.

The ability of the soil to support machinery and the hooves of grazing cattle is related to both texture and water content. Soils with high contents of organic matter and clay have low load-bearing capacity when wet.

Table 3-2

Relationship of Texture and Average Crop Yields or Soil Productivity

Dominant Texture of the Upper 3 Feet of Soil Profile	Yield, per Acre[a]			
	Corn, bushels	Potatoes, cwt	Pine, cords	Oak, cords
Clay	80	—	0.3	1.3
Loam	100	250	0.6	1.5
Sandy loam	80	300	1.3	1.3
Loamy sand	60	250	0.8	0.8
Sand	120 (irrigated)	300 (irrigated)	0.3	—

[a] Corn and potato yields with good management; yield of cord wood is annual sustained yield. Based on data from various state and federal agencies operating in Michigan.

Relation of Texture to Forestry

Soils affect trees importantly through air and water relations. The ability of the soil to store water between rains determines the seasonal supply of soil moisture and commonly determines what species grow in a forest and growth rate. On the glaciated uplands of southern Michigan, oaks are common on sands, hickory and oaks on sandy loams, and maple and beech on loams and clay loams. Ash and elm grow on soils that are wet and poorly aerated because of topography.

The soil affects tree growth and the presence of the forest, in turn, influences the growth of trees. The presence of trees modifies sun and wind conditions, which modify the effects of soils on tree growth. Where reforestation occurs after a fire, conditions for establishment of seedlings are severe. Seedling survival is highly dependent on water supply and is closely related to soil texture. Some recommendations made by Wilde for reforestation in Wisconsin in relation to texture are given in Table 3-3.

The recommendations in Table 3-3 assume uniform texture with increasing soil depth, which is not the common case in the field. Species more demanding of water and nutrients can be planted if the underlying soil horizons have finer texture. In some cases an underlying soil layer inhibits root penetration, thus creating a "shallow" soil with limited water and nutrient supplies. The effect of a plastic clay layer on tree growth is illustrated in Fig. 3-5. The effect of an inhibitory layer on an alfalfa taproot is shown in Fig. 2-3.

Nature and Evolution of Argillic Horizons

Clay particles are moved by percolating water from A horizons and deposited in B horizons. The result is polypedons with horizons that have different textures—a feature that is common throughout the

Table 3-3

Recommendations for Reforestation on Upland Soils in Wisconsin in Relation to Soil Texture

Percent Silt Plus Clay		Species Recommended
Less than 5		(Only for wind erosion control)
5–10		Jack pine, red cedar
10–15	increasing	Red pine, Scotch pine, jack pine
15–25	demand for	All pines
25–35	water and nutrients	White pine, European larch, yellow birch, white elm, red oak, shagbark hickory, black locust
Over 35		White pruce, Norway spruce, white cedar, white ash, basswood, hard maple, white oak, black walnut

S. A. Wilde, "The Significance of Soil Texture in Forestry, and Its Determination by a Rapid Field Method," *Journal of Forestry* (1935), *33*:503–508. Permission for use by *Journal of Forestry*.

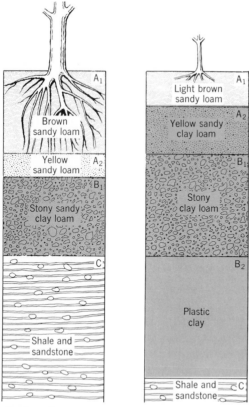

Fig. 3-5 Graphic relation between texture of upper 40 inches of soil and growth of yellow-poplar. Plastic clay layer at depth of 22 inches restricted root penetration and tree growth in the soil on the right. (Adapted from Auten and Plair, 1949.)

particles migrate downward in the water. When the percolating water encounters dry soil, the water is withdrawn into dry soil and the clay is deposited on the walls of pore spaces. Repeated cycles of wetting and drying build up layers of oriented clay particles termed *clay films* or *clay skins*. Clay skins are common in argillic horizons and can be observed in the field with a 10 power hand lens.

Thousands of years and alternating periods of wetting and drying are required to develop argillic horizons. These conditions are well satisfied in the eastern side of the Central Valley of California. Summers are dry and winters are rainy for clay formation and downward movement of clay. Great differences exist in the age of land surfaces, including alluvial floodplains less than 1000 years old and alluvial fans of varying age, some over 650,000 years old. Argillic horizon development as a function of time or soil age is presented in Fig. 3-6.

Soils 1000 years or younger (Hanford) show little or no evidence of clay translocation. Argillic horizons, however, had formed in 10,000 years (Greenfield). Maximum clay content in the argillic horizon increased with time, and the depth to maximum clay content decreased with time. The ratio of the maximum clay content of the argillic horizon and the clay content of the surface horizon increased with time to a maximum of 3.8 in the Redding soil. Furthermore, the San Joaquin and Redding soils had developed an iron and silica cemented layer impermeable to roots and water below the argillic horizon.

Miami soils have been developing for 10,000 to 15,000 years and have argillic horizons (Fig. 3-7). Similar percentages of clay exist in the A1 and A2 horizons. The maximum clay content in the Bt or argillic horizon is about two times greater than the clay content in the A horizon. This clay

world. Clay accumulation by movement is indicated by the subscript *t*, as in *Bt*. The symbol *t* comes from the German "ton," meaning clay. When the Bt horizon has at least 20 percent more clay than the A horizon above, the horizon qualifies as an *argillic horizon*. For our purposes we can use Bt and argillic horizon interchangeably.

The formation of argillic horizons requires parent material that contains clay or that weathers to form clay. Alternating periods of wetting and drying seem necessary. Some clay particles are believed to disperse when dry soil is wetted and the clay

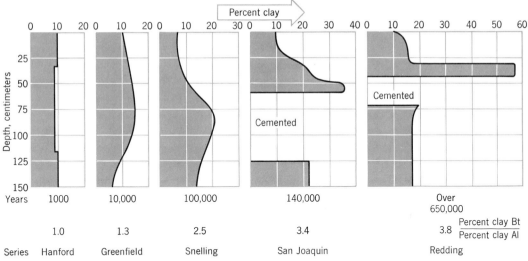

Fig. 3-6 Development of argillic horizons (and cemented horizons) as a function of time in some California soils developed from granitic parent materials. (From Arkley, 1964.)

distribution is similar to the Snelling and typical of many mature soils. Soils with different texture in the various horizons, as in the case of soils with argillic horizons, have a *texture profile*.

Influence of Argillic Horizon on Plant Growth

The presence of an argillic horizon can be beneficial or detrimental, depending on the degree to which it has developed. Up to a certain point, an increase in the amount of clay in the subsoil is desirable. It can increase the amount of water and nutrients stored in that zone. By slightly reducing the rate of water movement through the soil, it will reduce the rate of nutrient loss through leaching. If, however, the accumulation of clays is great, as in the case of a clay-pan soil, it will severely restrict the movement of air and water and the penetration of roots in the Bt horizon. It will also tend to increase the amount of water from rainfall or irrigation that will occur as runoff on sloping land.

The development of argillic horizons does not restrict root penetration in the Greenfield and Snelling soils of California. Roots penetrate over 6 feet deep and are able to use 8 to 10 inches of water stored in the root zone. Rooting depth in the San Joaquin and Redding soils is 20 inches or less, and only 3 to 4 inches of water is in the root zone available to plants. Root-zones become saturated and oxygen-deficient in wet years and soils are droughty in dry years. San Joaquin and Redding soils have an agricultural rating (Storie Index) of 30 and 25, respectively, compared to 95 or 100 (maximum rating) for the Hanford, Greenfield, and Snelling soils.

The Texture Profile and Deep Plowing for Wind Erosion Control

Deep plowing has been used to control wind erosion where sandy-surface soils are underlain by Bt horizons containing 20 to 40 percent clay. Scientists of the United States Department of Agriculture and the Agricultural Experiment Stations of

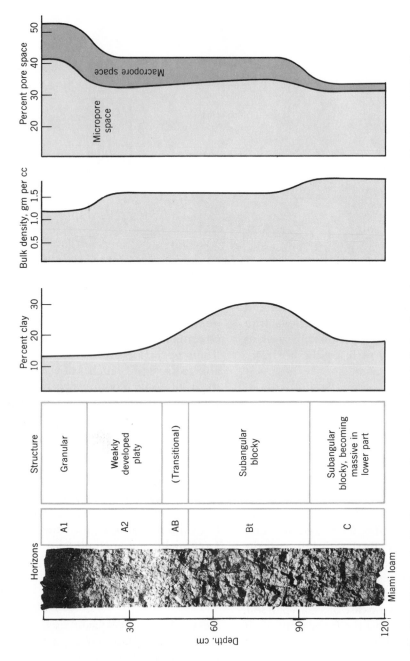

Fig. 3-7. Horizon designations, structure, clay content, bulk density, and percentage pore space of horizons of Miami loam (Alfisol) (Wascher, 1960).

Kansas and Texas observed that at least 1 inch of subsoil should be plowed up for every 2 inches of surface soil thickness in order to control wind erosion. In many cases, the soil must be plowed to a depth of 20 to 24 inches. This kind of plowing increased the clay content of the surface soil 5 to 12 percent in some cases. Since it was found that about 27 percent clay in the surface soil was required to halt soil blowing, deep plowing by itself resulted in only partial and temporary control of wind erosion. When deep plowing was not accompanied by other erosion control practices, wind erosion removed the clay from the surface soil and the effects of deep plowing were shortlived.

Alteration of Soil Texture

Fabrication of soils to meet certain textural specifications is common for golf greens, tree nurseries, and greenhouse use. Alteration of texture in the field is only occasionally attempted because of the enormous weight of such large quantities of soil. Deep plowing is used in some cases to break up root-inhibiting layers and to control wind erosion.

An interesting case of texture alteration occurs in the Netherlands. The low load-bearing capacity of peat lands limits the number of cattle that can be grazed in pastures. Cattle hooves make deep holes in the peat soil, destroying the sod. Where these peat lands have sand within 1 meter of the surface, giant plows are used to bring up the sand to form a surface with greater load-bearing potential and thus increase the number of cattle that can be pastured per area. The high value of this land for pasture also makes it profitable to use a machine that augers sand to the surface from depths as great as 10 or more feet or 3 meters.

Engineering Properties of Soils and Texture

The *shrink-swell potential* is one of the most important engineering properties of soils and is related to clay content. The shrink-swell potential is a measure of the volume change with wetting and drying; it is termed the *coefficient of linear extensibility* (COLE) and is defined as:

$$COLE = \frac{\text{length of moist sample}}{\text{length of dry sample}} - 1$$

If the COLE value exceeds 0.09 a significant shrink-swell potential can be expected. Soils with high COLE values cave in basement walls, displace and break gas and water pipelines, and crack masonry walls (see Fig. 3-8).

Fig. 3-8 This house was built on soil with a high shrink-swell potential. The soil's swelling and shrinking with changes in water content made the wall crack.

The load-bearing potential is also related to texture. The American Association of State Highway Officials has classified soil materials into seven groups for suitability as a road base. Group A-1 consists of gravelly soils with high load-bearing potential when wet. Group A-7 consists of high clay content soils with low load-bearing potential when wet.

Organic soil layers have low-bearing potential and are commonly removed and replaced with more stable soil material like sand and gravel.

Soil Structure

The term *texture* is used in reference to the size of soil particles but, when the arrangement of the particles is being considered, the term *structure* is used. Structure refers to the *aggregation of primary soil particles (sand, silt, clay) into compound particles*, or clusters of primary particles, which are separated from the adjoining aggregates by surfaces of weakness. The structure of the different horizons of a soil profile is an essential characteristic of the soil just as are color, texture, or chemical composition.

Structure *modifies* the influence of texture in regard to moisture and air relationships, availability of plant nutrients, action of microorganisms, and root growth. A good example of this occurs in the blacklands of Alabama and Texas, where the content of highly plastic clay is as high as 60 percent. These soils would be of limited value for crop production if they did not have a well-developed granular structure, which facilitates aeration and water movement.

Types of Soil Structure and Importance

Soil aggregates or peds are classified on the basis of shape as spheroidal, platelike, blocklike, or prismlike. These four basic shapes give rise to seven commonly recognized types, as listed in Table 3-4. A brief description and the horizon in which they are commonly found is also given in the table. The types of structure in the Miami loam soil are shown in Fig. 3-7.

The macroscopic size of most peds results in the existence of interped spaces that are much larger than those that can exist between adjacent sand, silt, and clay particles. It is this effect of structure on the pore space relationships that makes structure so important. Movement of air and water is facilitated. The interped spaces also serve as corridors for root extension and pathways for small animals. The effect of structure on the gross morphology of oat roots is presented in Fig. 3-9.

Structure is also described in terms of size and grade (distinctness) of peds. Those in the Bt horizon of the Miami loam shown in Fig. 3-7 are distinct and the grade is strong. In the A2 horizon the structure is indistinct, and the grade is called weak. Careful observation is required to see the peds in the A2 horizon; furthermore, only part of the soil mass is aggregated. Where no observable aggregation or definite orderly arrangement of peds is present, the grade is *structureless*.

Two structureless grades worthy of mention are the *single-grained* and the *massive*. If each particle in a soil functions as an individual, that is, if it is not attached to other particles, the condition is called single-grained. Such a structural condition may exist in very sandy soils. If soils containing considerable amounts of clay are tilled or are tramped by grazing animals when too wet, a portion of the colloidal material, including the clay, tends to fill the pore spaces. This makes the soil more dense and, upon drying, it forms large clods. The condition is called massive and is commonly referred to as *puddled*.

Table 3-4
Diagrammatic Definition and Location of Various Types of Soil Structure

Structure Type	Aggregate Description	Diagrammatic Aggregate	Common Horizon Location
Grangular	Relatively nonporous, small and spheroidal peds; not fitted to adjoining aggregates		A horizon
Crumb	Relatively porous, small and spheroidal peds; not fitted to adjoining aggregates		A horizon
Platy	Aggregates are platelike. Plates often overlap and impair permeability		A2 horizon in forest and claypan soils
Blocky	Blocklike peds bounded by other aggregates whose sharp angular faces form the cast for the ped. The aggregates often break into smaller blocky peds		Bt horizon
Subangular blocky	Blocklike peds bounded by other aggregates whose roinded subangular faces form the cast for the ped		Bt horizon
Prismatic	Columnlike peds without rounded caps. Other prismatic aggregates form the cast for the ped. Some prismatic aggregates break into smaller blocky peds		Bt horizon
Columnar	Columnlike peds with rounded caps bounded laterally by other columnar aggregates that form the cast for the peds		Bt horizon

Adapted from *Soils Laboratory Exercise Source Book,* Am. Soc. of Agron., 1964.

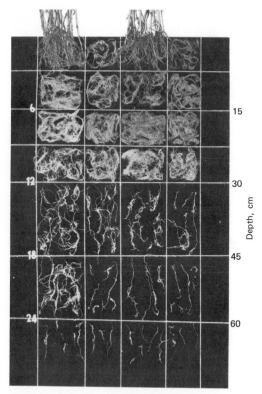

Fig. 3-9 Oat roots recovered from a slab of soil 4 inches (10 centimeters) thick. Note the marked change in amount and kind of roots at a depth of 12 inches (30 centimeters). This is the depth where the granular structure of the plow layer changes to the blocky structure of the Bt horizon. Roots in the B horizon were fewer in number, larger in diameter, and were growing primarily in the vertical spaces between structural units, old root channels, and in earthworm channels.

Formation of Soil Aggregates

An understanding of the causes for the development of structure in soils is of practical importance because structure has a great influence on plant growth and is also altered by tillage operations and traffic.

Structure develops from either a single-grained or massive condition. To produce aggregates there must be some mechanism that groups particles together into clusters and also some means by which they are firmly bound so that the structural forms persist. The colloidal fraction is the active constituent since, without its presence, single-grained structure prevails. Three groups of colloidal matter important as cementing material in aggregate formation are: (1) clay minerals, (2) colloidal oxides of iron and manganese, and (3) colloidal organic matter, including microbial gums.

There are several theories concerning the process by which aggregation is brought about, but the following is probably as widely accepted as any. Because colloidal particles are charged bodies, dipolar water molecules attach themselves readily and firmly to them. As water molecules and colloidal nuclei carry both positive and negative charges, oriented water molecules connect colloidal particles. The linkage may include cations because aggregation seems to be affected by the cations adsorbed (see Fig. 3-10).

Such water molecules are held very tightly and, as water evaporates from the soil, the length of each linkage becomes shorter and stronger and so pulls the colloidal particles close together. As more water is lost and the colloidal material becomes further dehydrated, the colloids stick or cement the particles into an aggregate. Thus it is seen that water, acting in conjunction with the colloids, constitutes the major force inducing granulation of soil and that colloidal material is the final binding agent.

Repeated cycles of wetting and drying occur in soils and the permanence of aggregates depends on two conditions. First, the soil along the ped faces must not disperse during rewetting or rehydration. Second, the colloids must be able to hold the particles within the peds together when the soil becomes wet. We consider next the importance of the various colloids in aggregate formation.

SOIL STRUCTURE

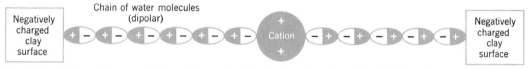

Fig. 3-10. Schematic presentation of the mechanism in structure formation. As the water evaporates, the clay plates are pulled closer together and, with cementing agents, structure develops.

Importance of Colloids in Aggregation

Wet-sieving is commonly used to measure aggregation. Dry aggregates are placed on a sieve that is lowered and raised in water. When the dry aggregates are immersed in water, the water moves into the aggregates from all directions and compresses the air in the pore spaces. Aggregates unable to withstand the pressure exerted by the entrapped air are disrupted and fall through the sieve along with the unaggregated soil. The aggregates remaining on the sieve after a standard period of time are considered water stable.

Researchers at the University of Wisconsin determined the aggregation of four soils in which they also determined the content of microbial gum, iron oxide, organic carbon (organic matter), and clay. The results are presented in Table 3-5. Microbial gum was the most important agent in producing aggregation in three of the four soils. Iron oxide was most important in one soil and second most important in three soils. Many red tropical soils are composed mainly of sand-sized aggregates that are very stable because of the high iron oxide content. For the four soils grouped together, the order of importance was microbial gum, iron oxide, organic carbon (organic matter), and clay. The great importance of microbial gum warrants consideration of the importance of microbial activity in aggregate formation.

Microbial Activity and Aggregate Formation

Various experimenters have found that the addition of plant residues to soil without microbial activity has little if any effect on aggregation. It is the microbial activity supported by the organic matter that is closely related to aggregate formation. There appears to be several factors involved.

Table 3-5

Order of Importance of Soil Constituents in Formation of Aggregates over 0.5 mm in Diameter in Four Wisconsin Soils

Soil Type	Order of Importance of Soil Constituents in Aggregate Formation
Parr silt loam	Microbial gum>clay>iron oxide>organic carbon
Almena silt loam	Microbial gum>iron oxide
Miami silt loam	Microbial gum>iron oxide>organic carbon
Kewaunee silt loam	Iron oxide>clay>microbial gum
All soils	Microbial gum>iron oxide>organic carbon>clay

Adapted from G. Chesters, O. J. Attoe, and O. N. Allen, "Soil Aggregation in Relation to Various Soil Constituents," Soil Science Society of America Proceedings (1957), Vol. 21, p. 276, by permission of the Soil Science Society of America.

Threadlike fungal mycelium bind soil particles together. Various substances synthesized by microorganisms also play an active role. The gums, along with other substances, act as cements that exhibit some resistance to rehydration. Fats and waxes also give aggregates the ability to resist wetting and the subsequent compression force of entrapped air that tends to disrupt aggregates. Last, some microbially produced substances have negative and positive charged groups that link soil particles in a manner analogous to that shown in Fig. 3-10. Many of these organic substances are decomposed within a year and more organic residues must be added to the soil to maintain the aggregation effects.

Aggregation Induced by Freezing and Thawing

As soil water freezes in a small crevice of the soil, it draws moisture from the surrounding soil as the ice crystals grow. Thus, under slow freezing, the ice crystals tend to break clods and at the same time exert a pressure on the surrounding soil particles that presses them together and stimulates the development of new clusters or granules. Furthermore, as moisture is drawn from the adjoining soil, the colloidal material is partially dehydrated and shrinks and tends to cement the particles into an aggregate. Repeated freezing and thawing from night to day accentuate these processes until they culminate in a thoroughly granulated soil.

Soil Aggregation and Crop Yields

From the previous paragraphs it is obvious that the aggregates in the soil are in a stage of continual change. Wetting and drying, freezing and thawing, tillage, and biological activities all contribute to the continual breaking down and building up of aggregates. The structure of the plow layer is affected by management practices, and where aeration and drainage limit plant growth, cropping systems that maintain greater aggregation are likely to result in higher crop yields. This was the case on the Paulding clay soils in Ohio where the percentage of the soil that was aggregated was positively correlated with the yield of corn (Fig. 3-11). When corn was grown in cropping systems that included a legume, the legume undoubtedly contributed to the higher yields because of an increase in the amount of available nitrogen. However, the fact that the use of fertilizer on the poorly aggregated plots resulted in yields of about 30 bushels, whereas yields from well-aggregated nonfertilized plots were about 70 bushels, indicates that the physical conditions of the poorly aggregated plots were limiting the yield of corn.

Figure 3-11 shows that as cropping systems went from continuous corn to a corn-oats system (corn and oats grown in alternate years) to a corn-oats-legume system (each crop grown every third year), there was a marked increase in aggregation. The greatest aggregation existed where cultivation was the least frequent and where the land was most often supporting an actively growing plant canopy. Where a no-till system is used and a mulch is maintained in a continuous corn program, the difference shown in Fig. 3-11 would be expected to be reduced.

A plant canopy absorbs the raindrop impact and lets the water fall gently on the soil without breaking the aggregates apart. Furthermore, an actively growing crop is associated with more microbial activity because there is a continual addition of organic matter to the soil, which serves as food for microorganisms. In addition, sev-

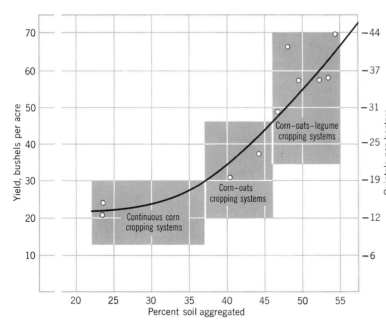

Fig. 3-11 Soil aggregation versus yield of corn on Paulding clay as affected by cropping system. Aggregation was determined at the time of corn harvest and is the average of 3 years. (Data from Page, 1946.)

eral means by which plant roots cause aggregation may be summarized as follows: (1) roots and root hairs penetrating the soil produce the lines of weakness along which the clod or soil mass may break into granules; (2) the pressure exerted by developing roots may induce aggregation; (3) root secretions may flocculate colloids, stabilize, or cement aggregates; and (4) use of moisture by roots may cause dehydration of colloids, thus resulting in shrinkage and finally in cementation.

Soil Consistence

Consistence is the resistance of the soil to deformation or rupture and is determined by the cohesive and adhesive properties of the entire soil mass. Whereas structure deals with the shape, size, and distinctness of natural soil aggregates, consistence deals with the strength and nature of the forces between particles. Consistence is important for tillage and traffic considerations. Dune sand exhibits minimal cohesive and adhesive properties and is so easily deformed that automobiles get stuck. Clay soils can be so sticky when wet as to make hoeing or plowing difficult.

Soil Consistence Terms

Consistence is described for three moisture levels: wet, moist, and dry. A given soil may be sticky when wet, firm when moist, and hard when dry. The terms used to describe consistence include:

1. Wet soil—nonsticky, sticky, nonplastic, plastic.
2. Moist soil—loose, friable, firm.
3. Dry soil—loose, soft, hard.

Cementation is also a type of consistence and is caused by cementing agents such as calcium carbonate, silica, or oxides of iron and aluminum. Cementation is little affected by moisture content. *Cemented* and *indurated* are used to describe cementation.

Indurated soil is so hard that a sharp blow of a hammer is required to break the soil apart and generally the hammer will ring as a result of the blow. Cemented horizons exist in San Joaquin and Redding soils (see Fig. 3-6) and rippers are used to break up the layers to improve rooting.

Weight, Pore Space, and Air Relationships

Soil texture and structure exert a large influence on weight and pore space. As the density of a soil increases, the volume of pore space decreases, and vice versa. We are interested in density and pore space relationships because air and water are stored in and move through pore spaces. Plant roots and other soil organisms also need space in which to live.

Bulk Density of Soil

The bulk density is the weight per unit volume of *oven dry* soil, commonly expressed as grams per cubic centimeter. Core samples used to determine bulk density are obtained with the equipment shown in Fig. 3-12. Care is exercised in the collection of cores so that the natural structure of the soil is preserved. Any change in the structure of the soil is likely to alter the amount of pore space and likewise the weight per unit volume. Four or more cores are usually obtained from each soil horizon to obtain a reliable average value.

Bulk density cores obtained in the field are brought to the laboratory for oven drying and weighing. The bulk density is calculated as follows:

Bulk density

$$= \frac{\text{weight of oven dry soil}}{\text{volume of oven dry soil}}$$

Fig. 3-12 Technique for obtaining bulk density cores. The light-colored metal core fits into the core sampler just to the left. This unit is then driven into the soil in pile-driver fashion using the handle and weight unit.

For 600 grams of oven dry soil that fills a 400 cubic centimeter core, the bulk density is 1.5 grams per cubic centimeter.

Bulk density may be expressed in several different kinds of units. A soil with a bulk density of 1.5 grams per cubic centimeter also has a bulk density of 93.6 pounds per cubic foot because water weighs 62.4 pounds per cubic foot and the soil is 1.5 times more dense than water (62.4 × 1.5 = 93.6).

When expressed in grams per cubic centimeter, the bulk density of granulated clay ·surface soils will commonly be in the range

1.0 to 1.3. Coarse-textured surface soils will usually be in the range 1.3 to 1.8. The greater development of structure in the fine-textured surface soils accounts for their lower bulk density as compared to more sandy soils.

The bulk densities of the various horizons of the Miami loam given in Fig. 3-7 show that the parent material is the densest layer. It has a bulk density of 1.8 grams per cubic centimeter. Formation of structure during soil development caused the overlying horizons to have lower bulk densities than the original parent material.

The Bt horizon in the Miami has a greater clay content than the A1 horizon. Its bulk density is greater than the A1 horizon and thus has a lower percentage of pore space. Clay deposition in the Bt horizon filled pore space and made the horizon more dense as the clay content increased. The general rule that fine-textured soils have more pore space and lower bulk densities than coarse-textured soils may hold when comparable structural conditions exist, as is the case when samples from plow layers are compared. This point should provide a basis for understanding that the greatest amount of available water may not necessarily be in the horizon with the highest clay content, since water is stored in the pore space.

Organic soils have very low bulk density compared to mineral soils. Considerable variation exists depending on the nature of the organic matter and the moisture content at the time of sampling to determine bulk density. Values ranging from 0.1 to 0.6 gram per cubic centimeter are common.

Weight per Acre Furrow Slice

Weight per acre furrow slice is the oven dry weight of the soil over 1 acre to a depth of 6 to 7 inches. A soil with a bulk density of 1.5 grams per cubic centimeter would have a density 1.5 times greater than water and would have an acre furrow slice weight equal to:

$$(1.5)\,(62.4) \times (43,560)\,(7/12)$$
(pounds per   (cubic feet in acre
cubic foot)   furrow slice)

$$= 2,378,376 \text{ pounds}$$

It is customary to consider that an average acre furrow slice has a bulk density of 1.3 grams per cubic centimeter and weighs about 2 million pounds or 1000 tons. On the basis an acre furrow slice that contains 1 percent organic matter on a weight basis would contain 20,000 pounds of organic matter. Soil losses are usually expressed as tons per acre. An average annual soil loss of 10 tons per acre would result in the removal of soil equal to the furrow slice about every 100 years. We also use the weight per acre furrow slice as the basis for calculating the amount of fertilizer and lime to apply per acre. From these examples it should be clear that the weight per acre furrow slice is a very useful soil property.

Particle Density of Soil

In determining the particle density of soil, consideration is given to the solid particles only. Thus, the particle density of any soil is a constant and does not vary with the amount of space between the particles. It is defined as the mass (weight) per unit volume of soil particles (soil solids) and is frequently expressed as grams per cubic centimeter. For many mineral soils the particle density will average about 2.6 grams per cubic centimeter. It does not vary a great deal for different soils unless there is considerable variation in content of organic matter or mineralogical composition.

Calculation of Total Pore Space

The percentage of pore space in a soil may be calculated from the bulk density and particle density if both are expressed in the same units of measurement. The following formula will give the percentage of the soil that is solid particles.

$$\left(\frac{\text{bulk density}}{\text{particle density}}\right)$$
$$\times 100 = \text{percent solids}$$

This percentage, taken from the total volume (100 percent), will give the percentage of pore space, hence the formula

$$100 \text{ percent} - \left(\frac{\text{bulk density}}{\text{particle density}}\right) \times 100$$
$$= \text{percent pore space}$$

Substitution in the formula of 1.3 for bulk density and 2.6 for particle density gives 50 percent total pore space, which is considered rather typical for medium-textured plow layers. Using the same formula, the C horizon of the Miami (Fig. 3-7) has only 32 percent total pore space. Soils research often requires an experimental determination of total pore space, as discussed in the next section.

Determination of Total Pore Space

The soil cores used for the determination of bulk density can also be used to determine total pore space. To determine the pore space, the cores are placed in a pan of water until completely saturated and then the cores are weighed. The difference in weight between saturated and oven dry cores represents a volume of water equal to the volume of the pore space in the soil. For a 400 cubic centimeter core that contained 200 grams (200 cubic centimeters) of water at saturation, the total pore space of the soil would be 50 percent.

Effect of Texture and Structure on Pore Space

Spherical particles in closest packing result in 26 percent total pore space and in open packing 48 percent pore space. This is true regardless of sphere size. Single-grained sands have a total pore space of about 40 percent. This suggests that sand particles are not perfect spheres, and that packing is not perfectly closed. The total pore space of single-grained sands is "low" and closely related to texture.

Fine-textured soils have a wide range in particle sizes and shapes. The particles are not in closest packing and the soil usually has peds. Soils with structural peds have pore space because of spaces between textural particles *and* between peds. A clay-textured A horizon with granular structure may have 60 percent total pore space resulting from the interactive effects of texture *and* structure. This is consistent with the fact that A horizons of clays commonly have low bulk density and sands have high bulk density. We have also noted that movement and deposition of clay in argillic or Bt horizons decreases pore space and increases bulk density.

Pore Size and Its Importance

In the discussion of bulk density, it was pointed out that sandy surface soils have a greater bulk density than clayey soils. This means that sandy soils have less volume occupied by pore space. Yet our everyday experiences tell us that water usually moves much faster through a sandy than through a clayey soil. The explanation for this seeming paradox lies in the size of the pores that are found in each soil.

The total pore space in a sandy soil may be low, but a large proportion of it is composed of large pores that are very efficient in the movement of water and air. The percentage of the volume occupied by small pores in sandy soils is low, which accounts for their low water-holding capacity. In contrast, the fine-textured surface soils have more total pore space, and a relatively large proportion of it is composed of small pores. The result is a soil with a higher water-holding capacity. Water and air move through the soil with difficulty because there are few large pores. Thus we see that the size of the pore spaces in the soil is as important as the total amount of pore space.

In a moist, well-drained soil the large pore spaces are usually filled with air; consequently they have been called *aeration pores* or *macropores*. The smaller pores usually tend to be filled with water and are commonly called *capillary* or *micropores*.

Distribution of Pore Space in the Soil

The pore space distribution in the profile of a mature soil is shown in Fig. 3-7. Development of granular structure in the A1 horizon results in high total porosity as well as favorable amounts of both micro- and macropore space. As pointed out earlier, clay deposition in the pores of the Bt horizon has been responsible for decreasing porosity. The Bt horizon of this soil has less of both types of pore space than does the A1 horizon.

Soil Permeability and the Percolation Rate

Water in a capillary tube does not move or drain out because the attraction between water and glass offers greater resistance than can be overcome by gravity. As a result, a substance can be very porous and yet be slowly permeable to water. In large (noncapillary size) tubes or pipes, water movement varies as the fourth power of the

radius. Thus, as the diameter of a pipe is doubled, the rate of water flow increases 16 times (2^4). Since water molecules are strongly adsorbed onto soil surfaces, as is the case with glass, *pore size* is of great importance in regard to the flow or movement of water into (infiltration) and through (percolation) soil. By contrast, the insignificant attraction between soil particles and air results in the air movement being primarily related to the *volume* of the vacant soil pores, not *size* of pores, and to the continuity of the pore spaces.

Permeability is the ability of the soil to transmit water or air. Permeability is commonly measured in terms of the rate of water flow through the soil in a given period of time and is commonly expressed as inches per hour. Soil permeability rates and classes are given in Table 3-6.

Table 3-6
Soil Permeability Classes and Rates

Classes	Permeability .n./hr[a]
Slow:	
1. Very slow	less than 0.05
2. Slow	0.05–0.20
Moderate:	
3. Moderately slow	0.20 to 0.80
4. Moderate	0.80 to 2.50
5. Moderately rapid	2.50 to 5.00
Rapid:	
6. Rapid	5.00 to 10.00
7. Very rapid	More than 10.00

[a] Rates through saturated soil cores under $\frac{1}{2}$ inch head of water.

Important decisions that are based on a knowledge of soil permeability include (1) determination of the distance between lines of drainage tile, (2) size of area of seepage beds for septic tank systems, (3) size of terrace ridges and the slope of terrace channels for erosion control, and (4) length and gradient of irrigation furrows. For example, note the characteristics of a claypan soil as presented in Fig. 3-13. The high clay content of the Bt horizon is associated with little aeration pore space and very little permeability. The claypan soil is unsuited for drainage by tile. If the soil existed on a slope, the claypan soil would be very susceptible to erosion resulting from a high water-runoff rate. During periods of rainy weather, the lower part of the A horizon would become saturated with water. The lack of oxygen in the saturated soil would inhibit root growth. Let us consider next the relationship of soil aeration and plant growth.

Soil Aeration and Plant Growth

The atmosphere contains about 79 percent nitrogen, 21 percent oxygen, and 0.03 percent carbon dioxide by volume. Respiration of roots and other organisms consumes oxygen and produces carbon dioxide; this causes the soil air to contain commonly 10 to 100 times greater concentration of carbon dioxide and slightly less oxygen than the atmosphere. (Nitrogen remains nearly constant at 79 percent in both soil air and atmosphere.) Differences in the partial pressures of the two gases are created that cause oxygen to diffuse from the atmosphere into the soil and carbon dioxide to diffuse from the soil into the atmosphere. Normally, this diffusion is sufficient to prevent oxygen deficiencies or carbon dioxide toxicities.

We have noted that gas diffusion is related to volume of gas filled pore space and not pore size. Poor soil aeration is caused more by the presence of water than the amount and size of pores. Clay soils are particularly susceptible to poor aeration when wet because most of the pore space is water filled and the spaces or avenues for gas diffusion become discontinuous.

The oxygen diffusion rate through water is about 10,000 times slower than through

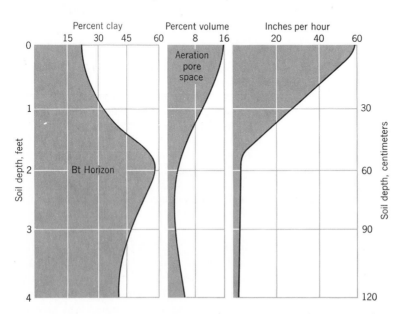

Fig. 3-13 Distribution with depth of clay, aeration pore space, and permeability in a claypan soil (Edina silt loam). During a prolonged rain, the infiltration of water into the surface of the soil will be limited by the permeability of the least-permeable horizon, the Bt horizon. (Data from Ulrich, 1950.)

Fig. 3-14 Both of these yews were planted along the west side of a new house. The failure to install a pipe on the right to carry away the down-spout water caused saturated soil and killed the plant.

air-filled space. As the water content of the soil increases, the diffusion path of oxygen to root surfaces increases in length, causing a decrease in availability of oxygen for root respiration. Experiments have shown that roots of many plants fail to penetrate soil when the oxygen diffusion rate is less than 20×10^{-8} grams per square centimeter per minute. Oxygen deficiencies are created when soils become saturated with water and plants commonly die (Fig. 3-14). That plants should respond to oxygen this way is not surprising because other forms of life are killed most quickly by suffocation, next by thirst, and slowest by starvation.

The data plotted in Fig. 3-15 show that the growth of peas and tomatoes was reduced by an oxygen deficiency that lasted only 24 hours. It is interesting to note that a deficiency early in the season was the most

detrimental for tomatoes, and for peas the greatest reduction in growth resulted from a deficiency occurring later in the season (near blossoming). Thus, two important

Fig. 3-15 The effect of a 24-hour, oxygen-deficient period on the growth of peas and tomatoes. Note that tomatoes were most sensitive in the early part of the season, whereas peas were injured the most near flowering time. (From data of Erickson and Van Doren, 1960.)

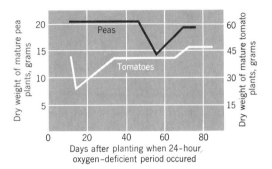

aspects of oxygen deficiency are the length of the oxygen-deficient period and the stage of plant growth when the oxygen deficiency occurs.

Tomato and pea plants are known for their susceptibility to oxygen deficiency. The reduced ability of tomato roots to absorb water from water-saturated soil and cause wilting (wet wilting) is shown in Fig. 2-2. Oxygen deficiency also impairs the ability of roots to absorb nutrients. Paddy rice, by contrast, is grown most of the season on water-saturated soil. Rice roots obtain atmospheric oxygen by the downward diffusion of oxygen through air spaces in the stems and roots. Many plants are intermediate and can make some adjustment to poor soil aeration. Even so, gas diffusion or aeration is one of the most important soil properties for plant growth.

Effects of Tillage and Traffic on Soils and Plant Growth

The beginning of agriculture marks the beginning of soil tillage. Crude sticks were probably the first tillage tools used to establish crops. Paintings on the walls of ancient Egyptian tombs, dating back 5000 years, depict oxen yoked together by the horns, drawing a plow made from a forked tree. The Greeks improved the Egyptian plow by adding a metal point. By Roman times tillage tools and techniques had advanced to the point where thorough tillage was a recommended practice for crop production.

Development and improvement of tillage tools and methods contributed to greater food production and an increase in the human population. Childe considered the discovery and development of the plow one of the nineteen most important discoveries or applications of science in the development of civilization.

The use of tillage tools requires power in the form of man, beast, or machine and produces traffic that has potential for compacting soils. As machines become larger, the potential for soil compaction increases. Recreational use of land has been marked by increases in foot and vehicular traffic. This section will discuss the need for tillage and the effects of tillage and traffic on soil properties and plant growth.

Definition of and Purposes of Tillage

Tillage is the mechanical manipulation of the soil for any purpose; but, in agriculture, it is usually restricted to the modifying of soil conditions for crop production.

There are three commonly accepted purposes of tillage: (1) to kill weeds, (2) to manage crop residues, and (3) to alter soil structure. Let us consider next the reasons for tillage and some modern tillage techniques and problems.

Tillage and Weed Control

Weeds compete with crop plants for nutrients, water, and light. If weeds are eliminated without tillage, can cultivation of row crops be eliminated? Data from many experiments support the conclusion that the major benefit of cultivating corn is weed control. On many soils herbicides have been used for weed control with good success. There are, however, some instances where cultivation during the growing season may be justified to improve soil aeration or to increase the infiltration of water.

Tillage and Management of Crop Residues

Crops are generally grown on land that contains the plant residues of a previous crop. The moldboard plow is widely used for burying such crop residues in the humid region. Fields free of trash permit precision placement of seed and fertilizer at planting and easy cultivation of the crop during the growing season. In the sub-humid and semiarid regions, by contrast, the need for wind erosion control and conservation of moisture have led to the development of machines that can successfully establish crops without plowing (Fig. 3-16). The plant tops remain on the surface and provide some protection from water and wind erosion. The plant residues left on the land over the winter may also cause snow to become lodged; this later melts and increases the water content of the soil.

Effect of Tillage on Soil Structure

All tillage operations change the structure of the soil. The lifting, twisting, and turning action of the moldboard plow leaves the soil in an aggregated and loose condition. Aggregate stability, however, remains unchanged. Cultivators, discs, and packers crush some of the soil aggregates. Cultivation of a field to kill weeds may have the immediate effect of loosening the soil, increasing soil aeration, and infiltration of

Fig. 3-16 A chisel planter designed to plant corn in the residues of the previous crop without plowing the land. In one operation the land is prepared, planted, fertilized, and sprayed with herbicide. (Photo USDA.)

water. The long-time (few weeks or months) effect of cultivation resulting from crushing of soil aggregates is a less well-aggregated and more compact soil. Exposed cultivated land also suffers from disruption of aggregates by raindrop impact in the absence of a vegetative cover. We have already observed in Fig. 3-11 that cropping systems with the least frequent tillage (corn-oats-legume) are associated with highest percentage of aggregated soil. The most frequent tillage occurred in continuous cropping systems (continuous corn) where soil aggregation and corn yield were the least.

Effect of Long-Time Cultivation on Pore Space

When forests or grasslands are used for agriculture, there is a decline in soil aggregation and soils become more compact. Compaction pushes aggregates and soil particles closer together. Obviously, the total volume of pore space decreases and the bulk density increases. Pushing parti-

cles together results in a decrease in the average pore size. Some of the macropore spaces are reduced in size to micropores. The result is an increase in the volume of micropore space. For a sand this could be desirable because the soil could retain more water. By contrast the increase in micropore space or water-filled space in fine-textured soils is generally detrimental because of reduced aeration and water movement. The decrease in total pore space with compaction results from a greater decrease in macropore space than the increase in micropore space.

An interesting study was made of the effects of growing cotton for 90 years on clay-textured Houston soils in the Texas blacklands. Soil samples from cultivated fields were compared with samples obtained from an adjacent area that had remained in grass. Long-time cultivation caused a significant decrease in soil aggregation and total pore space and an increase in bulk density. These changes are shown in Fig. 3-17. The most striking and important change was the decrease in macropore

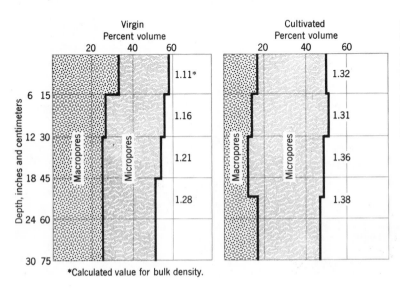

Fig. 3-17 Effects of 90 years of cultivation on pore space and bulk density of Houston soils (Vertisols). Macropore and total pore space decreased and micropore space increased. (Data from Laws and Evans, 1949.)

*Calculated value for bulk density.

space, which was reduced to about half of that of the uncultivated soil. Note that these changes occurred to a depth of 30 inches or 75 centimeters. Changes in total pore space parallel the changes in bulk density.

Summary of Compaction Effects on Soil Pore Space

The changes in soil pore space resulting from compaction are the same whether compaction is the result of long-time cropping or traffic involving tractor tires, animal hooves, or shoes. These changes are important enough to warrant their summarization as an aid to further discussion of effects of tillage and traffic on soils.

Soil compaction results in:

1. Increase in micropore space.
2. Decrease in macropore space.
3. Decrease in total pore space.

The decrease in total pore space is associated with an increase in bulk density.

Compaction Layers or Pressure Pans Produced by Tillage

Compaction at the bottom of the furrow frequently occurs during plowing because of the running of tractor wheels in the furrows made by the previous pass over the field. Subsequent tillage is usually too shallow to break up the compacted soil and compaction increases with time. This type of compaction produces a layer with high bulk density at the bottom of the plow layer and is appropriately termed a *plow sole* or *plow pan.*

Plow pans are a problem on sandy soils with insufficient clay content to cause enough shrinking and swelling with wetting and drying to naturally break up the compacted layer. Bulk densities as high as

1.9 grams per cubic centimeter have been found in plow pans of sandy coastal plain soils of the southeastern United States. Cotton root extension was inhibited when the bulk density exceeded 1.6 grams per cubic centimeter.

The Minimum Tillage Concept

It is obvious that plants grow without tillage of the soil. Sooner or later it was inevitable that the extent to which tillage was necessary would be questioned in the search for ways to maintain the soil in good physical condition and to produce high yields at minimum cost. Even though plants may grow very well in experiments without tillage, production of most crops will generally require at least some tillage. For the production of sugar beets it is interesting to observe in Table 3-7 that on an experimental field the yields were lowest when the land was not worked at all between plowing and planting and when the land was worked four or more times. The highest yields occurred where the land was worked only one or two times between plowing and planting. Clearly, this experiment shows that it is easy to overtill the soil. Some tillage is necessary for the practical production of crops, but the concept of "thorough tillage" is declining. The result has been considerable progress in the development of minimum tillage systems.

Minimum tillage systems employ fewer operations to produce crops. Machinery like that in Fig. 3-16 prepares the land, plants the seed, and applies fertilizer and herbicide in one trip over the field. Under favorable conditions, no further tillage will be required during the growing season. Press wheels behind the planter shoes pack soil only where seed is placed; leaving most of the soil surface in a state very conducive

Table 3-7
Beet Yields as Affected by the Number of Times a Field Was
Worked Prior to Planting on Lake Plain Soils

Times Worked	Tons per Acre	Metric tons per Hectare
None	14.0	31.4
One	16.8	37.6
Two	16.7	37.4
Three	15.2	34.0
Four	14.8	33.2
Five	14.2	31.8
Six	14.3	32.0

Adapted from Cook et al., 1959.

to water infiltration. Experimental evidence shows that minimum tillage reduces water erosion on sloping land. It also makes weed control easier because planting immediately follows plowing. Weed seeds left in the loose soil are at a maximum disadvantage. Crops yields are similar to those where more tillage is used, but the use of minimum tillage in crop production reduces costs.

Within the past 10 years much research has been directed toward the development of *no-till* systems of row crop production. In no-till a narrow slot is made in the untilled soil so that seed can be planted where moisture is adequate for germination. Weed control is entirely by herbicides, and this cost partially offsets the savings of fewer tillage operations. It has been estimated that corn grown in the United States each year could be grown with 221 million gallons less diesel fuel with no-till than with conventional tillage methods.

From this discussion of tillage it should be obvious that tillage is not a requirement of plant growth. If you have a garden you do not have to plow or engage in tillage operations that require a tractor or plow.

As long as weeds are controlled to reduce competition for nutrients, light, and water, you can grow a good garden with little tillage.

Tillage Operations of the Future

The minimum tillage concept represents a revolution in tillage practices. Rapidly expanding human populations and food requirements will probably intensify the use of the best agricultural lands and encourage another revolution in tillage techniques. Some futuristic planners foresee the day when many fields will be leveled, fields will be 10 miles long, and all production operations will be carried out with machines that run on tracks. Obviously, there would be no soil compaction. Such systems or machines would permit precise tillage operations and precise placement of seeds, fertilizers, and pesticides, and exact application of irrigation water and harvesting. Although these changes are being anticipated in the world where advanced technology exists, 70 percent of the world's farmers have only hoes or wooden plows as their sole tillage tool.

EFFECTS OF TILLAGE AND TRAFFIC ON SOILS AND PLANT GROWTH

Effect of Logging Traffic on Soils and Tree Growth

There has been a rapid increase in the use of tractors to skid logs because of their maneuverability, speed, and economy (Fig. 3-18). About 20 to 30 percent of the land area may be affected by logging. Tractor traffic can disturb or break shallow tree roots feeding just under the surface organic layers (refer back to Fig. 2-4).

Permeability of soil to water was found to be 65 and 8 percent as great on cutover and logging road areas, respectively, as compared to undisturbed areas in southwestern Washington. The reductions in permeability increase water runoff and erosion and reduce the amount of water available for the growth of transplanted seedlings or any remaining trees. Reduced growth and survival of chlorotic Douglas-fir seedlings on tractor roads in western Oregon was considered to be caused by poor soil aeration and low nitrogen supply.

Changes in bulk density and pore space of forest soils because of logging are similar to changes produced by long-time cultivation. After logging and in the absence of further traffic, however, forest soils are gradually restored to their former condition. Extrapolation of 5 years' data suggests that restoration of logging trails takes 8 years and in wheel-ruts requires 12 years in northern Mississippi.

Effect of Wheel Traffic on Soils and Crops

Wheel traffic compacts soils, resulting in reduced soil aeration and increased mechanical impedance of plant roots. Germination and emergence of seedlings may be delayed and or reduced. Poor plant vigor and reduced yields result from restricted root systems (Fig. 3-19).

Up to 75 percent of the entire area of an alfalfa field may be run over by machinery wheels in a single harvest operation. Plant

Fig. 3-18 Skidding logs with maneuverable rubber-tired tractor results in soil compaction. (Photo Courtesy B. P. Dickerson, Forest Service, USDA.)

Fig. 3-19 Beans on left grew in soil that had a compacted layer at the 3-inch (7.5-centimeters) depth. Beans on the right grew in uncompacted soil.

injury and soil compaction potential is great because 10 to 12 harvests per year are made where the crop is grown all year with irrigation. Wheel traffic damages plant crowns and plants are weakened and more susceptible to disease infection. Root development is restricted. Alfalfa stands (plant density) and yields have been reduced by wheel traffic.

Potatoes are traditionally grown on sandy soils that permit "easy" development of well-shaped tubers. Mechanical impedance in compacted soils not only reduces tuber yield but increases the amount of deformed tubers, which have reduced market value. In studies where bulk density was used as a measure of soil compaction,

potato tuber yield was negatively correlated with bulk density and the amount of deformed tubers was positively correlated with bulk density (Fig. 3-20).

Most potatoes are grown in the western states with irrigation. Soil compaction has also been observed in potato fields where irrigation water was allowed to reach excessive heights on the soil.

Effects of Recreational Traffic

Maintenance of plant cover in recreation areas is important to preserve the natural beauty and prevent the undesirable consequences of water runoff and erosion. Three campsites in the montane zone of the Rocky Mountain National Park in Colorado were studied to determine the effect of camping activities on soil properties. Soil in the areas where tents were pitched and where the fireplaces and tables were located had a bulk density of 1.60 grams per cubic centimeter compared to 1.03 grams per cubic centimeter in areas with little soil use at the Glacier Basin campground. This data suggests that camping activities can compact soil as much as tractors or other heavy machinery.

Snowmobile traffic on alfalfa fields significantly reduced the yield of forage at one out of four locations in a Wisconsin study. It appears that injury to alfalfa plants and reduced yields are importantly related to snow depth, there being less effect as the depth of snow increases. The large cross-sectional area of the snowmobile threads makes it likely that soil compaction will be minimal or nonexistent. More study is needed to determine the full effects of heavy traffic on public trails. It seems that moderate traffic on fields with a good snow cover has little effect on soil properties or dormant plants buried in the snow.

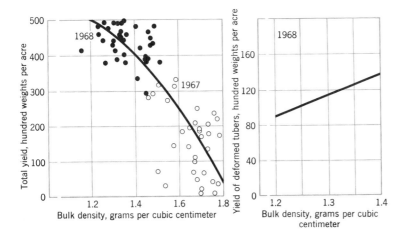

Fig. 3-20 Relation of total potato tuber yield and yield of deformed tubers to soil bulk density (or soil compaction). (Adapted from Grimes, 1971. Courtesy of *The American Potato Journal* (1971), Vol. 48.)

Heavy human traffic compacts soil on golf courses and lawns. Aeration and water infiltration can be increased by using coring aerators. Numerous small cores about 1 inch in diameter and 4 inches long are removed by a revolving drum and left lying on the ground (Fig. 3-21).

Soil Color

The color of soil serves layman, farmer, engineer, and soil scientist, provided that they understand the causes of the various colors and are able to interpret them in terms of soil properties. Organic matter content, drainage condition, and aeration

Fig. 3-21 Core aerator used to remove soil plugs to increase soil aeration and water infiltration of compacted soil on campus at Davis, Calif.

are soil properties related to color that are of interest to farmers. The investigator uses color as an aid in soil classification and draws from the color of the different horizons information about conditions pertaining to and forces active during soil formation.

Factors Affecting Soil Color

The minerals occurring in appreciable quantities in most temperature region soils are light in color. As a result, soils would be of a light gray color if composed of crushed minerals that had undergone little chemical change. Accordingly, for an explanation of the brown, red, and yellow colors in soils, we must look to chemical changes in the constituents (especially iron) of the minerals and to the addition of organic matter. There are instances, however, in which the proportion of colored minerals is sufficient to give the soil a decided color. The red color of many tropical soils is due to an abundance of iron oxide minerals.

The dark color of temperate region soils is generally due to the highly decayed organic matter they contain; in fact, with some practice, the percentage of organic matter in many soils may be judged with reasonable accuracy from their colors. Organic matter imparts a gray, dark gray, or dark brown color to soils unless some other constituent such as iron oxide or an accumulation of salts modifies the color. Many tropical soils with high content of iron oxide (hematite), however, are red, even with large amounts of organic matter.

If soils are poorly drained there is usually a greater accumulation of organic matter in the surface layers, thus giving a very dark color. The lower soil layers, which contain very little organic matter, on the other hand, are of a light gray color, indicating the poorly drained condition. If drainage is intermediate, the gray of the subsoil is likely to be broken by flecks of yellow.

When drainage permits aeration, and moisture and temperature conditions are favorable for chemical activity, the iron in soil minerals is oxidized and hydrated into red and yellow compounds. Highly hydrated iron oxides are yellow but, as hydration diminishes, reds replace the yellows. Accordingly, we find shades of red in soils extending from the southern deserts across the semiarid and subhumid states of the Southeast. The red and yellow colors of the subsoils in the southeastern states immediately catch the eye of the traveler. The low organic matter content of many soils in this area leaves undulled the brilliant colors of the iron oxides. With an appreciably humus content the red colors are converted into mahogany colors.

Measurement of Soil Color

The color of light can be accurately described by measuring its three principal properties, hue, value, and chroma. Hue refers to the dominant wavelength or color of the light. Value, sometimes called brilliance, refers to the total quantity of light. It increases from dark to light colors. Chroma is the relative purity of the dominant wavelength of light. It increases with decreasing proportions of white light.

The Munsell notation of color is a systematic numerical and letter designation of each of the three variable properties of color. The three properties are always given in the order hue, value, and chroma. For example, in the Munsell notation 10YR 6/4, 10YR is the hue, 6 is the value, and 4 is the chroma.

The Munsell notation for a given soil sample can be quickly determined by com-

Fig. 3-22 Color is determined by comparing a sample of soil with the color chips in the Munsell color book as shown on the right. These two soil scientists are writing a profile description. Other properties that will be recorded include texture, structure, horizon thickness, and pH. The latter is being determined on the left.

parison of the sample with a standard set of color chips (Fig. 3-22). The chips are mounted in a notebook with all the colors of a given hue on one page. The pages are arranged in the order of increasing or decreasing wavelength of the dominant color in order to facilitate the matching of the unknown soil color with the color of the standards.

Many soil horizons have a single dominant color. Horizons that are dry part of the year and wet part of the year tend to exhibit a mixture of two or more colors. These colors are intermediate between those of well-drained and poorly drained soils. When several colors are present in a spotted or variegated pattern, the word *mottled* is used to describe the condition. In these cases several of the dominant colors may be recorded.

Soil Temperature

Below freezing there is no biological activity, water does not move through the soil as a liquid and, unless there is frost heaving, time stands still for the soil. Germination of seeds and root growth hardly occur in the range 0 to 5°C. A horizon as cold as 5°C acts as a *thermal pan* to most plant roots. Each plant species has its own temperature requirements. The chemical processes and activities of microorganisms that convert plant nutrients into available forms are also materially influenced by temperature. Freezing and thawing play a role in rock weathering, structure formation, and heaving of plant roots. Temperature is thus seen to be an important soil property.

Heat Balance of a Soil

The heat balance of a soil consists of the gains and losses of heat energy. Solar radiation received at the soil surface is partly reflected back into the atmosphere and partly absorbed by the surface of the soil. A dark-colored soil and a light-colored quartz sand may absorb, respectively, about 80 and 30 percent of the incoming solar radiation. Of the total solar radiation available for the earth, about 34 percent is reflected back into space, 19 percent is absorbed by the atmosphere, and 47 percent is absorbed by the earth.

Absorbed heat is lost from the soil by (1) evaporation of water, (2) reradiation back into the atmosphere as longwave radiation, (3) heating of air above the soil, and (4) heating of the soil. In the long run the gains and losses balance each other. For the short run considered as daytime or summer, heat gains exceed heat losses and soil temperatures increase. During the night and winter, the reverse is true.

Heat Capacity of Soil

Mineral particles require a comparatively small amount of heat to raise their temperature. The quantity of heat required to raise the temperature of a gram of soil particles 1°C is only about one fifth as much as is required to warm a gram of water the same amount. In other words, the specific heat of dry soil particles is 0.2, and it is evident from this fact that moisture content is an important factor in determining soil temperature. Soil high in water content will warm up slowly in the spring and will cool down slowly in the fall. Drainage therefore exerts a major influence on soil temperature.

Location and Temperature

In the northern hemisphere, soils located on southern and southeastern slopes warm up more rapidly in the morning than those located on the level or on the northern slopes. The reason is that they are more nearly perpendicular to the sun's rays, and

hence a maximum amount of radiant energy strikes a given area. Soils with a southern or southeastern exposure often are selected for the growing of early vegetables and fruits.

Temperature decreases with an increase in elevation. Within the 40 miles from Fort Collins, Colorado to the top of the Rocky Mountains, the average July temperature changes as much as from San Antonio, Texas to Devil's Lake, North Dakota. The presence of bodies of water also modifies temperature; this is well illustrated by the location of the Hawaiian Islands in the Pacific Ocean. Temperature adjacent to the Great Lakes is modified so that the land warms slowly in the spring, delaying the blossoming of fruit trees. Delayed blossoming and an earlier date for the last killing frost in the spring makes land near the lakes suited for fruit production (Fig. 3-23). A delay in the date of the first killing frost in the fall also lengthens the growing season.

One of the most interesting cases of temperature and location is in the Salinas Valley of California. The valley is oriented northwest-southeast between two mountain ranges and is open to the Pacific Ocean. Cool winds off the ocean travel up the valley, producing cloudy weather and moderating the temperature. Conditions are ideal for production of head lettuce, enabling this small valley to be a major producer of lettuce in the United States.

Fluctuations in Soil Temperature

Soil temperatures vary in a characteristic manner on a daily and seasonal basis. Both fluctuations are greatest at the soil surface and decrease with increasing depth. A well-defined seasonal time lag occurs as the seasonal soil temperature is slow to change. Below about 10 feet or 3 meters the temper-

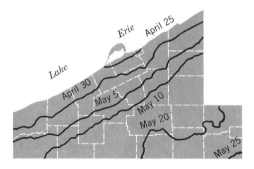 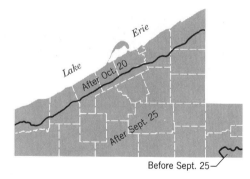

Fig. 3-23 Dates for last killing frost in spring and first killing frost in the fall at top and bottom, respectively, for Erie County, Pa. The longer growing season along Lake Erie favors use of land for fruits and vegetables. (Taylor, 1960.)

ature remains quite constant. Temperature fluctuations are greater at the soil-air interface than in the air above or soil below. Lethal temperature for plant seedlings may occur at midday. Below 6 inches or 15 centimeters there is little daily variation in soil temperature.

Snow acts as an insulator. A snow cover in the winter results in higher soil temperatures and a decrease in the depth of frost penetration. The insulation effect of a snow cover and a layer of organic matter at the top of the soil in forests of northern Ohio and southern Michigan has resulted in little if any soil freezing in some years. Melting of the snow in the spring delays the warming up of the soil because of the high specific heat of the melt water. A layer of organic matter on top of the soil and a snow cover are thus seen to reduce fluctuations in soil temperature.

Control of Soil Temperature

As has been pointed out, removal of excess water from a soil will facilitate changes in soil temperature. By providing drainage, man may exert some influence on the temperature relations of soils that are so situated that they hold excessive quantities of water. By use of mulches and various shading devices, the amount of solar radiation absorbed by the soil, loss of heat energy from the soil by radiation, infiltration of water, and loss of water by evaporation can be altered.

Light-colored organic matter mulches (1) reflect a large part of solar radiation, (2) retard heat loss by radiation, (3) increase infiltration of water, and (4) reduce evaporation of water from the surface of the soil. The net effect of a light-colored organic matter mulch is to reduce soil temperature. In regions where summers are cool the reduced soil temperature has been found to reduce crop yields. Dark-colored plastic mulches (1) absorb most of the solar radiation, (2) reduce heat loss from the soil by radiation, and (3) reduce the evaporation of water from the surface of the soil. The net effect of black plastic mulches is to increase soil temperature in the soil under the mulch when used (Fig. 3-24). The higher soil temperature increases crop yields in regions with cool summers. In Michigan plastic mulches are used to increase the growth rate of muskmelons, which results in an earlier harvest.

Fig. 3-24 A black plastic mulch increases soil temperature and hastens the beginning of the harvest period for muskmelons. The opaque mulch also aids weed control.

The earlier-harvested melons have a sufficient price advantage to make the mulching practice profitable. Clear plastic mulches increase soil temperature because of "greenhouse effect" and have proved more successful than opaque mulches when used with herbicides.

Cold air, being more dense than warm air, moves down slope and collects in low areas or depressions. Such areas have greater frost hazard in the spring. Landforming and large electrically driven fans (wind machines) have been used to improve air drainage and reduce frost hazard for fruit crops (Fig. 3-25).

Soil Temperature Regimes

Soil temperature, being an important soil property, is used to classify soils. Soil temperature classes or regimes are defined according to the average annual soil temperature in the root zone (arbitrarily set at 5 to 100 centimeters). Regime definitions and some characteristics and land uses associated with the regimes are given in

Fig. 3-25 Fruit land that has been reshaped to permit drainage of cold air onto the lake in the background. Spring frost hazard is reduced, and the smooth land results in more efficient operation of machinery.

Table 3-8
Definitions and Features of Soil Temperature Regimes

Temperature Regime	Mean Annual Temperature in Root Zone (5 to 100 cm)		Characteristics and Some Locations
	C	F	
Pergelic	0	32	Permafrost and ice wedges common. Tundra of northern Alaska and Canada and high elevations of the mid and northern Rocky Mountains
Cryic	0–8	32–47	Cool to cold soils of the northern Great Plains of the United States and southern Canada where spring wheat is dominant crop. Forested regions of eastern Canada and New England
Mesic	8–15	47–59	Midwestern and Great Plains regions where corn and winter wheat are common crops
Thermic	15–22	59–72	Coastal plain of southeastern United States where temperatures are warm enough for cotton. Central valley of California
Hyperthermic	Over 22	Over 72	Citrus areas of Florida peninsula, Rio Grande Valley of Texas, and southern California, and at low elevations in Puerto Rico and Hawaii. Tropical climates, and crops

Table 3-8. The use of soils for agriculture and forestry is importantly related to soil temperature because of the specific temperature requirements of plants.

References

Arkley, Rodney J., "Soil Survey of the Easter Stanislaus Area. California," *U.S.D.A. and Cal. Agr. Exp. Sta.*, 1964.

Auten, John T., and T. B. Plair, "Forests and Soils," *Trees*, USDA Yearbook, pp. 114–119, 1949.

Bouyoucos, George John, "Hydrometer Method Improved For Making Particle Size Analyses of Soils," *Agron. Jour.*, 54:464–465, 1962.

Camp, C. R., and L. F. Lund, "Effect of Soil Compaction on Cotton Roots," *Crops and Soils*, November 1964.

Chesters, G., O. J. Attoe, and O. N. Allen, "Soil Aggregation in Relation to Various Soil Constituents," *Soil Sci. Soc. Am. Proc.*, 21:272–277, 1957.

Childe, V. E., *Man Makes Himself*, Mentor, New York, 1951.

Cook, R. L., J. F. Davis, and M. G. Frakes, "An Analysis of Production Practices of Sugar Beet Farmers in Michigan—1958," *Ag. Exp. Sta. Quart. Bul.*, 42:401–420, 1959.

Dickerson, B. P., "Soil Changes Resulting From Tree-Length Skidding," *Soil Sci. Soc. Am. Proc.*, *40*:965–966, 1976.

Dotzenko, A. D., N. T. Papamichos, and D. S. Romine, "Effect of Recreational Use on Soil and Moisture Conditions in Rocky Mountain National Park," *Jour. Soil Water Con.*, *22*:196–197, 1967.

Erickson, A. E., and D. M. VanDoren, "The Relation of Plant Growth and Yield to Soil Oxygen Availability," *7th Int. Cong. Soil Sci.*, *3*:428–434, 1960.

Grimes, Donald W., and James C. Bishop, "The Influence of Some Soil Physical Properties on Potato Yields and Grade Distribution," *Am. Pot. Jour.*, *48*:414–422, 1971.

Laws, W. Derby, and D. D. Evans, "The Effects of Long-Time Cultivation on Some Physical and Chemical Properties of Two Rendzina Soils," *Soil Sci. Soc. Am. Proc.*, *14*:15–19, 1949.

Page, J. B., and C. J. Willard, "Cropping Systems and Soil Properties," *Soil Sci. Soc. Am. Proc.*, *11*:81–88, 1946.

Retzer, J. L., "Soil Development in the Rocky Mountains," *Soil Sci. Soc. Am. Proc.*, *13*:446–448, 1948.

Soil Survey Staff, *Soil Survey Manual*, USDA Handbook 18, Washington, D.C., 1951.

Soil Survey Staff, *Soil Taxonomy*, Agr. Handbook 436, USDA, Washington, D.C., 1975.

Steinbrenner, E. C., and S. P. Gessel, "The Effect of Tractor Logging on Physical Properties of Some Forest Soils in Southwestern Washington," *Soil Sci. Soc. Am. Proc.*, *19*:372–376, 1955.

Taylor, David C., "Soil Survey of Erie County, Pennsylvania," USDA and Pennsylvania State University, 1960.

Ulrich, Rudolph, "Some Chemical Changes Accompanying Profile Formation of the Nearly Level Soils Developed from Peorian Loess in Southwestern Iowa," *Soil Sci. Soc. Am. Proc.*, *15*:324–329, 1950.

Van Doren, D. M., and G. B. Triplett, Jr., "Mulch and Tillage Relationships in Corn Culture," *Soil Sci. Soc. Am. Proc.*, *37*:766–769, 1973.

Walejko, R. N., et al., "Effect of Snowmobile Traffic on Alfalfa," *Jour. Soil Water Con.*, *28*:272–273, 1973.

Wascher, Herman L., et al., "Characteristics of Soils Associated with Glacial Tills in Northeastern Illinois," *Univ. Ill. Agr. Exp. Sta. Bul. 665*, 1960.

Wilde, S. A., "The Significance of Texture in Forestry and Its Determination by a Rapid Field Method," *Jour. For.*, *33*:503–508, 1935.

Wimer, D. C. and M. B. Harland, "The Cultivation of Corn," Univ. Ill. *Agr. Exp. Sta. Bul. 259*, 1925.

Yaalon, D. H., " 'Calgon' No Longer Suitable," *Soil Sci. Soc. Am. Proc.*, *40*:333, 1976.

Youngberg, C. T., "The Influence of Soil Conditions, Following Tractor Logging on the Growth of Planted Douglas-Fir Seedlings," *Soil Sci. Soc. Am. Proc.*, *23*:76–78, 1959.

4
SOIL WATER

Water is the most common substance on the earth and is necessary for all life. The supply of fresh water on a long-time sustained basis is equal to the annual precipitation, which averages 26 inches for the world's land surface. The soil, located at the atmosphere-lithosphere interface plays an important role in determining the amount of precipitation that runs off the land and the amount that enters the soil for storage and future use. Approximately 70 percent of the precipitation in the United States is evapotranspired and returned to the atmosphere as vapor, with the soil playing a key role in water retention and storage. The remaining 30 percent of the precipitation represents the long-time annual supply of fresh water for use in homes, industry, and irrigated agriculture. About one fourth of this 30 percent is currently being used in the United States.

Fortunately, water is not easily destroyed. The earth has as much water now as it did thousands of years ago. However, the water is unevenly distributed by rainfall, changes form, moves from place to place, and can be polluted. For students interested in pollution of the environment, resource use, and plant science, this chapter contains important concepts and principles that are essential for gaining an understanding of the soil's role in the intelligent management of water resources.

Energy Concept of Soil Water

As water cascades over a dam, the energy content (ability to do work) of the water decreases. If water than has gone over a dam is returned to the reservoir, work will be required to lift the water back up into the reservoir, and the energy level of the water will be restored. Water movement in soils and from soils into plant roots, like water cascading over a dam, is from regions of higher-energy water to regions of lower-energy water. Thus, "water runs down hill." For this reason it is necessary to consider the forces that determine the physical state or energy content of water in order to understand the behavior of water in soils and plants.

Surface Forces Control the Physical State of Soil Water

The change of water from a vapor to a liquid (condensation) is accompanied by a great reduction in the movement of the molecules and their energy content. Energy is released as heat when water changes from vapor to liquid. The liberation of heat from the formation of raindrops is a major source of energy for storm systems. When raindrops fall on dry soil and are adsorbed on the surface of soil particles, a further reduction occurs in the motion and energy content of the water molecules. The adsorbed water may still be in a liquid state, but the tendency of the water molecules to move has been further reduced. This change in energy content can be explained by considering the forces operating between soil particles and water molecules.

Water molecules (H_2O) are electrically neutral; however, the electrical charge within the molecule is asymmetrically distributed. As a result, water molecules are strongly polar and attract each other by H bonding (see Fig. 4-1). Soil particles are also charged and have negative and positive charged sites. The strong attraction of the soil for the water molecules (adhesion) results in a spreading of the water over the surface of the soil particles as a film when liquid water comes in contact with dry soil particles. The adsorption of water on the surface of soil particles produces (1) a reduction in the motion of the water molecules, (2) a reduction in the energy content of the water, and (3) the release of heat associated with the transformation of water to a lower energy level. You can observe the release of heat, called *heat of wetting*, by adding water to oven-dry clay soil and observe the increase in soil temperature.

Fig. 4-1 Schematic diagram of two water molecules. The sharing of electrons by the oxygen and hydrogen produces water molecules that are negatively charged on the oxygen side and positively charged on the side where the proton of the hydrogen sticks out. Molecules attract each other by H bonding, the attraction of the proton of one water molecule for the negatively charged oxygen side of an adjacent water molecule.

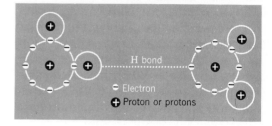

Several layers of water molecules are strongly adsorbed to the soil particles because of these strong adhesive forces (see Fig. 4–2). This water is called *adhesion water*. Adhesion water moves little, if at all, and some scientists believe that the innermost layers of water molecules exist in a crystalline state similar to the structure of ice. Adhesion water is not available to plants and is always present in the normal soil (even in the dust of the air), but the adhesion water can be removed by drying the soil in an oven.

Beyond the sphere of strong attraction of the soil particles, water molecules are held in the water film by cohesion (H bonding between water molecules). This outer film water is called *cohesion water*. Molecules of cohesion water, compared to adhesion water, are in greater motion, have a higher energy level, and move more readily. The water film (including adhesion and cohesion water) in soils may be as much as 15 to 20 molecular layers thick. The outer approximately two thirds of the film can be considered available to plants and constitutes the major source of water for plant growth (see Fig. 4-2).

Fig. 4-2 Schematic drawing showing the relationship of adhesion and cohesion water with respect to soil particles and air-filled macropores.

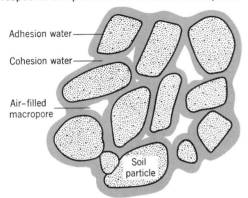

Adhesion water

Cohesion water

Air–filled macropore

Soil particle

You can demonstrate the existence of the forces important in holding water in soils by placing two clean microscope slides in water and bringing them together so that their flat sides are flush against each other. Then, try to pull them apart. Your failure to pull the slides apart demonstrates the existence of attractive forces between glass and water molecules. You can also observe that the forces are effective only over a very short distance, since the glass slides must be very close together before a strong attraction develops between the slides.

Energy and Pressure Relationships

We have just noted that water exists in the soil over a range of energy contents. This energy content of water can be expressed in terms of water pressure. Because it is much easier to determine the pressure of water, as contrasted to the energy content or level, we usually categorize water on the basis of pressure. For this reason we must understand the relationship between the energy content and the water pressure.

The hull of a submarine must be sturdily built to withstand the great pressure encountered in a dive far below the ocean's surface. The head of water above the submarine exerts pressure on the hull of the submarine. If the water pressure exerted against the submarine was directed against the blades of a turbine, the gravitational energy in the water could be used to generate electricity. The greater the water pressure, the greater is the tendency of water to move and do work and *the greater is the energy content of the water.* Thus, the existence of a relationship between the pressure of water and the energy of water is established. Our next consideration will be that of the water pressure relationships in saturated soils, which is analogous to the water pressure relationships in a beaker of water or any body of water.

Water Pressure in a Beaker or in Saturated Soil

The beaker in Fig. 4-3 has a bottom with an area of 100 square centimeters. The height of the water in the beaker is 20 centimeters. The water has a volume of 2000 cubic centimeters, weighs 2000 grams, and exerts a total force on the bottom of the beaker equal to 2000 grams. The pressure of the water at the bottom of the beaker is equal to:

$$P = \frac{\text{force}}{\text{area}} = \frac{2000 \text{ grams}}{100 \text{ square centimeters}}$$

$$= 20 \text{ grams per square centimeter}$$

The water pressure at the bottom of the beaker could also be expressed simply as equivalent to a column of water 20 centimeters high.

Fig. 4-3 A beaker with a cross-sectional area of 100 square centimeters contains 2000 grams of water when the water is 20 centimeters deep. At the water surface, the water pressure is 0 and the water pressure increases with depth to 20 grams per square centimeter at the bottom.

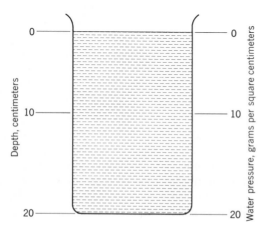

At the 10-centimeter depth, the water pressure is half of that at the 20-centimeter depth and is, therefore, 10 grams per square centimeter. The *water pressure* decreases with distance toward the surface and becomes zero at the free water surface (see Fig. 4-3).

To apply the considerations of water in a beaker to soil, we must consider soil that is *saturated* with water. At the top of the water table or at the top of the soil saturated with water, the water pressure is zero. The water pressure increases with increasing depth below a water table as it does in a beaker or any body of water. The water existing in the macro- or aeration pores of a *saturated* soil has a positive pressure determined by the distance below the surface of the saturated zone. Water in the macropores of saturated soil is under pressure and will freely flow through macropores from regions of higher pressure to regions of lower pressure (basically from higher elevations to lower elevations). The water that "freely" flows or drains out of the soil is called *gravitational* water. Where gravitational water exists in soil adjacent to the walls of the basement of a building, water pressure may force water through the cracks of walls and floor. If a soil remains saturated with water, one can assume that an impermeable layer is inhibiting the downward flow of gravitational water from the soil. Drainage ditches and tile drains are used to remove gravitational water from saturated soils.

After the gravitational water has drained out of the soil, the soil becomes unsaturated and the remaining water is held by the attractive forces between soil particle surfaces and water molecules (adhesion) and between water molecules (cohesion). A clear differentiation between saturated and unsaturated soils is needed to gain a meaningful understanding of the behavior of soil water. We will consider next the water pressure (energy) relationships in unsaturated soil.

Water Pressure in a Capillary Tube or in Unsaturated Soil

If the tip of a capillary tube is inserted in a beaker of water, the attraction between glass and water molecules (adhesion) causes water molecules to migrate up the interior wall of the capillary. The cohesive force between water molecules causes other water molecules to be drawn up the capillary. Now we need to ask, "What are the pressure relationships in the capillary tube?" and "Why is this knowledge important?"

We have already seen that the water pressure decreases from the bottom to the top of a beaker and, at the top of the water surface, the water pressure is zero. Beginning at the water surface of the beaker and moving upward into the capillary tube, the water pressure continues to decrease. Thus, the water pressure in a capillary tube is less than zero or is *negative*. From Fig. 4-4 one can observe that the water pressure decreases in a capillary tube with height above the water in a beaker. At a height of 20 centimeters above the water surface in a beaker, the water pressure in a capillary tube is equal to −20 grams per square centimeter. Figure 4-4 also shows that at this same height above the water surface the water pressure in *unsaturated* soil is the same; −20 grams per square centimeter at the 20 centimeter height. The two pressures are the same at any one height and there is no net flow of water from the soil or

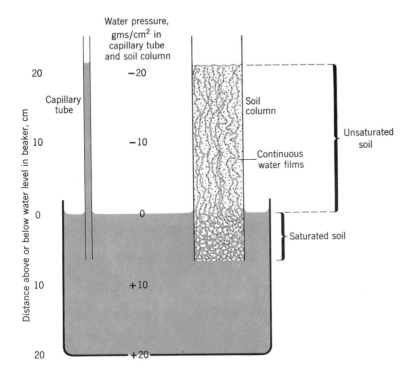

Fig. 4-4 Water pressure in a capillary tube decreases with increasing distance above the surface of the water in the beaker and is −20 grams per square centimeter at a height 20 centimeters above the water surface. Since the water column of the capillary tube is continuous through the beaker and up into the soil column, the water pressure in the soil 20 centimeters above the water level of the beaker is also 20 grams per square centimeter.

capillary after an equilibrium condition is established.

Applying the considerations of water in a capillary tube to an unsaturated soil, we can make the following statements: (1) water in unsaturated soil has a *negative* pressure or is under *tension*; (2) the water pressure in unsaturated soil *decreases* with *increasing distance* above the surface of a water table; and (3) water in unsaturated soil, compared to saturated soil, has a *lower pressure* and *lower energy* level. We consider next the importance of the nature of water in unsaturated soil in relation to water movement.

Water Movement in Unsaturated Soil

Several important consequences follow from the nature of water in unsaturated soils. Water in unsaturated soil exhibits very little tendency to move. Movement is very slow and mainly by adjustment of the thickness of water films on soil particles. We can visualize the water as occurring as surface films and as wedges in the angles of adjoining soil particles (Fig. 4-5). There is a tendency for spherical droplets to form at points A and B in Fig. 4-5. However, droplet formation is prevented because the two soil particle surfaces are acting against each other. Now, if the curvature of one film is greatly increased through the removal of moisture by a root, as shown at A, its pulling power will be proportionately increased, since the pull exerted by such a curved film is inversely proportional to the radius of curvature. As a result, water will be drawn toward A until the curvature of films A and B is equal. The equation for capillary movement may be written: p (pull) $= 2T/r$, in which T is surface tension and r the radius of curvature. If it is assumed, for example, that the r of A is 1 and the r of B is

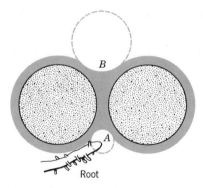

Fig. 4-5 As the root absorbs moisture from the accumulation between two soil particles, the film curvature increases, as is shown by the projected circles. Since the force tending to draw water into a given portion of the film varies inversely with the radius of the curvature ($p = 2T/r$), it follows that moisture will move to the feeding point of the root.

2, the pull exerted at the two points will be $p = 2T/1$ compared to $p = 2T/2$. Since the liquid in each case is water, T will be equal, and so the pull exerted by the film A will be twice as great as that exerted by film B. By film adjustment we can explain the movement of water in unsaturated soils in the direction of drier soil as a result of absorption of water by plant roots. Significant water movement occurs over distances of the order of less than an inch. As stressed in Chapter 2, the limited mobility of water in well-aerated soils (unsaturated soils) requires extension of plant roots into all soil horizons from which water is absorbed.

When moist unsaturated soil is adjacent to the basement walls of a building, water may slowly move by capillarity through small pores or cracks in the walls or floor. The edges of cracks may appear moist but pools of water will not collect on a basement floor if water is moving by capillarity. Can you explain this on the basis of energy and pressure relationships?

Capillary Water Movement in Stratified Soil

In the Columbian Basin of the northwestern United States some of the best agricultural soils consist of 2 feet of fine sandy loam soil with sand and gravel underneath. These soils are noted for their great water-holding capacity. One's first reaction is that soils underlain by sand and gravel would be very droughty. But this is readily understood if one realizes that when a downward moving water front (from rainfall or irriga-

tion) encounters a coarse-textured layer, water movement will stop, at least temporarily. Water movement by capillarity through the upper 2 feet of fine sandy loam soil will stop when the sand and gravel layer is encounted (Fig. 4-6). The sand layer has "large" pores and very low suction capacity. Water will not be pulled down into the sand layer until the water content of the overlying finer-textured layer is increased so that the suction capacity of the pores in the sand has been attained, as shown in the lower

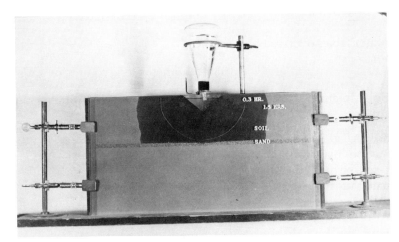

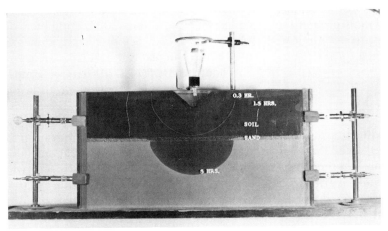

Fig. 4-6 Photographs illustrating water movement in stratified soil where the water, under conditions of unsaturated flow, is moving from a finer-textured soil into a coarser-textured layer. In the upper photo it can be seen that the downward movement of water is inhibited by the sand layer. The lower photo shows that sometime after an elapsed time of 1.5 and 5.0 hours, the SMT just above the sand layer was lowered sufficiently so that the pores in the sand layer could pull the water downward by capillarity. (Photos courtesy W. H. Gardner of Washington State University, Pullman, Wash.)

part of Fig. 4-6. The soil just above the sand interface will become nearly saturated before water continues to move down into the sand. This explains the high water-holding capacity of the Columbian Basin soils. A similar case occurs on terraces of glaciated regions where loamy soils are underlain by sands and gravels.

When a moving water front encounters a finer-textured layer or horizon, as a Bt horizon, the water front continues to move. The finer pores pull the water downward. The low transport capacity of the finer pores, however, may cause a buildup of water above the finer-textured layer. During wet seasons, this commonly produces water-saturated soil or perched water tables overlying well-developed Bt horizons.

Water Pressure Versus Soil Moisture Tension (SMT)

A few crops, like paddy rice, are grown on saturated soils. In most situations plants are growing on unsaturated soils where the water is under tension analogous to water in a capillary tube. The water pressure is a negative quantity and expression of the pressure of soil water is more conveniently done with positive tension values. Thus, a negative pressure of 20 grams per square centimeter becomes a soil moisture tension (SMT) of 20 grams per square centimeter or:

(Water pressure) $\times (-1)$

= soil moisture tension (SMT)

Since it is the weight or downward pull of a 20-centimeter column of water that produces 20 grams per square centimeter of tension (see Fig. 4-4), the SMT as grams per square centimeter can be directly converted into equivalent lengths of water columns.

That is, a tension of 20 grams per square centimeter is equivalent to the tension produced by a 20-centimeter-long water column. It is analogous to saying a pressure of 14.7 pounds per square inch is equal to 1 atmosphere.

Soil moisture tension is commonly expressed in atmospheres. As a first step to express SMT in atmospheres consider the following equivalents.

1 atmosphere

= 33-foot-long water column

= 14.7 pounds per square inch

1 atmosphere

= 1036-centimeter-long column of water

= 1036 grams per square centimeter

To convert the SMT at the top of the soil column in Fig. 4-4 to atmospheres, the 20 must be divided by 1036.

$$\frac{20 \text{ grams per square centimeter}}{1036 \text{ grams per square centimeter}}$$

= 0.019 atmospheres

The bar is a metric unit and is nearly equal to 1 atmosphere (1023 centimeters of water). Since they are so similar, they are used interchangeably.

When the SMT is 0.019 bars or atmospheres, an opposing tension or drawing force (suction) greater than 0.019 atmospheres must be exerted by a plant root before water will move from the soil into the root. Shortly, in the consideration of the biological aspects of soil water, it will become clear that this tension (0.019 atmosphere) is very low and that water at this tension is very available to plants.

Measurement of Soil Moisture Tension

A vacuum gauge-type tensiometer, shown in Fig. 4-7, consists of a porous, fired clay cup, which is attached to a vacuum gauge by a water-filled pipe. If the porous cup is buried in soil where the water has a tension greater than zero, water will move from the cup into the soil. At equilibrium the tension created within the tensiometer and registered by the vacuum gauge is the SMT. This means that a force or pull (suction) *greater* than this tension must be exerted to remove some water from the soil. Wetting the soil releases the tension, and water then moves from the soil into the porous cup of the tensiometer. Tensiometers work up to the range of about 0.8 atmospheres. This range is important in terms of plant growth, making the tensiometer a useful instrument to determine when to apply irrigation water.

Measurement of SMT at higher tensions is made with the pressure chamber in the range of 1 to 50 atmospheres. A membrane with very fine pores that are water filled rests on a screen in the bottom of the pressure chamber. The soil is placed on the membrane, the chamber is sealed, and air pressure is applied. Water is forced out of the soil and flows through the pores of the membrane. At equilibrium the SMT equals the air pressure that was applied. The pressure chamber is an invaluable research tool and is used to obtain the kind of data plotted in Fig. 4-8 that relate SMT to water content of soils.

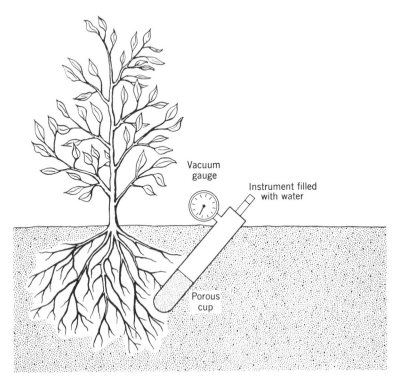

Fig. 4-7 Vacuum-gauge tensiometer used to measure soil moisture tension. Tensiometer gauges can be wired to turn on and off irrigation systems automatically in response to changes in SMT. (Drawn after Richards. Reprinted by permission, from C. A. Black, *Soil-Plant Relationships,* 2nd ed, p. 77. Copyright © 1968 by John Wiley & Sons Inc.)

Vacuum gauge

Instrument filled with water

Porous cup

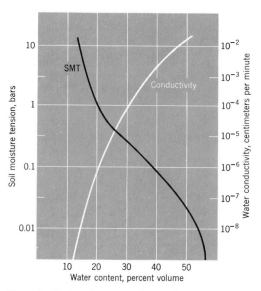

Fig. 4-8 Soil moisture tension and water conductivity as related to water content of silt loam soil. Soil moisture tension increases and conductivity decreases very rapidly as the soil dries. (Data from Kunze, et al. 1968.)

Water Content of Soils Versus SMT and Conductivity

Soil moisture tension and rate of water movement or conductivity are importantly related to water content of soils. A soil sample that is saturated has a SMT of zero. As water is removed from saturated or nearly saturated soil, the SMT increases rapidly, as shown in Fig. 4-8. This has two important consequences. First, as the soil dries or the water content decreases, plant roots and microorganisms must exert more energy to absorb water or the availability of soil water decreases. Second, water (hydraulic) conductivity or rate of water movement decreases sharply as the soil dries out (Fig. 4-8). Note that the vertical scales of Fig. 4-8 are logarithmic, meaning that each unit change in SMT or conductivity repre-

sents a tenfold change. The change in SMT from 0.01 to 10 atmospheres is a 1000-fold increase, which is entirely within the range important to plant growth. The corresponding decrease in conductivity, however, is of the order of 1 million. This fact highlights the relative immobility of water in unsaturated soils and has important consequences for water movement to plant roots and loss of water from soils by evaporation.

The Soil Water Potential

The energy content or free energy of soil water is also expressed as the *water potential*. The water potential has three components or subpotentials. The *gravitational potential* or component is due to the position of the soil water in a gravitational field. The gravitational potential is important in saturated soils and is shown by the tendency of water to flow to a lower elevation. The *matric potential* is the result of the adhesive and cohesive forces associated with the particle network of the soil or the soil matrix. The potential is expressed relative to pure water; thus, as soils dry and the energy content of water decreases, the matric potential decreases. The matric potential varies inversely as the SMT and is the controlling factor in water movement in unsaturated soils and usually in the movement of water from the soil into plant roots and microorganisms. The *osmotic potential* is due mainly to the attraction of water molecules for ions produced by soluble salt. Normally in leached soils the osmotic potential is small and is a minor factor in water absorption. The osmotic potential of saline soils, by contrast, reduces the ease that water moves into plant roots and microorganisms and will be considered in Chapter 5.

Summary Statement

Adhesion water in soils:

1. Is held by strong surface (electrical) forces existing between soil particles and water molecules.
2. Is mostly "crystalline," exhibits little or no movement, and possesses a low energy content.
3. Exists as a film on the surface of soil particles, several molecular layers thick.
4. Is unavailable to plants.
5. Exists on surfaces of dry soil particles that occur as dust in the air, but can be removed by drying the soil in an oven.

Cohesion water in soils:

1. Is held by the attraction of water molecules for each other through hydrogen bonding.
2. Exists in the liquid state in the water films around soil particles and in micropores and approximates the soil solution.
3. Is the major source of water for plant growth.
4. Has a higher energy content than adhesion water.
5. Moves very slowly by film adjustment from areas where the films are thick and SMT is low to areas where the water films are thin and the SMT high.

Gravitational water in soils:

1. Exists in the macropores.
2. Is either free water or under very low tension.
3. Moves freely through macropore spaces in response to very small water pressure differences or gravitation.

In saturated soils water movement is rapid in macropore spaces in response to gravity and the soil lacks oxygen for biological activities. In unsaturated soils the capillary forces govern water movement, which occurs very slowly over very small distances, and the macropore spaces contain air. As the water is removed by plants in unsaturated soil, the SMT increases and this results in greater resistance to movement of water and uptake by plant roots. The role of SMT in utilization of soil water by plants will be examined in the next section.

Plant-Soil Water Relations

In the course of a summer day, it is not unusual for a plant to transpire an amount of water equal to many times its weight. Since the water that plants absorb from soils is not free flowing but diffuses slowly into plant roots by osmosis, an enormous area of contact between roots and soil particles is required. The total length of the root system of a corn plant could easily equal the distance from New York to San Francisco and back. Our concern in this section centers around the ability of plants to satisfy their water requirements from the soil.

Gravitational Water and Field Capacity

When rain falls on the soil, water is pulled or sucked down into the soil by capillary action. If free water builds up on the surface of the soil, water may flow freely down through large macropores. The macropores, however, *must be open at the top of the soil*. Where a typical loamy soil is unprotected by vegetation, rain drop impact will break apart soil aggregates and fine soil particles will float over the top of the soil creating a "dense" layer that will be composed of micropores (capillary-sized openings). Then, the water moves into the soil by capillarity. We will be more correct if we visualize water being pulled into the soil surface during a rain by capillarity instead of thinking that water is running into the soil through "large" holes. The situation at the soil surface is the *reverse* of a capillary tube standing in a pan of water, where the tube fills from the bottom to the top. By contrast, capillary-sized pores in the surface of the soil pull water downward into the soil during a rain or irrigation.

Water in excess of the ability of the soil to retain adhesion and cohesion water exists in the large or noncapillary pore space and moves downward in response to *gravity* and the *suction* or *pull* of the underlying soil pores. This "excess" or gravitation water moves downward and moistens drier soil below. We see, then, that water that was considered gravitational at one level becomes capillary (nongravitational) water at a lower level in drier soil. Under these conditions, the water moves downward as a front (Fig. 4-9). A sharp line of demarcation is formed between the moist upper layer and the drier lower layer, and this line may persist for days. The upper moist soil layer is at *field capacity*, which is the water content of soil in the field after the gravitational water has drained out.

Two questions present themselves. First, "Why doesn't the water continue to move with 'reasonable' speed from the moist upper layer into the drier lower layer?" Second, "What is the soil moisture tension in the moist soil (at field capacity)?" (Fig. 4-9).

The answers to these two questions are related. Consider, again, the capillary tube. What determines how high water rises in a capillary? Obviously, it is the size or diameter of the capillary. The same sitation is true of the soil in regard to the ability of soil

pores to pull water downward. The question then becomes, "What is a reasonable or typical capacity of soil pore spaces to pull water into the soil surface or of the soil pores to pull water out of a moist soil layer?", as shown in Fig. 4-9. On a theoretical basis it has been calculated that water may be lifted from a free water surface to a height of $1\frac{1}{2}$ feet by coarse sand and to 150 feet by fine silt. Soils vary greatly in their kinds and sizes of pores, causing the SMT at field capacity to be highest in clays and lowest in sands. In sands the SMT at field capacity may be as low as 0.01 atmosphere. It has become customary, however, to consider a value of $\frac{1}{3}$ atmosphere (equivalent to a column of water about 11 feet high) to be equal to the typical pulling power of capillaries in well-moistened soil. Thus, the SMT at field capacity is generally considered to be $\frac{1}{3}$ atmospheres or 346 centimeters of water (see Fig. 4-10). This means a suction or tension *greater* than $\frac{1}{3}$ atmosphere must be applied to remove water or to cause water to move in moist soil.

The water in the moist layer in Fig. 4-9, as we have seen, exhibits very little tendency to move downward into the underlying drier soil. More than capillarity is involved, and we must draw on one additional factor to account for this phenomenon. This factor is the lack of water contacts or "bridges" between soil particles. As seen in Fig. 4-10, the moistened soil layer at field capacity has about half of the pore spaces filled with air (macropores) and only half of the pore space is filled with water (film water plus water in capillary pores). Although gravity and capillary-sized pores in the underlying dry soil exert a pull to encourage the downward movement of water out of the moist layer, water must move along films on soil particles to reach the underlying capillaries in the dry soil. The downward movement

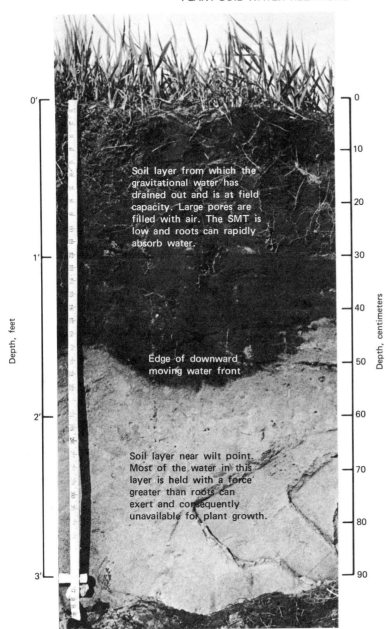

Soil layer from which the gravitational water has drained out and is at field capacity. Large pores are filled with air. The SMT is low and roots can rapidly absorb water.

Edge of downward moving water front

Soil layer near wilt point. Most of the water in this layer is held with a force greater than roots can exert and consequently unavailable for plant growth.

Depth, feet

Depth, centimeters

Fig. 4-9 Moisture relationships 1 day after a rain when the soil was near the wilt point to a depth of 3 feet (1 meter) or more. The sharp line of demarcation between the moist upper and drier lower layer should be observed.

of water is inhibited by the discontinuity of water films or absence of water *bridges* between many soil particles. As an analogy, the water movement at the wetting front a day or two after a rain can be compared to a car traveling down a highway. If the car is moving rapidly and comes to a bridge that is out, the car comes to an abrupt halt. In

Soil moisture classification	Tension		Approximate % pore space occupied by water
	Atmospheres or bars	Centimeters of water	
— Oven dry —	10,000	10,000,000	0
Hygroscopic water (unavailable to plants)			
— Hygroscopic coefficient —	31	31,600	15
Capillary water (unavailable to plants)			
— Wilt point —	15	15,800	25
Capillary water (available to plants)			
— Field capacity —	1/3	346	50
Gravitational water (subject to drainage)			
— Saturation —	0	0	100

Fig. 4-10 Diagram showing soil moisture classification, soil moisture tension equivalents, and the approximate percentage of the soil pore space occupied by water at various tensions. The soil is assumed to be a well-aggregated, medium-textured plow layer.

energy terms the water in films is held more tightly and has a lower energy content than water in capillary pore spaces; thus, water stops moving downward (for all practical purposes) into the drier soil because the energy gradient prevents the flow of film water into soil capillaries.

The water between saturation and field capacity (gravitational water) has a low tension and would be easily absorbed by plant roots. However, the gravitational water is of little value in most soils because it drains downward rather quickly. In addition, the presence of gravitational water excludes air that is needed for root respiration and many other biological activities. The downward movement of gravitational water pulls fresh air into the soil.

Water Absorption and the Wilt Point

At field capacity the SMT is low and plant roots can easily absorb water. As roots absorb water by osmosis, water near the roots will move slowly in the direction of the root by film adjustment (see Fig. 4-5). We have already observed that as the soil becomes drier, the SMT increases and the movement of water becomes slower. Eventually, if no additional water is added to the soil, the plant will absorb water slower than water is lost by transpiration. A water deficit is developed inside the plant and, eventually, wilting occurs (unless the plant has some special adaptation such as found in many desert plants that can stop transpiration loss). To determine the wilt point, plants such as sunflowers or wheat are grown on the soil until the plants wilt and are unable to regain turgor when placed in a saturated atmosphere. A soil moisture tension of 15 atmospheres has been found to correspond generally with the wilt point.

The water between field capacity and wilt point is considered the available water for plants (see Fig. 4-11). Assuming field capacity of $\frac{1}{3}$ atmosphere for each of the soils, plants would have to exert the same amount of energy to remove water from each soil, even though the water content of the clay loam is about three times greater than that of the fine sandy loam. This point brings out a most important fact: the ability of plants to remove water from soils is primarily related to SMT and not water content. It is the SMT and not the water content that indicates when to irrigate, and this shows the importance of tensiometers (and other SMT measuring devices) in irrigation agriculture. The wilt point, like the field capacity, is not a precise value and varies with soil, plant, and environmental condi-

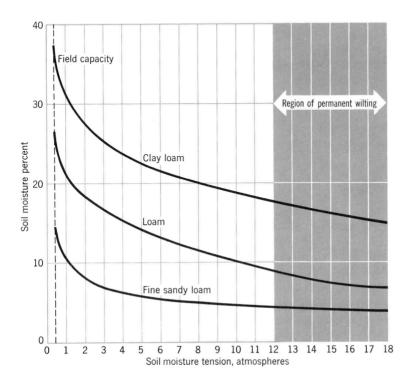

Fig. 4-11. Soil moisture characteristic curves for three soils. Water between field capacity and wilt point is available to plants. The sandy loam holds a small amount of available water at low tension compared to the clay loam that holds much more available water over a wide range of tensions.

tions. High temperatures and strong winds could cause some plants to wilt with SMT as low as 2 atmospheres. The minimum moisture content in subsoils for wheat on the Great Plains has an SMT greater than 26 atmospheres. For most soils and plants the wilting range appears to be about 10 to 60 atmospheres. By the time the water content of the soil has been reduced to the wilting range, soil moisture tension increases rapidly, with little change in moisture content of soils. Thus, a plant that wilts on a *fine sandy loam* at 10 atmospheres has extracted about as much water from the soil as a plant that wilts at 50 atmospheres (Fig. 4-11).

Other Soil Moisture Coefficients

The soil still contains water at the wilt point that is considered to be unavailable to plants (see Fig. 4-10). To remove the remaining water (excluding water of hydration), the soil is dried in an oven for 24 hours at 110°C. The soil is then brought to the *oven-dry* state. If oven-dry soil is placed in a water-saturated atmosphere, water will be adsorbed by the soil. At equilibrium the soil will contain an amount of water described as the *hygroscopic* coefficient or *air dry*. The hygroscopic coefficient has little relevancy for plant growth but is a qualitative measure of the surface area in the soil.

SMT and Plant Growth

We have already noted that at very low SMT the lack of air may limit plant growth. The rate of plant growth is at or near a maximum at field capacity because there is adequae oxygen accompanied with low SMT for rapid water absorption. As soil moisture is absorbed, the moisture films become thinner, the SMT increases, and the rate of water absorption decreases.

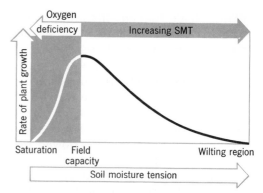

Fig. 4-12 Generalized relationship between soil moisture tension (SMT) and plant growth.

Generally, increasing SMT between field capacity and the wilt point is associated with a reduced rate of photosynthesis and growth. Thus, two important facets of plant growth are associated with SMT: lack of oxygen at low SMT and slow rate of water absorption at high SMT (Fig. 4-12). Forest tree growth is commonly limited by low water supply. It is interesting to note that maximum growth of sugar maple in the north central states was found to occur when SMT was in the range of $\frac{1}{3}$ to 3 atmospheres.

Effect of Texture on Available Water

The capacity of the soil to hold water is related to both surface area and pore-space volume. Water-holding capacity is therefore related to structure as well as to texture. It can be seen in Fig. 4-13 that fine-textured soils have the maximum *total* water-holding capacity but that maximum *available* water is held in medium-textured soils. Research has shown that available water in many soils is closely correlated with the content of silt and very fine sand.

It is generally known that sandy soils are more droughty than clayey soils. One

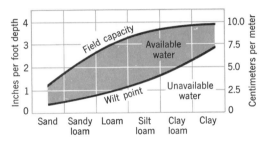

Fig. 4-13 Typical water-holding capacities of different textured soils. Note that maximum *available* water-holding capacity occurs in the silt loam soil. (Adapted from USDA Yearbook, 1955, p. 120.)

reason is that the finer-textured soils are able to retain more available water, as shown in Fig. 4-13. The soil moisture curves in Fig. 4-11 show a clay loam soil that can hold 18 percent available water compared to 8 percent for a fine sandy loam. Note that the clay loam contains more water at the wilt point than the fine sandy loam soil contains at field capacity.

Another important difference between sands and clayey soils is related to the differences in the slope of the soil moisture curves in Fig. 4-11. The flatness of the curve for the fine sandy loam at tensions greater than 4 atmospheres means that most of the available water is held at low tension. Consequently, the available water can be rapidly used by transpiring plants and, unless there is frequent rain or irrigation, plants are likely to suffer moisture stress. The steeper curve for the clay loam at tensions greater than 4 atmospheres indicates that considerable available water is held at "high" tension. Uptake of water by plants occurs more slowly and the water is conserved. A given amount of water was found to be more efficiently used or produced greater yields by wheat growing on clay soil than on sandy loam soil in Saskatchewan (Fig. 4-14). The researchers

assumed that the generally greater moisture stress in the early season of plants growing on the clay soil resulted in less early growth and transpiration so that more moisture was available late in the season for production of the grain. If wheat runs out of water before the grain matures, yields are greatly reduced.

Many red tropical soils (Oxisols) with high iron oxide contents are rich in clay; however, the soils exhibit moisture characteristics of sands. That is, a modest amount of water is held at tensions less than 0.3 atmospheres. Recall from Chapter 3 that such soils may be composed of very stable sand-sized aggregates. The aggregates act as sand particles for water retention. Moisture within the aggregates is held in a matrix high in clay and is mostly held at tensions greater than wilt point.

Fig. 4-14 Effect of texture on wheat yields in Saskatchewan. A given amount of evapotranspiration produced more wheat on the clay soil. (Lehane and Staple, 1965. Reproduced courtesy of the Canadian Journal of Soil Science, June 1975, p. 213.)

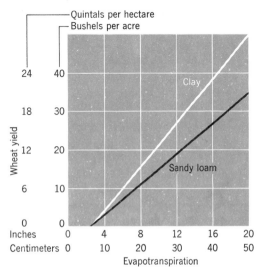

Pattern of Water Removal from Soils by Plants

When the SMT throughout the root zone is low or near field capacity, roots will absorb water most rapidly from the upper part of the soil where oxygen is the most abundant and near the base of the plant because less resistance will be encountered in translocating the water through the roots to the stem. As the soil dries and the SMT in the surface soil layers increases, water uptake will shift to deeper soil layers where the oxygen supply is less but the soil is moist and the SMT is low. In this way the root zone is progressively depleted of available soil moisture (in the absence of rain or irrigation water) (see Fig. 4-15). When the upper soil layers are rewetted by rain or irrigation water, water absorption shifts back toward the surface soil layers near the base of the plant. This pattern of water utilization results in: (1) more deeply penetrating roots in dry years than in wet years, and (2) a greater use of water from the upper soil layers than from the lower soil layers (Fig. 4-16). Crops that have a long growing season and a deep root system, like alfalfa, absorb a greater proportion of water below the 1 foot depth (30 centimeters).

Forests exert considerable influence on how water is utilized. Much of the precipitation is intercepted by the canopy. Pine and hardwood forests of the southeast can intercept as much as $\frac{1}{4}$ inch of water, which is lost by evaporation. A similar amount may be retained in the unincorporated organic matter on the forest floor so that rains must exceed about 0.5 inch before the upper mineral soil is wetted. Frequent rain in humid regions contributes to a large proportion of the soil water being absorbed by roots near the surface of the soil. This is related to the extensive root growth in forests in the upper soil horizons, as shown in Fig. 2-4. Use of water by deep roots is minor but critical to enable forest trees to survive drought.

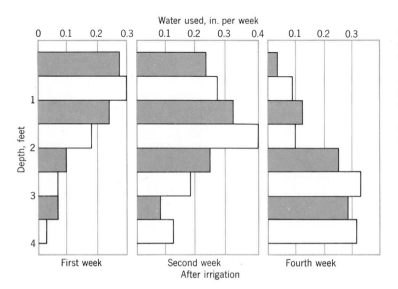

Fig. 4-15 Pattern of water used from soil by sugar beets during a 4-week period following irrigation. Water in the upper soil layers was used first. Then water was removed from increasing depths with time since irrigation. (From Taylor, 1957.)

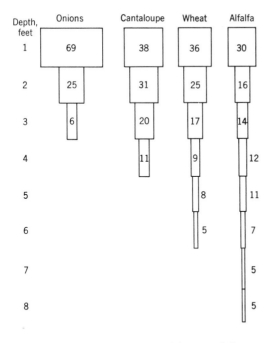

Fig. 4-16 Percent of water used from each foot (30-centimeter layer) of soil when produced with irrigation in Arizona. Onions are a shallow rooted annual crop that were grown in the winter and used a total of 17.5 inches (44 centimeters) of water. Alfalfa, by contrast, is a deep-rooted perennial crop that grew the entire year and used a total of 74.3 inches (186 centimeters) of water. (Data from *Arizona Agr. Exp. Sta. Tech. Bul.* 169, 1965.)

Studies of beech trees in Ohio showed that considerable rainfall moved down along the branches and trunk. About five times more water was estimated to reach the soil near the trunk as compared to soil further away (450 versus 90 centimeters annually). The soil near the trunk was also more leached and acid. The channeling of water from summer rains into the soil near the trunk results in deeper penetration of water into the soil and more efficient use of water by the trees.

Water Loss by Transpiration and Water Uptake

The amount of water transpired to produce 1 pound of dry matter was studied intensely by Briggs and Shantz shortly after the turn of this century. They grew plants in large galvanized pots that had tight-fitting covers with openings only for the stems of the plants. They measured the amount of water utilized by the plants by weighing the pots and harvested the plants to determine the quantity of dry matter produced. Some of their findings, presented in Table 4-1, show that plants commonly transpire 500 pounds or more of

Table 4-1
Water Transpired by Plants

Pounds of Water Transpired per Pound of Dry Plant Tissue Produced					
Crop	1911	1912	1913	Greatest Variation	Average
Wheat	468	394	496	102	452.7
Oats	615	423	617	194	551.7
Corn	368	280	399	119	349.0
Sorghum	298	237	296	61	277.0
Alfalfa	1068	657	834	411	853.0

From Briggs and Shantz, 1914.

water for each pound of dry matter produced. Furthermore, differences existed in the amount transpired between crops.

Because transpiration is simply the evaporation of moisture from plant surfaces, it is influenced by the same factors that affect the evaporation of water from any moist surface; exposure to direct sunlight, air temperature, humidity, wind movement, and atmospheric pressure are among the most important. Since these are variable from year to year, so should the amount of water transpired vary (Table 4-1). Perhaps you have reduced the transpiration of plant cuttings by placing them in a plastic bag to prevent wilting and enabling the cuttings to get established.

Loss of water by transpiration creates the major driving force for water uptake by roots of transpiring plants. Tensions created in leaves by loss of transpiration water are transmitted to the xylem (water-conducting vessels) of the stem and eventually to the roots. When the water tension in the roots is greater than the tension holding water in the soil, water moves into the roots. Tensions of 4 to 5 bars are common in the xylem of trees in damp forests and are sufficient to pull water to the top of tall trees. Tensions over 80 bars have been found in desert plants. As pointed out previously, however, this high tension may not enable desert plants to remove a significantly greater amount of water from some soils.

Consumptive Use—Amount of Water Used to Produce a Crop

Consumptive water use is that amount of water lost by evaporation from the soil and plants during the time a crop is grown. The quantity varies widely from less than 10 inches for a quickly maturing crop in a humid region to more than 70 inches for a long-season crop in an arid environment. Soybeans commonly require about 13 to 23 inches per season, which averages to about 0.14 to 0.18 inch per day. During periods of maximum use, the rate commonly rises to a fourth or third of an inch per day.

Evaporation of water requires energy, making the climatic environment the major factor in determining the amount of water used. This is shown in Table 4-1, where the amount of water transpired to produce 1 pound of dry matter varied greatly between years. In fact, studies made with an experimental setup like that shown in Fig. 4-17 reveal that a green, actively growing crop that completely covers the ground may lose water more rapidly than a free water surface (open pan), if the soil moisture tension is low.

On any given day a given amount of radiation is available to evaporate water. Therefore, many different kinds of crops require essentially the same amount of water when the cover is complete, they are green, and the soil moisture tension is the same. Great differences, however, exist in the total amount of water crops utilize because they have different maturation periods or they grow at different seasons of the year (Fig. 4-18).

Role of Water in Nutrient Absorption

Plant roots do not engulf and absorb the soil solution containing nutrients as animals drink water containing soluble material. Instead, the water enters the roots as pure water without regard to the intake of any of the materials dissolved in it. The entrance of dissolved substances is entirely a separate process. Nutrients dissolved in the soil solution move with it and so, when moisture flows by capillarity toward the roots to

Fig. 4-17 Experimental setup for studying the effect of various factors on consumptive water use. When rain starts to fall, the sheds automatically cover the plots; they return to their former position when rain ceases. This permits control of the amount of water added to the plots.

Fig. 4-18 The seasonal distribution of water use by alfalfa and corn. The arrows indicate harvest dates. (From "Agricultural Water Use," D. E. Angus, *Adv. in Agronomy* (1959), *II*: 20. Used by permission.)

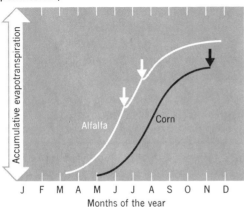

replace that which has been taken up by plants, a supply of nutrients is moved near the roots. Although this action takes place through short distances only, the net result in the course of a growing season may add materially to the nutrient supply of plants.

Soil Moisture Regimes

The potential available water supply for plants is governed mainly by the climate. The continental United States receives 30 inches of precipitation annually. About 5.5 inches or 20 percent runs off directly as overland flow to streams and 24.5 inches or 80 percent infiltrates the soil. An average of 3 inches percolates through the soil to the water table or underground water reservoir with the bulk of the water, 21.5 inches or 70 percent, being retained by the soil and later returned to the atmosphere by transpiration and evaporation (evapotranspiration).

The actual amount of water available to plants is affected by soil as well as by climate. For example, wet soils exist in deserts where soils have impermeable layers and receive runon water from surrouding higher land or springs. Gravelly soils in humid regions may be droughty because little water is retained. The soil property that expresses the order of soil moisture changes over time is the *soil moisture regime*. In addition to the supply of water for plants, the soil moisture regime expresses the availability of water for weathering and leaching and whether the root zone lacks oxygen because of water saturation.

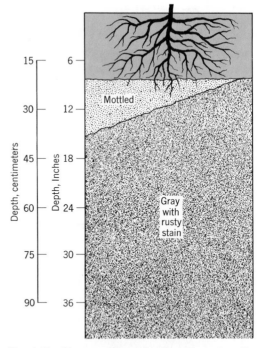

Fig. 4-19 Characteristics of soils with aquic soil moisture regime.

The Soil Moisture Control Section

Soil moisture regimes are based on moisture conditions in the *soil moisture control section*. The upper boundary of the moisture control section is the depth to which 1 inch (2.5 centimeters) of water will moisten dry soil (tension over 15 bars but not air dry) in 24 hours. The lower boundary is the depth of penetration of 3 inches (7.5 centimeters) of water in dry soil in 48 hours. These depths are exclusive of large cracks open at the soil surface. For many loamy soils the moisture control section is located between depths of 8 inches and 2 feet (20 to 60 centimeters).

Aquic Soil Moisture Regime

Soils with aquic (L. *aqua*, water) moisture regimes are wet and dissolved oxygen is virtually absent because the soil is satu-

rated. Very commonly the level of ground water will fluctuate with the season. In some cases, as in tidal marshes, the water table is at or close to the soil surface all the time. Drainage is needed to grow plants that require an aerated root zone. Subsoil colors in mineral soils are frequently gray, indicating reducing conditions, or mottled, indicating alternating reducing and oxidizing conditions (Fig. 4-19).

Aridic Soil Moisture Regime

The driest soils have aridic (L. *aridus*, dry) soil moisture regimes (also called *torric* meaning hot and dry). The moisture control section is dry in all parts more than half the growing season and is not moist (SMT less than 15 bars) in some part for as long as 90 consecutive days during a growing season in most years. Most soils with aridic moisture regimes are in arid or desert regions with widely spaced shrubs and cacti (Fig. 4-20). A crop cannot be matured without irrigation.

Some climatic data and the soil water balance typical of an aridic soil moisture regime are given in Fig. 4-21. Rainfall is low most months and the potential evapotranspiration is very high in summer relative to the precipitation. Therefore, little soil moisture recharge or water storage occurs and the little water that is stored is quickly used in late winter or early spring. The result is a lack of water for plant growth most of the time and a large water deficit. Many of the native plants have an unusual capacity to endure a high degree of desiccation without serious injury. Grazing is the dominant land use, and crop production requires irrigation.

Weathering occurs when soils are moist but there is little or no leaching. Soluble salts commonly accumulate in a zone, marking the average depth of moisture penetration. Some soils with aridic moisture regimes exist in the semiarid regions if they are shallow over rock or have very low infiltration rates.

Udic Soil Moisture Regime

Udic (L. *udus*, humid) means humid. Soils with udic moisture regimes have a moisture control section that is not dry in any part as much as 90 cumulative days in most years. Udic soil moisture regimes are common in soils of humid climates that have well-distributed rainfall. The amount of summer rainfall plus stored soil water is approximately equal to or exceeds the amount of evapotranspiration (see Fig. 4-21). Forests and tall grass prairies are typical vegetation on udic soils. If precipitation exceeds the amount of evapotranspiration each month, the moisture regime is called *perudic*.

In most years there is surplus water and leaching occurs. Data in Table 4.2 from a location in Ohio show that about one sixth

Fig. 4-20. Typical vegetation on soils with aridic soil moisture regime in Sonoran Desert, Ariz.

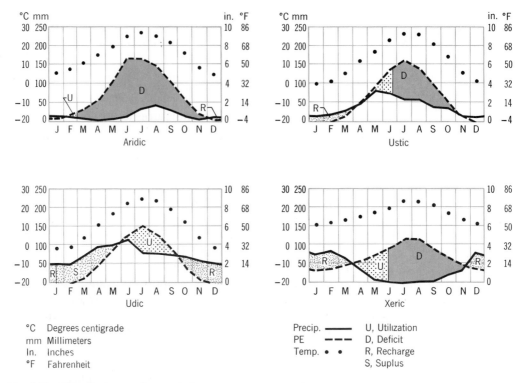

Fig. 4-21 Climatic data and water balance representing aridic, udic, ustic, and xeric soil moisture regimes. (From *Soil Taxonomy*, 1975.)

of the precipitation percolated through the soil. Significant amounts of plant nutrients were removed from the soil and the soils developed acidity. Leaching losses of calcium and magnesium in Ohio each year were more than the amount required to produce an average crop of wheat but less than the amount contained in a 3-ton crop of alfalfa. Sufficient leaching to produce soil acidity and low natural fertility is typical of forest soils with udic moisture regimes.

Short droughts occur occasionally. Irrigation is not widespread and is used for specialty crops and for special benefits as frost control. Udic soils are common from the east coast to the western boundaries of

Minnesota, Iowa, Missouri, Arkansas, and Louisiana.

Ustic Soil Moisture Regime

The ustic (L. *ustis*, burnt, implying dryness) soil moisture regime is intermediate between aridic and udic regimes. The concept is one of limited available soil moisture, but soil moisture is available for significant plant growth when other conditions are favorable for growth. Compared to the aridic regime, significantly more water storage occurs between fall and spring, and summer rainfall is greater (see Fig. 4-21). The water deficit is much less than in the

Table 4-2
Plant Nutrient Losses in Lysimeter Percolates on Kenne Silt Loam During a Period of High Precipitation (1950) and Low Precipitation (1953), as Compared with the 16-Year Average (1940–1955), by Practice

Practice and Period	Total Precipitation, in.	Percolation, in.	Nutrients Percolated Per Acre lb					
			Ca	Mg	K	N	Mn	S
Conservation								
1950 (high precipitation)	47.28	12.61	51.79	30.18	11.5	5.62	0.64	82.21
1953 (low precipitation)	28.20	3.54	5.88	0.74	2.77	0.64	0.13	8.23
16-year average (1940–1955)	37.16	6.29	29.35	17.61	9.74	3.47	0.30	31.42
Poor								
1950 (high precipitation)	47.28	13.40	30.81	18.69	20.56	2.87	0.94	51.18
1953 (low precipitation)	28.20	5.04	4.59	0.68	3.61	1.18	0.19	12.51
16-year average (1940–1955)	37.16	7.20	21.04	12.29	13.41	4.13	0.37	20.89

From Harrold and Dreibelbis, 1958.

aridic regime and much greater than in the udic regime. Ustic soil moisture regimes are common on the Great Plains where water is available for wheat production in winter, spring, and early summer in most years. Droughts are not uncommon. Sorghum is grown because it can interrupt growth when water is lacking and grow again if more rainfall occurs. Corn requires irrigation. Wheat and sorghum are the major dry land crops and grazing is important. Surplus water is rare and soils are unleached.

Fig. 4-22 Grape production on soils with xeric soil moisture regime in the Central Valley of California. Irrigation water in this case is being applied in March because the deep root zone was not completely recharged by winter rains.

Native vegetation was mainly mid, short, and bunch grasses.

Xeric Soil Moisture Regime

Soils in areas with Mediterranean climates typically have xeric (Gr. *xeros*, dry) soil moisture regimes. Winters are cool and moist and summers are hot and dry. The precipitation occurs in the cool months when evapotranspiration is low and is very effective for weathering and leaching (see Fig. 4-21). Surplus water may occur. Pastures and crops are well supplied with moisture in the winter and the landscape is green. A large water deficit occurs in summer and hilly grasslands are brown. Xeric soils of the Central Valley of California are used for a wide variety of crops including vines, fruits, nuts, vegetables, seeds, and agricultural crops (Fig. 4-22). Xeric soils of the Palouse in Washington and Oregon are used for winter wheat.

References

Angus, D. E., "Agricultural Water Use," in *Advances in Agronomy*, Vol. 11, Academic, New York, 1959, pp. 19–35.

Briggs, L. J., and H. L. Shantz, "Relative Water Requirements of Plants," *Agr. Res., 3*:1–65, 1914.

Erie, L. J., Orrin F. French, and Karl Harris, "Consumptive Use of Water by Crops in Arizona," *Arizona Agr. Exp. Sta. Tech. Bull. 169*, 1965.

Gardner, Walter H., "How Water Moves in the Soil," in *Crops and Soils*, pp. 7–12, November 1968.

Gersper, P. L., and N. Halowaychuk, "Effects of Stemflow Water on a Miami Soil Under a Beech Tree: I. Morphological and Physical Properties," *Soil Sci. Soc. Am. Proc., 34*:779–786, 1970.

Harrold, Lloyd L., and F. R. Dreibelbis, "Evaluation of Agricultural Hydrology by Monolith Lysimeters 1944–1955," *USDA Tech. Bull.*, 1179, 1958.

Heninger, R. L., *A Field Study of the Soil-Nutrient Status for Sugar Maple*, MS Thesis, Michigan Technical University, 1969.

Jacobs, H. S., and L. V. Withee, "Soil Moisture Tension: Presentation for Beginning Soil Students," *Agron. Jour., 57*:639–642, 1965.

Kunze, R. J., G. Uehara, and K. Graham, "Factors Important in the Calculation of Hydraulic Conductivity," *Soil Sci. Soc. Am. Proc., 32*:760–765, 1968.

Lehane, J. J., and W. J. Staple, "Influence of Soil Texture, Depth of Soil Moisture Storage, and Rainfall Distribution on Wheat Yields in Southwestern Saskatchewan," *Can. Jour. Soil Sci., 45*:207–219, 1965.

Musgrave, G. W., "How Much of the Soil?", in *Water*, USDA Yearbook, Washington, D.C., 1955, pp. 151–159.

Scholander, P. F., H. T. Hammel, Edda D. Bradstreet, and E. A. Hemmingsen, "Sap Pressure in Vascular Plants," *Science, 148*:339–346, 1965.

Sharma, M. L., and G. Uehara, "Influence of Soil Structure on Water Relations in Low Humic Latosols: I. Water Retention," *Soil Sci. Soc. Am. Proc., 32*:765–770, 1968.

Soil Survey Staff, *Soil Taxonomy*, Agr. Handbook 436, USDA, Washington, D.C., 1975.

Taylor, Sterling A., "Use of Moisture by Plants," in *Soil*, USDA Yearbook, Washington, D.C., 1957, pp. 61–66.

Voight, G. K., "Distribution of Rainfall Under Forest Stands," *Forest Science, 6*:2–10, 1960.

United States Department of Agriculture, *Water*, USDA Yearbook, Washington, D.C., 1955.

5
SOIL WATER MANAGEMENT

In a very real sense the story of water is the story of man. Civilization and cities emerged along the rivers of the Near East. The world's oldest known dam in Egypt is over 5000 years old and was used to store water for drinking and irrigation and perhaps to control flood waters. As the world's population increases and the need for food and fiber production increases, water management becomes more important. The rapidly increasing need of water in some urban areas is causing serious competition for available water between agriculture and industry.

Basically there are three approaches in water management to increase food and forest production: (1) conservation of natural precipitation in subhumid and arid regions; (2) removal of excess water from wet lands; and (3) adding water to supplement the amount of natural precipitation.

Water Conservation

Water conservation is important where large water deficits occur in soils with aridic, ustic, and xeric soil moisture regimes. Most of the wheat is produced on these soils. Techniques for water conservation aim at increasing the amount of water that enters the soil and making better use of this water.

Effect on Surface Soil Conditions on Infiltration

Infiltration is the movement of water into the soil surface. The nature of the pores and water content are the most important factors determining the amount of precipitation that infiltrates and the amount that runs off. High infiltration rates, therefore, not only increase the amount of water stored in the soil for plant use but also reduce flood threats and erosion resulting from runoff.

Raindrop impact on bare soil (Fig. 5-1) breaks up soil aggregates and causes the average pore size in the surface soil to decrease; this decreases infiltration. The presence of a vegetative cover or mulch that absorbs raindrop impact is effective in maintaining a high infiltration rate in a given situation. In Fig. 5-2 there is a comparison of infiltration rates for a meadow protected by grass and bare soil in a corn field in Illinois. When dry, both soils had high and similar infiltration rates, which decreased rapidly as the soil became wet. The infiltration rate for the bare corn field soil, however, decreased the most rapidly and remained below that of the meadow soil for at least 5 hours. Infiltration is also decreased by overgrazing, deforestation, and soil compaction resulting from traffic.

The presence of crop residues or other organic matter on the soil surface has an effect on infiltration similar to that of living plants. At Hays, Kansas, the infiltration during a storm was 1.16 inches where stubble of the previous wheat crop had been left on the land compared to 0.71 inches where

Fig. 5-1 Raindrops falling on bare soil. *(a)* A drop just before striking the soil. *(b)* Just after the drop has struck the soil. *(c)* Scattering of soil particles in all directions. The scattered particles form a compact layer that reduces infiltration. (Photo USDA.)

the stubble had been burned. Standing stubble is also effective for trapping snow. Efforts to trap snow are used extensively in

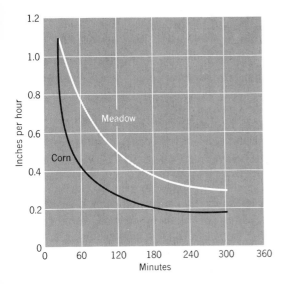

Fig. 5-2 Infiltration under bluegrass pasture or meadow and corn showing greater infiltration in the meadow where vegetation absorbed raindrop impact.

the newly developed Virgin Lands of Kazakhstan in the Soviet Union where natural precipitation is adequate for about one good crop of wheat every 4 years. Narrow strips of plants are grown perpendicular to the winds to trap snow.

Many farmers modify the soil surface with contour tillage and terraces that hold water on the land for a longer time to produce greater opportunity for infiltration (Fig. 5-3).

Effect of Internal Soil Properties on Infiltration

Water that infiltrates the soil surface moves downward. The downward movement is affected by nature of soil pores, aggregate stability, texture, depth to impermeable layers, and presence or absence of swelling clays. Soils with a high content of expanding or swelling clays, as in the Texas blacklands, develop large cracks in the dry season that permit water from intense storms to enter the dry soil without runoff. When these soils become wet in the rainy season, however, infiltration approaches zero and nearly all the rainfall runs off.

Many Great Plains soils have well-developed argillic horizons that limit the downward movement of water. It would seem that the use of deep tillage to disrupt these horizons would increase infiltration. Numerous studies were conducted from 1909 to 1916 at 12 different locations.

Fig. 5-3 Lister furrows on the contour hold water on the land to increase infiltration on this Oklahoma field. Runoff and erosion are reduced. (Photo USDA.)

Results at most locations showed that there was no increase in wheat yields as a result of deep tillage. Deep tillage may be effective in some cases, but this is uncommon. When one considers the extra cost, it is unlikely that any increases in yield would be able to pay the extra cost for deep tillage on the Great Plains.

Flooding Related to Infiltration Rates

Many storms have intensities that exceed infiltration rates of many soils. Although floods are frequently caused by intense storms of low duration, many floods are caused by low-intensity storms of long duration. Flooding is frequent in the spring when soils are wet and infiltration rates are low and when temperatures further reduce infiltration rates by lowering the viscosity of the water (Fig. 5-4). Flood water on many rivers is impounded in reservoirs and conserved for use as irrigation water and recreation.

Summer Fallowing for Water Conservation

Many soils have ustic soil moisture regimes in the wheat-growing areas of the western United States and Canada and receive too little water to produce a profitable crop every year. Summer fallowing is used to increase soil water storage so that a profitable crop can be produced every other year. After wheat harvest the land is left fallow (no crop is grown) to accumulate soil moisture. Weeds are controlled by cultivation to prevent water loss by transpiration (Fig. 5-5). The land in these areas has a characteristic pattern of alternating strips of bare fallow land and land used for wheat production. An explanation of how water storage occurs during fallowing in dry regions depends on a consideration of evaporation of water from the soil surface and water movement within the soil.

Any rain on dry soil immediately saturates the soil surface, and excess water moves downward by capillarity. The depth

Fig. 5-4 Flooding is common on many rivers in the spring when low infiltration rates exist because soils are at field capacity and low temperature reduces viscosity of the water.

Fig. 5-5 Fallowing on the Great Plains where the annual rainfall is about 15 inches (37 centimeters). The fallow strip is being cultivated to kill weeds to maximize water storage. The strip will be used the next year for wheat production.

of water penetration depends on texture and amount of water that infiltrates. Each inch of water that infiltrates the soil will wet about 1 foot of dry sandy loam soil to field capacity. Water will evaporate from the soil surface after the rain and some water from below in moist soil will migrate upward as capillary flow of liquid water to keep the surface moist. This phase of soil drying is characterized by rapid water loss and is represented by the horizontal part of the curve shown in Fig. 5-6. As water below the surface migrates upward and is lost by evaporation, soil moisture tension increases and hydraulic conductivity decreases rapidly. The loss of water bridges between soil particles brings the upward movement of water by capillarity to a near halt, and the soil surface dries. A sharp boundary is created between the dry surface soil layer and moist soil below similar to the boundary shown in Fig. 4-9. Drying of the soil surface causes a sharp decline in water loss from the soil (Fig. 5-6).

Once the surface soil dries, water can move from the underlying moist soil to the dry soil surface only in the vapor phase. This vapor movement is so slow as to be largely discounted (unless cracks exist). This produces a "capping" feature, which has great significance for water conservation. Once the surface of the soil has become dry after a rain (with or without cultivation), the water in the underlying moist layers is largely protected from loss by evaporation when the time considered is a few weeks or months. If another rain

Fig. 5-6 Evaporation of water from a bare soil surface decreases sharply when the surface soil becomes dry, thus conserving the moisture remaining in the soil. (Adapted from Evans and Lemon, 1957.)

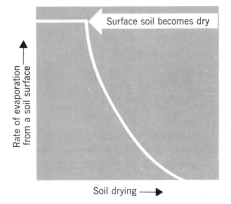

occurs the soil may be moistened to a greater depth but, when the surface of the soil again becomes dry, the water is again "trapped" in the soil. Repetition of this sequence of events during the fallow year progressively increases the amount of moisture stored in the soil. Many studies have shown a close correlation between water stored at planting time and grain yield.

The fallowing system is not 100 percent efficient because the self-mulching effect does not work perfectly and some runoff and evaporation occur. A good estimate is that about 25 percent of the rainfall during the fallow period will become stored in the soil for use in crop production. This extra quantity of water, however, has a great effect on yields (Table 5-1). Yields over 20 or 40 bushels were significantly increased by summer fallowing. Fallowing was also more effective at Pendleton, where maximum rainfall occurs in winter (xeric soil moisture regime) as compared to Akron and Hays, where maximum rainfall occurs in the summer (ustic soil moisture regime).

Stored soil moisture has been shown to be as effective as precipitation during the growing season for wheat production (Fig. 5-7).

Effect of Fertilizers on Water Use Efficiency

Plants growing in a medium containing relatively small quantities of nutrients appear to grow slower and transpire more water per pound of plant tissue produced than those growing in a medium containing an abundance of plant-nutrient materials. Since it has been shown that water loss is mainly dependent on the environment, any management practice that increases the rate of plant growth will tend to result in more dry matter produced per pound of water used. The data in Table 5-2 show that the yield of oats was increased from 2.4 to 4.0 bushels for each inch of water utilized by the addition of fertilizer. In the humid regions it has been commonly observed that fertilized crops are more drought resistant.

Table 5-1

Percentage distribution of wheat yield categories at two locations in the Great Plains and one location in the Columbia River Basin

Yield Category per Acre	Wheat after Wheat %			Wheat on Fallowed Land %		
	Akron, Colo.	Hays, Kans.	Pendle- ton, Oreg.	Akron, Colo.	Hays, Kans.	Pendle- ton, Oreg.
Under 5 bushels	44	33	0	15	17	0
Under 20 bushels	88	62	100	61	36	0
Over 20 bushels	12	38	0	39	64	100
Over 40 bushels	0	5	0	5	12	83

From Mathews, O. R., "The Place of Summer Fallow in the Agriculture of the Western States," USDA Circular 886, 1951.

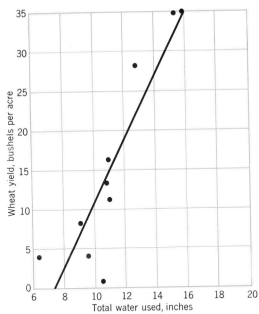

Fig. 5-7 Wheat yield and moisture use (stored soil water plus precipitation) at Edgely, N.D. At least 7 inches (17 centimeters) of water was needed for some grain production, and each additional inch (2.5 centimeters) of water used increased grain yield 4 bushels per acre. (Cole and Mathews, 1940.)

Table 5-2

Effect of Nitrogen Fertilizer on Water Utilization

| Year | Bushels of Oats per Inch of Water Utilized | |
	Low Nitrogen	High Nitrogen
1949	2.1	4.4
1950	2.7	3.7
Average	2.4	4.0

From Hanks and Tanner, 1952.

This may be explained on the basis that increased top growth results in increased root growth and penetration so that the total water consumed is greater (Fig. 5-8).

Fertilizer used in a subhumid region may occasionally decrease yields. If fertilizer causes a crop to grow faster early in the season, the greater leaf area at an earlier date results in greater loss of water by trans-

Fig. 5-8 Fertilizer increased corn yield nearly fourfold on the Cisne soil of southern Illinois. Fertilizer use resulted in a deeper rooting depth and greater use of soil water (and nutrients). (Photo courtesy Dr. Joe Fehrenbacher, University of Illinois.)

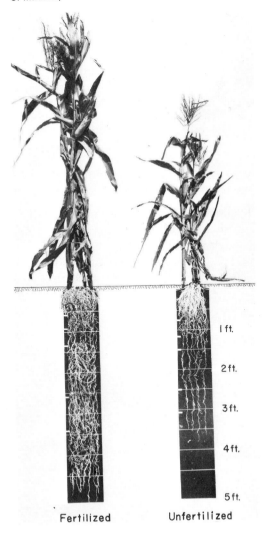

Fertilized   Unfertilized

piration. If no rains occur or no irrigation water is applied, plants could run out of water near harvest and a serious reduction in grain yield could occur. Work in Kansas has shown this to be the case for grain sorghum. Forage, when fertilized, does not depend on a supply of water to the end of the growing period, and can be produced, the same as in the humid region, when fertilized, with less water per pound. The forage will stop growing when the water is exhausted after having used the water "efficiently" as long as it lasted.

This concept may have far-reaching consequences in the future for crop production. Allocation of water to those situations resulting in the most efficient use of the water, in terms of crop yield per unit of water consumed, may be required if supplies become sufficiently limited.

Asphalt Barriers to Increase Retention of Available Water

Deep sand soils retain little water for plant growth and allow lots of water to percolate through the soil profile, resulting in inefficient use of natural precipitation. Michigan State University scientists theorized that installing a subsurface barrier would increase the retention of available water for plant growth. Asphalt barriers were installed in soils at a depth of about 2 feet with the experimental machine (Fig. 5-9). A close-up view of the barrier is shown in Fig. 5-10.

The barrier holds up rain or irrigation water and, in effect, creates a free water surface with soil moisture tension approaching zero at the immediate top of the barrier. Water films from water underneath the barrier are broken by the barrier and cannot exert a downward pull to cause water to move downward from soil above the barrier. Thus, at the barrier surface, the soil is saturated after the soil has been thoroughly wetted and excess water drains off laterally from the barrier. Above the barrier the water content of the soil is a function of height above the barrier (Fig. 5-11). The effect has been to double the available water retention capacity of the soil above the barrier. This means that in the spring of the growing season the barriered soils contain two times more water in the soil above the barrier. It also means that with each significant rain, such as 1 inch or more, more water will be retained throughout the growing season. The effect on the

Fig. 5-9 Installation of asphalt barrier 2 feet (60 centimeters) under the soil surface. (Photo Michigan Agr. Exp. Sta.)

Fig. 5-10 Close up view of exposed asphalt layer. (Photo Michigan Agr. Exp. Sta.)

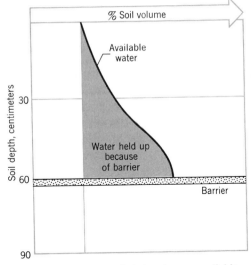

Fig. 5-11 Effect of asphalt barrier on *available* water retention capacity of sand soil. Available water retention capacity was about doubled in the soil above the barrier.

yields of vegetables has been sufficient to pay for the cost of the barrier in about 3 years. The life of the barrier is expected to be at least 15 years.

Soil Drainage

About 1200 A.D. farmers in the Netherlands became engaged in an interesting water control problem. Small patches of fertile soil affected by tide and flood waters were enclosed by dikes and drained. Windmills provided the energy to lift gravitational water from drainage ditches into higher canals for return to the sea. Now, 800 years later, about one third of the Netherlands is protected by dikes and kept dry by pumps and canals. The use of drainage systems to remove water from soil is highly important throughout the world where the water table is close enough to the surface so that saturated soil occurs within the root zone of plants, foundations of engineering structures, or septic tank drain fields. We will discuss the principles underlying these problems in this section.

Properties and Distribution of Aquic Soils

Aquic soils have aquic soil moisture regimes and develop under the influence of water saturation. As a result, aquic soils have dark-colored surface horizons high in organic matter content, are commonly on level land with minimal erosion hazard, and are frequently fine textured with high native fertility (see Fig. 4-19). Aquic soils comprise some of the best agricultural soils after drainage.

Small areas of wet or aquic soils are widely distributed locally. Extensive areas occur on the coastal plain of the southeastern United States because of high precipitation and elevation near sea level. Other major areas include the Mississippi River Valley, the lake plains in the Midwest, the recently glaciated nearly level plains of northern Iowa and southern Minnesota, and the Red River Valley between Minnesota and North Dakota.

Effect of the Water Table on Air and Water Content of Soil

Soil is water saturated at the surface of the water table and below. Water migrates upward by capillarity from the water table and creates a capillary fringe where the water content is greater than field capacity (Fig. 5-12). Large pores pull water up only a short distance and small pores pull water to a greater distance above the water table. This causes a decreasing water content and increasing air content in the capillary fringe with increasing distance above the water table (Fig. 5-12).

The thickness of the capillary fringe varies from a few inches or less in sandy and gravelly soils to a maximum of about 10 feet

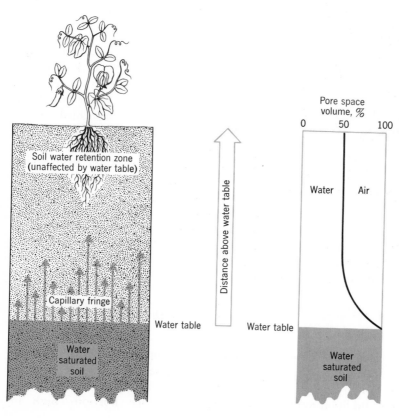

Fig. 5-12 Generalized relationship of water table to soil moisture zones and the air and water content of soil above the water table. Drawn to represent field capacity in soil above the capillary fringe and where root zone is unaffected by water table.

(3 meters) in fine-textured soils. For all practical purposes the upward movement of water at distances greater than 5 feet above the water table is so slow that roots must be within 5 feet or less of the water table to use a significant amount of water supplied by the water table. This means that most plants, even in humid regions, do not use water from the water table. In most cases the soil above the capillary fringe will have a water content governed by the balance between infiltration and the evapo-transpiration and percolation. Soils with aridic moisture regimes will have a permanently dry zone between the root zone and the water table.

Benefits of Soil Drainage

Root tips are regions of rapid cell division and elongation; they have a high oxygen requirement. Typically, roots of most crop plants do not penetrate water-saturated soil because of oxygen deficiency. The response of roots to the depth of the water table is shown in Fig. 5-13. The major purpose of

drainage in agriculture and forestry is lowering the water table to increase the depth of rooting.

Some of the most phenomenal increases in growth resulting from drainage of aquic soils occur in forests. Drainage on the Atlantic Coastal Plain increased the growth of pine from 80 percent to 1300 percent (Table 5-3). There are few, if any, management practices as effective as drainage for increasing tree growth. Additional benefits of drainage in forests are easier logging, less soil disturbance during logging, and easier site preparation for the next crop.

Drainage of wet soils also increases the length of the growing season in regions where low soil temperature and frost hazard restrict plant growth. Drainage lowers the water content in the spring, causing soils to warm more rapidly. The higher early soil temperature is associated with drier soil that can be tilled earlier so that crops can be established sooner. Seed germination is more rapid and roots grow faster. The net result is greater potential for plant growth.

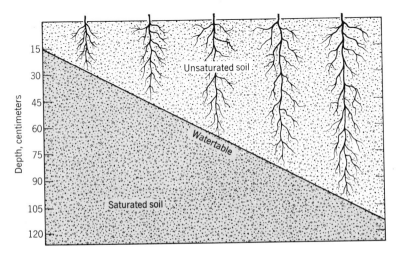

Fig. 5-13 The effect of depth to water table on the length of millet roots. Roots did not enter the saturated soil. (Based on data from Williamson et al., 1969.)

Table 5-3
Effect of Drainage on Mean Annual Growth of Pine on Aquic Soils of the Atlantic Coastal Plain

Species	Ages	Mean Annual Growth[a]		Percent Increase Over Undrained
		Drained	Undrained	
Planted loblolly	0–17	17.9	1.28	1,298
Natural pond	0–22	0.74	0.41	80
Natural slash	19–22	9.0	4.9	84
Planted slash	0–5	14.6	5.7	156
Planted slash	0–5	13.3	5.7	133
Planted loblolly	0–5	7.6	1.3	585
Planted loblolly	0–5	4.9	1.3	277
Planted loblolly	0–13	4.3	0.54	696

Adapted from Terry and Hughes, 1975.
[c] Cubic meters per hectare.

Surface Drainage

Surface drainage is the collection and removal of water from the surface of the soil. Two conditions favoring the use of surface drainage are low areas that receive water from surrounding higher land and impermeable soils that have insufficient capacity to dispose of the excess water by movement downward through the soil profile. Surface drainage is a good choice on the Sharkey clay soils along the lower Mississippi River. The soil is very slowly permeable and receives water from surrounding higher land. Sugar cane is widely grown and is sensitive to poor aeration. The ridges on which the sugar cane is planted also provide improved soil aeration for roots.

It is difficult to irrigate without applying excess water in many instances. Surface drainage ditches are used to dispose of excess water to prevent saturation of soil on the low end of irrigated fields. Surface drainage is also widely used along highways and in urban areas for water control. Rapid removal of surface water from low areas of golf courses is essential to permit golfing soon after a rain.

Subsurface Drainage

Ditches can be quickly and inexpensively made to remove gravitational water. Drainage ditches, however, require periodic cleaning and are inconvenient for the use of machinery. There are also many situations where ditches are not satisfactory, as in the case of water removal around the walls of the basement of a building (Fig. 5-14). The drainage tiles shown in Fig. 5-14 are made of fired clay and are laid side by side with a small crack between adjacent tiles. When the soil surrounding the tile is saturated with water, water seeps into the tile laid on a grade and the water eventually reaches an outlet where it is disposed.

Drain tile are installed in fields with trenching machines (Fig. 5-15). Perforated

Drainage tile

Fig. 5-14 Clay tile lain at the base of the foundation of the walls of a basement. Cracks between adjacent tile are covered with a durable black material to keep soil from entering the tile.

plastic tubing for small diameter laterals is less expensive than ceramic tile and is gaining in popularity. The plastic tubing is desired for draining organic (muck) soils where the low supporting capacity of the soil results in unequal settling and misalignment of short clay tile sections.

Unless some unusual condition prevails, tile should be laid at least $2\frac{1}{2}$ feet deep. The depth of the outlet has some bearing on the depth at which the drains may be placed and still provide a satisfactory fall. A fall of at least $\frac{1}{10}$ foot per 100 feet is considered necessary, and more is desirable. The water

Fig. 5-15 Installation of 4-inch drainage tile in a field. (Photo USDA.)

level is lowered by the lines (Fig. 5-16). In order, therefore, to have a sufficiently low water table between tile lines to permit crops to develop an adequate root system, it is essential that the tile be $2\frac{1}{2}$ feet deep or more.

The freedom with which water moves through the soil determines the distance apart that drains may be placed and still afford adequate drainage. If the soil through which the water must percolate to reach the tile contains a high percentage of clay and the structural condition is such as to make water movement slow, the tile lines should not be more than 70 feet apart. In more permeable soil a spacing of 80 feet is permissible, and in soil through which water drains rapidly the drains may be placed at 100 feet intervals. Spacing in sandy loam soils with rapid permeability should vary from 100 to 300 feet.

Drainage in the Soil of Container-Grown Plants

Most of us grow plants in containers that we use to decorate rooms. One of the most common problems of growing indoor plants is poor soil aeration caused by over-watering. Perhaps you wonder how this could be true, since most flower pots have a large drainage hole in the bottom. We can

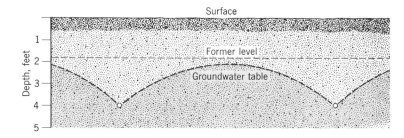

Fig. 5-16 The effect of tile lines in lowering the water table or groundwater level. The benefit of the drainage is first evident immediately over the tile lines and gradually spreads to the soil area between them.

use our knowledge of changes in air and water content of soil above the water table to help us understand this problem.

The smallest soil pores can be completely water filled from the bottom to the top of a flower pot (Fig. 5-17). If all soil pores were very small the soil in a flower pot would be completely water saturated after watering and drainage. A good soil for container-grown plants has many large pores that cannot hold up capillary water or can hold capillary water up only a very short distance. After wetting and drainage, the large

pores are air filled. Since soils are usually composed of pores of varying size, there will be an increase in soil moisture tension and air content from the bottom to the top (Fig. 5-17). Roots in tall containers will also have a better aerated soil environment than roots growing in short containers. Since there is no capillary pull on soil at the very bottom of the container, the soil will be water saturated at the bottom.

Special attention must be given to the soil mix used for container-grown plants. A mix of half fine sand and half sphagnum peat

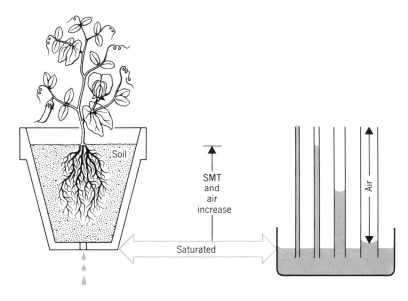

Fig. 5-17 Illustration of conditions in a flower pot after a thorough wetting. At the bottom of the pot, the soil is saturated. Soil moisture tension and air content increase (water content decreases) with increasing height in the soil.

moss by volume is recommended by the University of California to produce a mix with excellent physical properties. If you want to prepare a mix using a garden or lawn soil, add only a small amount, perhaps 10 percent soil by volume, to some medium or fine sand. When equal amounts of soil and sand are mixed together, the small particles fill the spaces between the sand particles, producing a dense mix with few large aeration pores.

Stratified soils interfere with water movement. Coarse material like gravel should not be placed in the bottom of holes dug for transplanting trees and shrubs because a water-saturated zone will be formed above the gravel layer in wet seasons. Blending of the soil in the container with the surrounding soil will encourage better water movement and root extension from the container-grown soil into the surrounding soil.

Drainage of Septic Tank Effluent Through Soils

Farmers who reside beyond the limits of municipal sewer lines have used septic tank sewage disposal systems for many years. The recent rapid expansion of rural residential areas and the development of summer homes near lakes and rivers has greatly increased the role of the soil in waste disposal. The sewage enters a septic tank where solid material is digested and the liquid effluent flows out of the top of the septic tank and into the tile lines of a filter field (Fig. 5-18). The seepage lines are laid in a bed of gravel from which the effluent seeps into and drains through the soil.

A major factor influencing the suitability of the soil for filter field use is permeability. Soil permeability of less than 1 inch per hour (or a percolation rate of more than 60 minutes per inch) is too slow. Sand may have percolation rates in excess of 10 inches

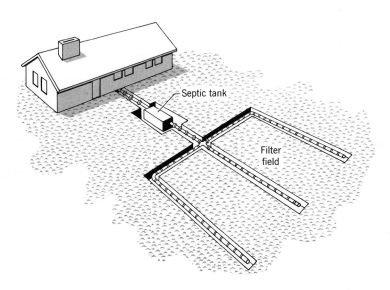

Fig. 5-18 The layout for a septic tank and filter field.

Septic tank

Filter field

per hour and may be too permeable. When sewage effluent moves too rapidly through soil, there is danger that shallow water supplies may become contaminated.

Septic tank filter fields should never be installed in poorly drained soils because the soil will become saturated with water and there is a danger that disease organisms in the effluent will contaminate work or play areas. The basements of homes constructed on poorly drained soils are likely to become flooded.

Irrigation

Irrigation is an ancient agricultural practice that was used 7000 years ago in Mesopotamia. Other ancient notable irrigation systems were located in Egypt, China, Mexico, and Peru. Today, about 11 percent of the world's cropland is irrigated. Some of the world's densest populations are supported by producing crops on irrigated land, as in the United Arab Republic (Egypt), where 100 percent of the cropland is irrigated. Other countries that have large percentages of irrigated cropland include Peru, 75 percent; Japan, 60 percent; Iraq, 45 percent; and Mexico, 41 percent. About 8 percent or 38 million acres of cropland are irrigated in the Unites States. "Nearly two thirds of the world's population lives in diet deficient countries having less than half of the world's arable land but with three quarters of the irrigated land."[1] Thus, we can see the great importance of irrigation in the world today and for the future.

[1] The White House, The World Food Problem, A Report of the President's Science Advisory Committee, May 1967, p. 442.

Water Sources

Most of the irrigation water is surface water resulting from rain and melting snow. Many rivers in the world have their headwaters in mountains and flow through arid or semiarid regions. Examples include the Indus River that starts in the Himalaya Mountains and the numerous rivers on the western slope of the Andes Mountains that flow through the desert of Peru to the Pacific Ocean. Much of the water in these rivers comes from the melting of snow in the high mountains. In fact, the extent of the snow pack is measured to obtain information on streamflow and the amount of water that will be available the next season for irrigation (see Fig. 5-19).

Water from melting snow and rainfall is collected in reservoirs and released as needed for irrigation. Rivers serve as avenues for the transportation of the water. Eventually, the water arrives at farms (Fig. 5-20). Many farmers have their own source of irrigation water because they use well water. About 20 percent of the water currently used in the United States comes from irrigation wells.

Selecting Land for Irrigation

In choosing land for irrigation, a careful examination should be made of the soil to determine (1) texture of soil to depth of several feet; (2) presence of an impermeable stratum or of gravel within a depth of 5 or 6 feet; (3) accumulation of soluble salts in injurious quantities; (4) slope and evenness of soil surface; and (5) behavior of the soil under irrigation. A desirable soil is readily permeable to water and yet is moisture retentive. Infiltration rates should be in the range of 0.1 to 3 inches per hour. It is well if

Fig. 5-19 Measuring the snow pack near Cooke City, Mont. to obtain information for predicting streamflow and the quantity of water available for irrigation and other uses. (Photo USDA-SCS by P. E. Farnes.)

the soil will absorb sufficient moisture in 24 hours to wet it to a depth of 2 or 3 feet. Some soils are so slowly permeable that they become wet to a depth of only 1 foot or less in 24 hours. Some soils are so coarse textures that water passes rapidly below the reach of plant roots, and little available moisture is retained. On the other hand, where fine- or medium-textured soil materials are underlain by coarser sands or gravels at a depth of several feet, the limited capacity of the coarser material to "pull" water downward by capillarity results in greater retention of water in the upper layer than if the soil was medium or fine textured throughout. The importance of soil properties in placing irrigated soils into land capability classes is illustrated in Table 5-4.

The land surface should be comparatively smooth, because the cost of leveling land is high. (Fig. 5-21). A uniform slope of 10 to 20 feet to the mile is desirable, although much steeper slopes are in use. Land cut up by ravines, gullies, or buffalo wallows or covered with sand dunes and hummocks should be avoided if possible.

Fig. 5-20 Concrete-lined canal bringing irrigation water to a farm in Utah. Although much of the water originates from the melting of the snow pack, it is collected in a reservoir and then discharged into canals for delivery to farms as needed. (Photo USDA-SCS.)

Table 5-4
Guide For Placing Irrigated Soils Into Land Capability Classes (The Guide Shows Soil Properties of Importance in Irrigation)

| Land Capability Class | Surface Texture | Coarse Fragments in Surface, % | Available Water-Holding Capacity | | Soil Permeability | Effective Depth, in. | Salinity or Sodium Hazard |
			Surface Foot, in.	Soil Profile to 60 Inches, in.			
I	Sandy loam to clay loam or silty clay loam	<15	>1.5	>7.5	Slow to medium	>40	None
II	Loamy sand to silty clay loam or clay loam	15–35	1.0–1.5	5.0–7.5	Slow to rapid	30–40	Slight
III	Sand to clay	>35	0.75–1.0	3.75–5.0	Slow to rapid	20–30	Moderate
IV	Sand to clay	No limits	<0.75	<3.75	Slow to rapid	<20	Severe
V, VI, VII VIII	Sand to clay	No limits	No limits	No limits	Slow to rapid	<20	No limits

Adapted from Soil Conservation Service, USDA, Phoenix, Arizona State Guide by Donald Post.

Fig. 5-21 Planned land with proper grade to give uniform distribution of water down the rows. Gated pipe is being used to distribute the water.

Methods of Applying Irrigation Water

Choice of the various methods of applying irrigation water is influenced by a consideration of (1) seasonal rainfall, (2) slope and general nature of the soil surface, (3) supply of water and how it is delivered, (4) crop rotation, and (5) permeability to water of the soil and subsoil. The methods of distributing water can be classified as surface, subsurface, sprinkler, and drip or trickle. A brief discussion of each of these methods is given.

Surface Irrigation. Surface irrigation distributes water down rows or into basins and similar areas that are surrounded by ridges or dikes. Flooding of basins and similar areas is used for pastures, orchards, and the like. Crops commonly irrigated by furrow irrigation include row crops such as potatoes, sugar beets, corn, grain sorghum, cotton, vegetables, and fruit trees. Furrows are made across the field, leading down the slope. Water is let into the upper end of the furrow from a "head ditch" or pipeline running across the end of the field. Siphon tubes are commonly used to transfer the water from the head ditch into the furrows (Fig. 5-22). Gated pipe is also used extensively.

Subirrigation. Subirrigation is irrigation by water movement upward from a free water surface some distance below the soil surface. In arid regions where almost all of the water used to grow crops is from irrigation, subirrigation would cause serious salt accumulation problems in the upper part of the soil. Subirrigation works best where natural rainfall removes any salts that may accumulate. In many poorly drained areas, as in the Sand Hills of Nebraska, there are areas where subirrigation is a natural occurrence. Artificial subirrigation is practiced in the Netherlands where tile drainage systems in polders are used in wet seasons for drainage and in periods of drought for subirrigation. Subirrigation, as compared to other systems, is inefficient in use of water and is adapted only for special situations.

Sprinkler Irrigation. Everyone is familiar with the sprinklers used to water or

Fig. 5-22 Use of siphon tubes to transfer water from head ditch into furrows. (Photo USDA.)

irrigate lawns. Sprinkler systems are versatile and have special advantages where high infiltration rates or topography prevents proper leveling of the land for surface distribution of water. The rate of application can also be carefully controlled. The portable nature of many sprinkler systems makes them ideally suited for use where irrigation water is used to supplement the natural rainfall. When connected to a soil moisture measuring apparatus, sprinkler systems can be made automatic.

Large self-propelled sprinkler systems have been developed in recent years. A self-propelled system that irrigates most of the land in a quarter-section (160 acres) is shown in Fig. 5-23. If you have flown over the western part of the United States you have probably seen large green irrigated areas produced by self-propelled sprinklers.

Sprinkler irrigation modifies the plant environment by completely wetting the soil and leaves. Reductions in relative humidity and temperature reduce water stress in plants. A very small amount of water applied with sprinklers has been observed to reduce midday surface soil temperature as much as 22°F (see Fig. 5-24). The high specific heat of water makes sprinkling an effective means to reduce frost hazard. Sprinkler irrigation has been used for frost protection in diverse situations such as strawberries in Michigan and grapes in California.

Drip Irrigation. Drip irrigation is the frequent or daily application of water drops to localized areas of the soil. Plastic hoses about $\frac{1}{2}$ inch in diameter with emitters are distributed down rows or around trees (Fig. 5-25). Only a small amount of the root zone

Fig. 5-23 Circular patterns made by self-propelled sprinklers in Holt County, Neb. A unit is operating in the front-left quarter-section. (Photo USDA-SCS.)

is wetted, but roots in the localized moistened areas absorb water rapidly with the low SMT. A major advantage of drip irrigation is the large reduction in water used. Other advantages include a more uniform SMT and water stress within plants during the growing season and adaptability to very steep land where other methods are unsuited.

Rate and Frequency of Irrigation

An ideal application of water would be a sufficient quantity to bring the soil, to the depth of the root zone of the crop, up to its field capacity. More water may result in the waterlogging of a portion of the subsoil or in the loss of water by drainage. If much water percolates below the root zone, it may accumulate under low areas, thus raising the water table unless suitable drainage

facilities are provided. On the other hand, unless enough water is applied to result in appreciable drainage, it is difficult to remove excess salts from soils in which there is a tendency for salt to accumulate.

A commonly accepted generalization is that it is time to irrigate when 60 percent of the available water in the root zone has been used. Knowing the daily evapotranspiration rate and available soil water storage, the timing of irrigation can be calculated. Tensiometers are used to measure SMT at various depths in the root zone. Wilting symptoms are also used but in most cases, crops should be irrigated before marked wilting occurs. Other factors affecting timing include affect of SMT on germination, plant development, and maturity. Corn is sensitive to high moisture stress at silking. Low moisture stress increases the rate of growth during periods when plants are

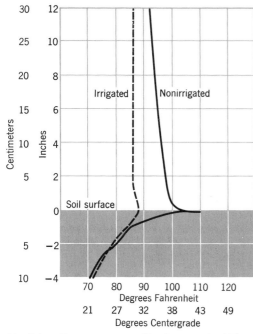

Fig. 5-24 Temperature profile above and within irrigated and nonirrigated muck soil at 2:00 P.M., June 9, 1964 at East Lansing, Mich. (From Brink and Carolus, 1965.)

growing the fastest. High soil moisture tension late in the season favors the development of flavors in fruits and the opening of cotton bolls.

Water Quality

Careful attention should be given to the nature of the water to be used before an irrigation system is constructed. Sometimes the available water carries so high a concentration of soluble salts that its use for irrigation is not advisable, particularly on land that already contains a considerable concentration of soluble salt. Sodium salts in the water are much more objectionable than are salts of calcium and magnesium because of the tendency for sodium to cause deflocculation of the colloidal fraction of the soil and to develop an undesirable structure. Some waters also contain sufficient boron to be toxic to plants with continued use. To make an estimate of the quality of irrigation water, the measure-

Fig. 5-25 A section of plastic tubing and emitter. Rate of water application can be adjusted from less than 1 gallon to about 4 gallons per hour per emitter.

ment of the following three characteristics is essential: (1) conductivity or total concentration of salts, (2) the concentration of boron, and (3) the sodium adsorption ratio (SAR). The SAR is calculated as follows.

$$SAR = \frac{Na^+}{\sqrt{\frac{(Ca^{2+} + Mg^{2+})}{2}}}$$

where the concentrations of the cations are given in milli-equivalents per liter. The diagram for the classification of irrigation waters is shown in Fig. 5-26 and is based on the electrical conductivity in micromhos per centimeter and the SAR.

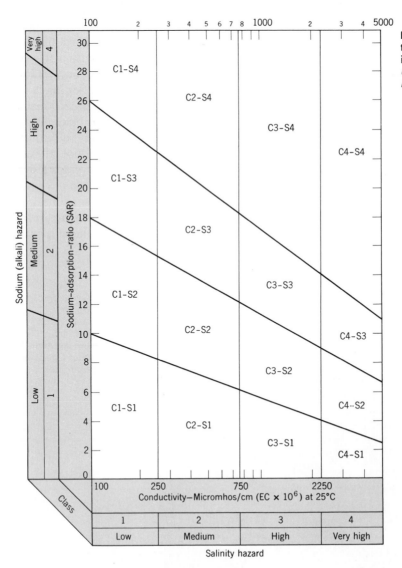

Fig. 5-26 Diagram for the classification of irrigation waters. (From *USDA Agriculture Handbook*, 60, 1954.)

Conductivity and Water Quality. Low-salinity water ($C1$) can be used for irrigation with most crops with little likelihood that soil salinity will develop. Some leaching is required, but this occurs under normal irrigation practices except in soils with extremely low permeability.

Medium-salinity water ($C2$) can be used if a moderate amount of leaching occurs. Plants with moderate salt tolerance can be grown in most cases without special practices for salinity control.

High-salinity water ($C3$) cannot be used on soils with restricted drainage. Even with adequate drainage, special management for salinity control may be required and plants with good salt tolerance should be selected.

Very high-salinity water ($C4$) is not suitable for irrigation under ordinary conditions, but may be used occasionally under very special circumstances. The soils must be permeable, drainage must be adequate, irrigation water must be applied in excess to provide considerable leaching, and very salt-tolerant crops should be selected.

SAR and Water Quality. The classification of irrigation waters with respect to SAR is based primarily on the effect of exchangeable sodium on the physical condition of the soil. Sodium-sensitive plants may, however, suffer injury as a result of sodium accumulation in plant tissues when exchangeable sodium values are lower than those effective in causing deterioration of the physical condition of the soil.

Low-sodium water ($S1$) can be used for irrigation on almost all soils with little danger of the development of harmful levels of exchangeable sodium. However, sodium-sensitive crops such as stone-fruit trees and avocados may accumulate injurious concentrations of sodium.

Medium-sodium water ($S2$) will present an appreciable sodium hazard in fine-textured soils having high cation-exchange capacity, especially under low-leaching conditions, unless gypsum is present in the soil. This water may be used on coarse-textured or organic soils with good permeability.

High-sodium water ($S3$) may produce harmful levels of exchangeable sodium in most soils and will require special soil management, good drainage, high leaching, and organic matter additions. Gypsiferous (high-calcium sulfate) soils may not develop harmful levels of exchangeable sodium from such waters. Chemical amendments may be required for replacement of exchangeable sodium, except that amendments may not be feasible with waters of very high salinity.

Very high-sodium water ($S4$) is generally unsatisfactory for irrigation purposes except at low and perhaps medium salinity, where the solution of calcium from the soil or use of gypsum or other amendments may make the use of these waters feasible.

Sometimes the irrigation water may dissolve sufficient calcium from calcareous soils to decrease the sodium hazard appreciably, and this should be taken into account in the use of $C1$–$S3$ and $C1$–$S4$ waters. For calcareous soils with high pH values or for noncalcareous soils, the sodium status of waters in classes $C1$–$S3$, $C1$–$S4$, and $C2$–$S4$ may be improved by the addition of gypsum to the water. Similarly, it may be beneficial to add gypsum to the soil periodically when $C2$–$S3$ and $C3$–$S2$ waters are used.

Effect of Boron on Water Quality. Boron is essential to normal plant growth, but the amount needed is very small. Boron is very toxic to certain plants, and the concentration that will injure these sensitive plants is often approximately the amount required for normal growth of very tolerant plants. The occurrence of boron in toxic concentrations in certain irrigation waters makes it necessary to consider this element in assessing the water quality. Permissible limit of boron in several classes of irrigation water is given in Table 5-5 and the relative tolerance of some plants is given in Table 5-6.

In making an estimate of the quality of water, the effect of the salts on both the soil and the plant must be considered. Various factors such as drainage, soil texture, and kind of clay minerals present influence the effects on the soil. The ultimate effect on the plant is the result of these two and other factors operating simultaneously. Consequently, there can be no method of interpretation that is absolutely accurate under all conditions. The schemes for interpretation are accordingly based chiefly on experience.

Salt Accumulation and Plant Response

All natural waters contain varying amounts of dissolved salt, with most irrigation waters containing 0.5 to 5 tons of salt per acre foot.

Since 3 to 5 acre feet of water are commonly applied annually, the potential for salt accumulation is great. Assume that 1 acre foot of soil weighs 4 million pounds and that 20 tons of salt would produce minimal salinization (0.2 percent salt) of soil to a depth of 5 feet. Irrigation of soils in the absence of natural leaching results in loss of water by evapotranspiration and the deposition of salt in the soil. It is obvious that a saline soil could be produced in less than a decade. No wonder that the prevention of or control of salt in soils is the prime concern for the maintenance of permanent agriculture in irrigated regions. No wonder that the demise of some early civilizations, as in Mesopotamia, coincided with salt accumulation in irrigated lands.

Saline soils contain sufficient salt to impair plant growth. The salts are mainly chlorides, carbonates and sulfates of sodium, potassium, calcium, and magnesium. The best method for assessing soil salinity is measurement of the conductivity of the saturated soil extract. The procedure involves preparing a saturated soil paste by stirring, during the addition of distilled water, until a characteristic endpoint is reached. A suction filter is used to obtain a sufficient ·amount of the extract for

Table 5-5

Permissible Limits of Boron for Several Classes of Irrigation Waters in Parts per Million

Boron Class	Sensitive Crops	Semitolerant Crops	Tolerant Crops
1	<0.33	<0.67	<1.00
2	0.33 to 0.67	0.67 to 1.33	1.00 to 2.00
3	0.67 to 1.00	1.33 to 2.00	2.00 to 3.00
4	1.00 to 1.25	2.00 to 2.50	3.00 to 3.75
5	>1.25	>2.50	>3.75

From *Agriculture Handbook* 60, USDA, 1954.

Table 5-6
Relative Tolerance of Plants to Boron (In each group, the plants first named are considered as being more tolerant and the last named more sensitive)

Tolerant	Semitolerant	Sensitive
Athel (*Tamarix aphylla*)	Sunflower (native)	Pecan
Asparagus	Potato	Black walnut
Palm (*Phoenix canariensis*)	Acala cotton	Persian (English) walnut
Date palm (*P. dactylifera*)	Pima cotton	Jerusalem artichoke
Sugar beet	Tomato	Navy bean
Mangel	Sweetpea	American elm
Garden beet	Radish	Plum
Alfalfa	Field pea	Pear
Gladiolus	Ragged Robin rose	Apple
Broadbean	Olive	Grape (Sultanina and
Onion	Barley	Malaga)
Turnip	Wheat	Kadota fig
Cabbage	Corn	Persimmon
Lettuce	Milo	Cherry
Carrot	Oat	Peach
	Zinnia	Apricot
	Pumpkin	Thornless blackberry
	Bell pepper	Orange
	Sweet potato	Avocado
	Lima bean	Grapefuit
		Lemon

From *Agriculture Handbook* 60, USDA, 1954.

making the conductivity measurement.

The saturated extract method has two advantages over other methods. First, the conductivity measurement of the saturated extract is directly related to the field moisture range that is important to plants. In the field the moisture content of the soil fluctuates between a lower limit represented by the permanent-wilting percentage and the upper, wet end of the available range, which is approximately two times the wilting percentage. Measurements on soils indicate that over a considerable textural range the saturation precentage (*SP*) is approximately equal to four times the 15-atmosphere percentage (*FAP*) which, in turn, closely approximates the wilting percentage. The soluble-salt concentration in the saturation extract, therefore, tends to be about one half of the concentration of the soil solution at the upper end of the field-moisture range and about one fourth the concentration that the soil solution would have at the lower, dry end of the field-moisture range. The salt-dilution effect that occurs in fine-textured soils, because of their higher moisture retention, is thus automatically taken into account. For this reason, the conductivity of the saturation extract (*EC*$_e$) can be used directly for appraising the effect of soil salinity on plant growth.

The second advantage of the saturated extract method is that the differences in the effectiveness of different ions in different salts is accounted for. If the salt content was expressed as tons of salt per acre, there would be no account of the fact that different ions have different atomic (or molecular) weights and different solubilities or activities.

Salinity effects on plants as measured by the conductivity of the saturated extract as milliohms per centimeter (EC_e) at 25°C are as follows.

1. 0–1: salinity effects mostly negligible.
2. 2–4: yields of very sensitive crops may be restricted.
3. 4–8: yields of many crops restricted.
4. 8–16: only tolerant crops yield satisfactorily.
5. Over 16: only a few very tolerant crops yield satisfactorily.

The salt-tolerance of crops in Table 5-7 are arranged according to major crop divisions. Within each group, the crops are listed in the order of decreasing salt tolerance, but a difference of two or three places in a column may not be significant. EC_e values given represent the salinity level at which a 10-, 25-, and 50-percent decrease in yield may be expected as compared to yields on nonsaline soil under comparable growing conditions.

Many studies of the effects of salt on plants included a number of varieties. Significant varietal differences have been found for cotton, barley, and smooth brome, while the varietal differences were of no consequence for green beans, lettuce, onions, and carrots.

The effect of the salts on plants is mainly indirect; that is, the effect of the salt on the osmotic pressure of the soil water (osmotic potential) and the resultant reduced uptake of water by germinating seeds and roots. Adsorption of water molecules on ions pro-

Table 5-7
Salt Tolerance of Crops Expressed as the EC_e at 25°C for Yield Reductions of 10, 25, and 50 Percent as Compared to Growth on Normal Soils.

Crop	10%	25%	50%
Field crops			
Barley	12	16	18
Sugar beet	10	13	16
Cotton	10	12	16
Safflower	8	11	14
Wheat	7	10	14
Sorghum	6	9	12
Soybean	5.5	7	9
Sesbania	4	5.5	9
Rice (paddy)	5	6	8
Corn	5	6	7
Broadbean	3.5	4.5	6.5
Flax	3	4.5	6.5
Beans	1.5	2	3.5

Table 5-7—*cont.*

Crop	10%	25%	50%
Vegetable crops			
Beet	8	10	12
Spinach	5.5	7	8
Tomato	4	6.5	8
Broccoli	4	6	8
Cabbage	2.5	4	7
Potato	2.5	4	6
Corn	2.5	4	6
Sweet Potato	2.5	3.5	6
Lettuce	2	3	5
Bell pepper	2	3	5
Onion	2	3.5	4
Carrot	1.5	2.5	4
Beans	1.5	2	3.5
Forage Crops			
Bermuda grass	13	16	18
Tall wheat grass	11	15	18
Crested wheat grass	6	11	18
Tall fescue	7	10.5	14.5
Barley hay	8	11	13.5
Perennial rye	8	10	13
Harding grass	8	10	13
Birdsfoot trefoil	6	8	10
Beardless wild rye	4	7	11
Alfalfa	3	5	8
Orchard grass	2.5	4.5	8
Meadow foxtail	2	3.5	6.5
Clovers, alsike and red	2	2.5	4
Fruit crops			
Date palm	8		16
Pomegranate ⎫			
Fig ⎬	4.6		9
Olive ⎭			
Grape (Thompson)	4		8
Muskmelon	3.5		No data
Orange, grapefruit, lemon	2.5		5
Apple, pear	2.5		5
Plum, prune, peach, apricot, almond	2.5		5
Boysenberry, blackberry, raspberry	1.5–2.5		4
Avocado	2		4
Strawberry	1.5		3

Adapted from Agr. Information Bull. Nos. 283 and 292 and Western Fertilizer Handbook.

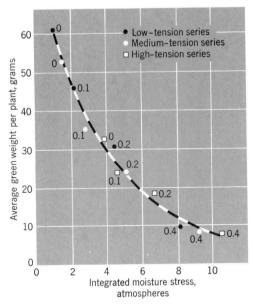

Fig. 5-27 Growth of bean plants as influenced by total soil moisture stress. Data shows the additive effects of SMT and osmotic suction on plant growth. (From *USDA Agr. Handbook*, 60, 1954.)

duced by the salt reduces the average motion of the water molecules and, consequently, the energy content of the water. Thus, an increase in salinity produces the same effect on water uptake as that produced by increased SMT. In fact, the effects appear to be additive. For example, if a plant wilts when the SMT is 20 bars and the osmotic pressure of the soil solution is one, the plant will wilt if the soil moisture tension is one and the osmotic pressure of the water is 20 bars. The additive effects of SMT and salinity on plant growth is illustrated in Fig. 5-27.

Salinity Control

For permanent agriculture there must be a favorable salt balance. The salt added to soils in irrigation water must be balanced by

the removal of salt by leaching. The fraction of the irrigation water that must be leached through the root zone to control soil salinity has been defined as the leaching requirement. The leaching requirement is directly related to the salt content of the irrigation water. A leaching requirement of 0.25 or 25 percent means that 25 percent of the water applied should leach through the root zone and carry with it the salt contained in the original irrigation water. Obviously the drainage water contains a higher concentration of salt than the irrigation water. As an illustration, calculate the concentration of salt in drainage water, assuming:

1. Leaching requirement of 0.25.
2. Irrigation water contains 2 tons salt per acre foot.
3. Four acre feet of irrigation water applied.
4. No leaching by natural rainfall.
5. The concentration of salt in the drainage water can be calculated with the equation:

$$\left[\begin{array}{c} \text{Tons salt per acre foot} \\ \text{drainage water} \end{array} \right] \times \left[\begin{array}{c} \text{acre} \\ \text{feet} \end{array} \right]$$

$$= \left[\begin{array}{c} \text{tons salt per acre foot} \\ \text{irrigation water} \end{array} \right] \times \left[\begin{array}{c} \text{acre} \\ \text{feet} \end{array} \right]$$

From the illustration the drainage water would contain 8 tons of salt per acre foot or four times greater concentration of salt than original irrigation water. The need for leaching soils for salt control highlights a major disadvantage of drip irrigation in that the method is not suited to application of large amounts of water.

Leaching of salts on upland soils, as on the High Plains of Texas, results in salt being deposited below the root zone. Most of the irrigated land in arid regions is in basins or alluvial valleys where long continued leaching of salts results in a buildup of water tables. When the water table rises to 3 or 4 feet of the surface, water moves upward by capillarity at a rate sufficient to deposit salt on top of the soil from the evaporation of water (Fig. 5-28). It is essential in these cases that tile drainage systems are installed with tile laid at least 5 feet or more below the surface to remove drainage water. Thus we see that salinity control and maintenance of permanent irrigated agriculture in arid regions is dependent on drainage for salt removal. Archeological studies in Iraq showed that the Sumerians grew about half wheat and half barley in 3500 B.C. One thousand years later only one sixth of the grain was the less salt-tolerant wheat and, by 1700 B.C., the production of wheat was abandoned. This decline in wheat production and increase in the more tolerant barley coincided with soil salinization. Records show widespread land abandonment and that salt accumulation in soils was a factor in the demise of the Sumerian civilization.

Even though good irrigation practices are followed, there can be important differences in salt content of soil immediately under the irrigation furrow and the beds. Downward movement of water below the furrow leaches the soil and produces low salinity. At the same time, water moves by capillarity to the top of beds carrying salt and depositing salt as water evaporates. Important differences in salt concentrations are produced (Fig. 5-29). Seeds are commonly planted on the shoulder of the beds between the furrow and high point of the beds to insure greater germination and growth. Research has shown that water uptake is greatest in the soil with the lowest salt content.

Effect of Irrigation on River Water Quality

About 60 percent of the diverted water for irrigation is evaporated or consumed. The remainder appears as irrigation return flow (includes drainage water) and is returned

Fig. 5-28 Salt accumulation at the soil surface caused by high water table in the Central Valley of California.

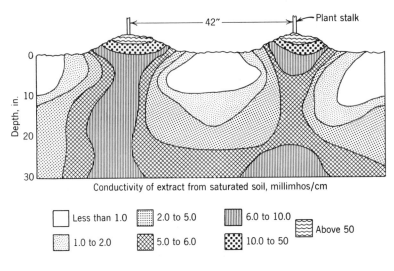

Fig. 5-29 Salt distribution under furrow-irrigated cotton for soil initially salinized to 0.2 percent salt and irrigated with water of medium salinity. (From *USDA Agriculture Handbook*, 60, 1954.)

to the rivers. As a consequence, the salt concentration of river waters is increased. A 2- to 7-fold increase in salt concentration is common for many rivers. Along the Rio Grande River in New Mexico, water diverted at Percha Dam for Rincon Valley irrigation had an average 8 milliequivalents of salt per liter from 1954 to 1963. Salt content increased to 9 at Leasburg, 13 at American Diversion Dam, and 30 at the lower end of El Paso Valley (Fig. 5-30). The Rio Grande River serves as a sink for the desposition of salt. The Salton Sea serves as a salt sink for the Imperial Valley in southern California.

Salts also appear in rivers from salting of roads, industries, and the like. In the rivers of the western United States, however, irrigation is responsible for large increases in salt. The Colorado, one of the largest rivers, experiences a 21-fold salt increase between Grand Lake in northwestern Colorado and the Imperial Dam. Large increases in salt concentrations have become harmful to freshwater fish. Salts decrease water quality for downstream users, and there is no inexpensive method to remove the salt at the present time.

Fig. 5-30 Total salt content of Rio Grande River water in milliequivalents per liter increased from 8 at Percha Dam to 30 at lower end of El Paso Valley. (Adapted from C. A. Bower, "Salinity of Drainage Waters," Agronomy Monograph # 17, p. 481 (1974), by permission of the American Society of Agronomy.

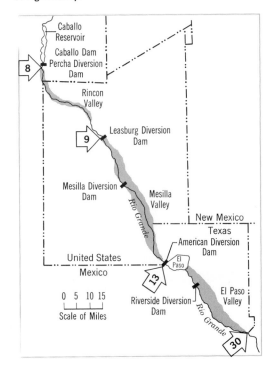

References

Allison, Lowell R., Salinity in Relation to Irrigation, *Advances of Agronomy*, *16*:139–180, 1964.

Baker, Kenneth F., Editor, "The U.C. System for Producing Healthy Container-Grown Plants," Manual 23, Cal. Agr. Exp. Station, Berkeley, 1957.

Bernstein, Leon, "Salt Tolerance of Plants," *Agr. Information Bull.*, No. 283, USDA, Washington, D.C., 1964.

Bernstein, Leon, "Salt Tolerance of Fruit Crops," *Agr. Information Bull.*, No. 292, USDA, Washington, D.C., 1965.

Bower, C. A., "Salinity of Drainage Waters," in *Drainage for Agriculture*, Jan Van Schilfgaarde, Editor, Am. Soc. Agronomy, Madison, Wis. 1974, pp. 471–487.

Brink, Ceel Van Den, and Robert L. Carolus, "Removal of Atmospheric Stresses from Plants by Overhead Sprinkler Irrigation," *Quart. Bull.*, Mich. Agr. Exp. Sta., *47*:358–363, 1965.

Cole, J. S., and O. R. Mathews, "Use of Water by Spring Wheat on the Great Plains," *USDA Bul.* 1004, 1923.

Criddle, Wayne D., and Howard R. Haise, "Irrigation in Arid Regions," in USDA Yearbook, *Soil*, Washington, D.C., pp. 359–367, 1957.

Edminister, T. W., and Ronald C. Reeve, "Drainage Problems and Methods," in USDA Yearbook, *Soil*, Washington, D.C., pp. 379–385, 1957.

Erickson, A. E., C. M. Hansen, and A. J. M. Smucker, "The Influence of Subsurface Asphalt Barriers on the Water Properties and the Productivity of Sand Soils," *9th Int. Cong. Soil Sci. Trans.*, *1*:331–337, 1968.

Evans, Chester E., and Edgar R. Lemon, "Conserving Soil Moisture," in USDA Yearbook, *Soil*, Washington, D.C., pp. 340–359, 1957.

Hanks, R. J., and C. B. Tanner, "Water Consumption by Plants as Influenced by Soil Fertility," *Agron. Jour.*, *44*:99, 1952.

Jacobsen, Thorkild, and Robert M. Adams, "Salt and Silt in Ancient Mesopotamian Agriculture," *Science*, *128*:1251–1258, 1958.

Law, James P., and Jack L. Witherow, "Irrigation Residues," *Jour. Soil and Water Con.*, *26*:54–56, 1971.

Mathews, O. R., "Place of Summer Fallow in the Agriculture of the Western States," *USDA Cir. 886*, 1951.

Richards, L. A., and S. J. Richards, "Soil Moisture," in USDA Yearbook, *Soil*, Washington, D.C. pp. 49–60, 1957.

Soil Improvement Comm. Cal. Fert. Assoc., "Western Fertilizer Handbook," 5th edition, Interstate, Dansville, Cal., 1975.

Staple, W. J., "Dryland Agriculture and Water Conservation," in H. L. Hamilton, Editor, *Research on Water*, ASA Spec. Pub. No. 4, Soil Sci. Soc. Am., pp. 15–30, 1964.

Taylor, Sterling A., "Use of Moisture by Plants," in USDA Yearbook, *Soil*, Washington, D.C., pp. 61–66, 1957.

Terry, T. A., and J. H. Hughes, "The Effects of Intensive Management on Planted Lobolly Pine (*Pinus taeda* L.) Growth on Poorly Drained Soils of the Atlantic Coastal Plain," Proc. Fourth North Am. Forest Soils Conf., University of Laval, Quebec, Canada, pp. 351–377, 1975.

The President's Science Advisory Committee Panel on World Food Supply, *The World Food Problem*, U.S. Govt. Printing Office, Washington, D.C., 1967.

Thorne, Wynne, and H. B. Peterson, "Salinity in United States Waters," in *Agriculture and the Quality of Our Environment*, Am. Assoc. Adv. Sci., Washington, D.C., 1967.

United States Salinity Laboratory Staff, "Diagnosis and Improvement of Saline and Alkali Soils," *USDA Agr. Handbook 60*, Washington, D.C., 1969.

Williamson, R. E., et al., "Effect of Water Table Depth and Flooding on Yield of Millet," *Agron. Jour.*, *61*:312, 1969.

6
SOIL ECOLOGY

The soil is the home of innumerable forms of plant, animal, and microbial life. Some of the fascination and mystery of this underworld has been described by Peter Farb.

We live on the rooftops of a hidden world. Beneath the soil surface lies a land of fascination, and also of mysteries, for much of man's wonder about life itself has been connected with the soil. It is populated by strange creatures who have found ways to survive in a world without sunlight, an empire whose boundaries are fixed by earthen walls.[1]

Life in the soil is amazingly diverse, ranging from microscopic single-celled organisms to large burrowing animals. As is the case with the organisms above the ground, there are well-defined food chains and competition for survival. The study of the relationships of these organisms in the soil environment is *soil ecology*. We will discuss diverse activities such as decomposition of organic matter, pesticide degradation by

[1] "Living Earth," Peter Farb, Harper and Brothers Publishers, 1959.

microorganisms, and predation and earth-moving activities of animals. A major theme of this chapter is the soil as a source of nutrients for living organisms and the role of soil organisms in nutrient cycling.

The Ecosystem

The sum total of life on earth together with the global environments constitutes the *ecosphere*. The ecosphere, in turn, is composed of numerous self-sustaining communities of organisms and their inorganic environments and resources, called *ecosystems*. Each ecosystem has its own unique combination of living organisms and abiotic resources that function to maintain a continuous flow of energy and nutrients. All ecosystems have two types of organisms based on carbon source. *Autotrophs* use inorganic carbon and are the *producers*. *Heterotrophs* use organic carbon and are the *consumers* and *decomposers*. The sun is the major source of energy to run the system.

Primary Producers

The major primary producers are vascular plants that use solar energy to fix carbon from carbon dioxide in photosynthesis. The tops of plants provide food for consumers and decomposers above the soil-atmosphere interface. Roots, tubers, and other underground organs provide food for consumers and decomposers within the soil. A very small amount of photosynthesis occurs at or near the surface of some soils by algae, which are the dominant plants in aquatic ecosystems. A small amount of inorganic carbon is fixed by chemotrophic bacteria by using the energy of chemical bonds. Thus, the productivity of a terrestrial ecosystem is basically a measure of the net photosynthesis (photosynthesis less respiration) of vascular plants.

Tropical rain forests have high productivity and deserts low. Agricultural ecosystems are unique in that people determine the primary producers. In the case of monoculture there is usually one major plant, as in the case of sugar cane, corn, or pineapples. Productivity of agricultural ecosystems is highly variable depending on soil, climate, and the inputs of labor, capital, and management. The biomass produced by the producers is the food for the consumers and decomposers, including humans.

Consumers and Decomposers

It is characteristic that ecosystems have many species but that only a few species are common. In a meadow, for example, several grasses may account for over 90 percent of the primary production. In turn, most of the primary production is consumed by a few animals, like mice and rabbits. The fox and a few birds may represent the bulk of secondary consumers feeding on the mice and rabbits (see Fig. 6-1).

About 1 pound of animal biomass is produced for every 10 pounds of plant material consumed. In the transformation of plant material into animal biomass, considerable carbon is returned to the atmosphere as carbon dioxide from respiration and some energy is dissipated as heat. Most of the original carbon and most of the nutrients, however, appear in the dung or feces. The result is that fecal material is a good source of nutrients and energy. It is not uncommon for primary consumers to recycle part of their feces. Rabbits eat feces when other food is in short supply. Some

Fig. 6-1 Examples of primary producers, consumers, and decomposers. Examples below the soil-atmosphere interface include roots, nematode consuming root, earthworm feeding on leaves, springtail feeding on decaying organic matter, bacteria decomposing organic matter, centipede eating a springtail, and mushroom fungi decomposing organic matter.

animals appear to recycle part of their feces instinctively as a survival mechanism or as a means to improve their nutrition. Recycling manure through animals is being studied as a means to use unwanted wastes.

The primary consumers become food for secondary consumers, and there may be additional levels of consumers. Eventually all the consumers die and are added to the soil along with the fecal material and unused primary production. These materials serve as food for soil dwelling consumers and decomposers (Fig. 6-1). Ultimately, all the carbon fixed in photosynthesis is returned to the atmosphere as CO_2 and the energy is lost as heat. The nutrients originally absorbed by the producers are released for use in another cycle of growth and decay. The result is that the major function of the soil organisms in the ecosystem centers around the flow or cycling of energy and nutrients.

We have just seen that the soil is the "dumping ground" for all the unused primary production and wastes associated with animal life. The great bulk of soil organic matter is dead, as shown in Table 6-1. Soil is the "stomach" of the earth and fulfills the dictum "Ashes to ashes and dust to dust." Without consumers and decomposers to release the fixed carbon, the atmosphere would be depleted of carbon dioxide, life would cease, and the cycle would stop.

Importance of Microorganisms as Decomposers

Life within the soil is analogous to life above the soil. Roots, tubers, and other underground plants are parts of primary producers. There are consumers and decomposers interrelated by food chains. Perhaps the major difference between the ecology above and below the soil-atmosphere interface is that above the interface animals play the dominant role as *consumers* and below the interface microorganisms play the dominant role as *decomposers*. These decomposers are mainly single-celled and microscopic and are called the *microbiota*. The dominant decomposer role of the microorganisms in the soil is complemented by the activity of many small animal consumers.

Some Characteristics of Microorganisms

Earliest life on earth consisted of microorganisms—both autotrophic and heterotrophic. The cycling of energy and nutrients was established before plants and animals evolved. No wonder that microorganisms play the major role as the ultimate decomposers. They secrete enzymes that *digest organic matter outside the cell* and absorb the soluble end products of digestion. The enzymes and processes are no different than those that occur in the digestive systems of animals.

At the root hair level the higher plants (vascular) and microorganisms have much in common. Both absorb soluble nutrients from the same soil solution and use energy to accumulate nutrients against a concentration gradient. Both must overcome the same soil moisture tension to absorb water, are inhibited by the same salts, and compete for soil oxygen. Vascular plants and microorganisms are competitors for the soil growth factors, but they also depend on each other in the continuous cycling of energy and nutrients.

The major distinguishing feature of microorganisms is their relatively simple biological organization. Many are unicellu-

Table 6-1

Estimates of Amount of Organic Matter and Proportions, Dry Weight, and Number of Living Organisms in a Hectare of Soil to a Depth of 15 Centimeters in a Humid Temperate Region.

Item	Dry Weight		Estimated Number of Individuals
	%	kg/ha	
Organic matter, live and dead	6	120,000	—
Dead organic matter	5.28	105,400	—
Roots of higher plants	0.5	10,000	—
Microorganisms (protists)		2,000	
Bacteria	0.10	2,600	2×10^{18}
Fungi	0.10	2,000	8×10^{16}
Actinomycetes	0.01	220	6×10^{17}
Algae	0.0005	10	3×10^{14}
Protozoa	0.005	100	7×10^{16}
Nonarthropod animals			
Nematodes	0.001	20	2.5×10^{9}
Earthworms (and potworms)	0.005	100	7×10^{3}
Arthropod animals			
Springtails (Collembola)	0.0001	2	4×10^{5}
Mites (Acarine)	0.0001	2	4×10^{5}
Millipedes and centipedes (Myriapoda)	0.001	20	1×10^{3}
Harvestman (Opiliones)	0.00005	1	2.5×10^{4}
Ants (Hymenoptera)	0.0002	5	5×10^{6}
Diplopoda, Chilopods, Symphyla	0.0011	25	3.8×10^{7}
Diptera, Coleoptera, Lepidoptera	0.0015	35	5×10^{7}
Crustacea (Isopods, crayfish)	0.0005	10	4×10^{17}
Vertebrate animals			
Mice, voles, moles	0.0005	10	4×10^{5}
Rabbits, squirrels, gophers	0.0006	12	10
Foxes, badgers, bear, deer	0.0005	10	<1
Birds	0.0005	10	100

Adapted and reprinted by permission from *Soil Genesis and Classification* by S. W. Buol, F. D. Hole, and R. J. McCracken, copyright 1972 by Iowa State University Press, Ames, Iowa 50010.

lar and even the multicellular organisms lack differentiation into cell types and tissues characteristic of plants and animals. They are members of the protist kingdom, and the words protists and microorganisms are used interchangeably.

Bacteria—The Most Abundant Soil Organisms

Bacteria are single-celled and exceed all other soil organisms in numbers and kinds. A gram of fertile topsoil may contain over 1 billion bacteria. The most common soil bacteria are rod-shaped, a micron (1/25,000 of an inch) or less in diameter, and up to a few microns long (see Fig. 6-2). Researchers have estimated that the live weight of bacteria per acre may exceed 2000 pounds or 2000 kilograms per hectare (Table 6-1).

Soil bacteria may be divided broadly into two large groups, based on their carbon source: (1) the *heterotrophic*, and (2) the *autotrophic*. In the autotrophic group are found organisms such as the nitrite formers, the nitrate formers, the sulfur-oxidizing bacteria, the iron oxidizers, and those that use hydrogen and its compounds.

Most of the soil bacteria require oxygen from the soil air and are classified as *aerobes*. Some aerobic bacteria can adapt to living in the presence or absence of oxygen; they are *facultative aerobes*. Other bacteria cannot live in the presence of oxygen and are *anaerobes*. The soil bacteria also differ considerably in their nutrition and in their response to environmental conditions. Consequently, the kinds and abundance of bacteria depend both on the available nutrients present and on the soil environmental conditions.

Bacteria normally reproduce by binary fission. Some divide as often as every 20 minutes and may multiple very rapidly under favorable conditions. It has been calculated that if a single bacterium divided every hour and every subsequent bacterium did the same, 17 million cells would be produced in a day. A mass the size of the earth would be produced in 6 days. Such rapid growth rates cannot be maintained for long because nutrients and other growth factors become exhausted. The

Fig. 6-2 (*Left*) Colony of bacteria on sand grain magnified about 6000 times. (*Right*) Rod-shaped bacteria magnified 20,000 times. (*Left* from "Stereoscan Electron Microscopy of Soil Microorganisms," T. R. G. Gray, *Science*, Vol. 155, p. 1668, Fig. 1, March 31, 1967. Copyright © 1967 by the American Association for the Advancement of Science. *Right* courtesy of Dr. Stanley Flegler of Michigan State University.)

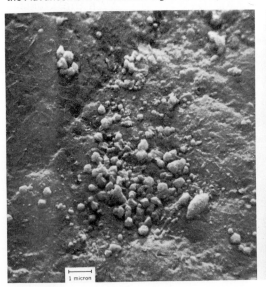
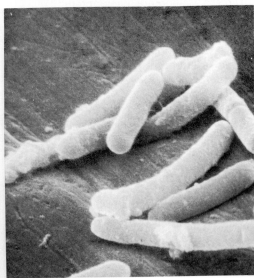

major limiting growth factor in nature is food or energy source.

Some bacteria form spores when conditions become unfavorable. Most bacteria are quite resistant to drying, and some have existed in air-dry soil for several years. Bacteria and fungi are the major decomposers; the fungi will be discussed next.

Fungi—The Effective Lignin Decomposers

Fungi are heterotrophs that vary greatly in size and structure from single-celled yeasts to molds and mushrooms. Fungi typically grow from spores by a threadlike structure that may or may not have crosswalls. Individual threads are *hypha*, and a mass of extensive threads is the *mycelium*. The mycelium is the working structure that absorbs nutrients, continues to grow, and eventually produces special hyphae that produce reproductive spores. The average diameter of hyphae is about 5 microns or about 5 to 10 times the diameter of a typical bacterium. Fungi have an advantage over bacteria in that fungi can invade and penetrate organic materials (Fig. 6-3).

It is difficult to determine accurately the number of fungi per gram of soil, since mycelium are easily fragmented. It has been observed that a gram of soil commonly contains 10 to 100 meters of hyphae per gram. On the basis of the amount of filament, researchers have concluded that live weight of fungal tissue exceeds or equals that of the bacterial tissue in most soils.

All of us have seen mold mycelia growing on bread, clothing, or leather goods. Some mold colonies growing on plant leaves produce a white-cottonish appearance called downy mildew disease. Many fungi have morphological features that resemble

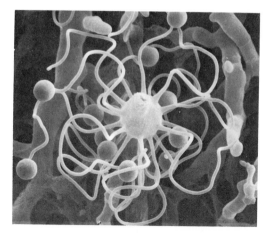

Fig. 6-3 A soil fungi showing mycelium and reproductive structure with spores. (Photo courtesy Michigan State University Pesticide Research Electron Microscope Laboratory.)

higher plants. *Rhizopus*, a common mold in soil and on bread, has rootlike absorptive structures called *rhizoids*, which penetrate the substrate on which the mold is growing. Hyphae elongating over the substrate are called *stolons*. Stalk or stemlike hyphae originating from the stolons bear spore cases. Unlike higher or vascular plants, however, fungi have no specialized xylem or phloem-conducting tissue.

Fungi are important in all soils and their tolerance of acidity makes them particularly important in acid forest soils. The woody residues of the forest floor provide an abundance of food for certain fungi that are effective decomposers of lignin (Fig. 6-4). In the Sierra Nevada Mountains there is a large mushroom fungus called the "train wrecker" fungus because it grows abundantly on railroad ties unless they are cresoted.

Yeasts are fungi and are found in soils only to a limited extent and are believed to be of no great importance in soil development or the growth of higher plants.

Fig. 6-4 Mushroom fungi (*Agaricales*) growing on soil, wood, and bark. The mushroom fruiting bodies form after the hyphae have accumulated sufficient food and the moisture and temperature are favorable.

Actinomycetes—the "Fungilike" Bacteria

The actinomycetes occupy a position between the bacteria and fungi from the morphological viewpoint. They are frequently spoken of as *ray fungi* or *thread bacteria*. The actinomyces resemble bacteria in that they have the same cell structure and are of about the same size in cross section. They resemble the filamentous fungi in that they produce a branched filamentous network. Many of these organisms reproduce by means of spores, and these spores appear very much like bacterial cells.

These organisms are present in great abundance in soil, making up as much as 50 percent of the colonies that develop on plates containing artificial media inoculated with a soil extract. The numbers of actinomyces may vary between 0.1 million and 36 million per gram of soil. In actual weight of live substance per acre, they may exceed bacteria but, as a rule, will not equal fungus tissue.

Algae—Chlorophyllous Protists

Algae exhibit great diversity in form and size, ranging from single-celled organisms, with a diameter about 5 to 10 times greater than that of a bacterium, to kelps of the ocean that are over 100 feet in length. Although algae are the most important plants living in water, algae are of only minor importance in most soils. The most common soil algae are single-celled or are small filaments. Algae are universally distributed in the surface layer of soils wherever moisture and light are favorable. A few algae are found below the soil surface in the absence of light and appear to function heterotrophically.

The common forms of algae inhabiting soils are (1) blue-green, (2) green, and (3) diatoms. The blue-green are the most abundant in soils and, to the extent that they fix carbon, they contribute to the organic matter content of soils. Their photosynthetic ability accounts for their growth on numerous exposed surfaces, including rocks and soils. Some algae grow in close association with fungi in a form known as *lichens*. In the initial weathering of rock exposures and the formation of soils from freshly exposed parent material, lichens play an important role in the early accumulation of organic matter. Furthermore, the ability of some blue-green algae to fix atmospheric nitrogen also helps plant communities become established on freshly exposed rocks and parent materials. The nitrogen fixed by algae living in the water of rice paddies is of great importance in rice production.

Protozoa—Aquatic Protists

Protozoa are single-celled protists that exhibit great diversity. Soil protozoa live in the films of water surrounding soil particles and, in a sense, are aquatic organisms. When the soil dries out, food supplies become short, or conditions are harmful, protozoa encyst; they become active again when conditions become favorable. Soil protozoa are largely predators, feeding on soil bacteria, although some protozoa also feed on fungi, algae, or dead organic matter. Although protozoa are very numerous in soils, they appear to have only a minor affect on organic matter decomposition and bacterial activity.

Vertical Distribution of Microbiota in the Soil

The surface of the soil is the interface between the lithosphere and the atmosphere. At or near this interface the quantity of living matter is greater than at any region above or below. As a consequence, the A horizon contains more organic debris that serves as food for the microorganisms than the B or C horizons. Although other factors influence the activity and numbers of microorganisms besides nutrient and energy supplies, the greatest number of microbes, as a rule, occurs in the A horizon or surface layers (Fig. 6-5). In forest and meadow soils the greatest number of microbes are likely to be in or very close to the surface of the

soil, although the number at the very surface of cultivated soils may be relatively low because of a lack of moisture and the germicidal action of sunlight.

Soil Animals as Consumers and Decomposers

Perhaps higher plants could grow and provide us with food and microorganisms could recycle all the nutrients without the aid of animals. Soil animals, however, are numerous (see Table 6-1) and play an important role in organic matter decomposition. Nematodes and white grubs are primary consumers that feed largely on the roots of primary producers (vascular plants). Secondary and tertiary consumers include predatory animals like mites, centipedes, spiders, ants, and moles. A food chain in the soil could consist of plant roots, nematodes, mites, and centipedes. Most of the soil animals are consumers that feed on dead and decaying refuse and include springtails, sowbugs, millipedes, mites, slugs, earthworms, and various soils insects. Soil animals can be considered both consumers and decomposers because they eat or ingest organic matter, and some decomposition of organic matter occurs in digestion.

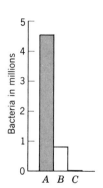

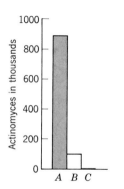

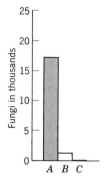

Fig. 6-5 Distribution of microorganisms in the A, B, and C horizons of a cultivated Prairie (Mollisol) soil. All values refer to the number of organisms per gram of air-dry soil. (Data from *Iowa Res. Bul.* 132.)

Interdependence of Microorganisms and Animals in Organic Matter Decomposition

The microorganisms and fauna work together as a team. Consider for a moment that a leaf falls on the forest floor. Both microorganisms and animals attack the leaf. Holes made in the leaf by springtails and mites facilitate the entrance of microorganisms inside the leaf. Soil animals ingest bacteria when they feed, and the bacteria continue to function in the digestive tract of the small animals. The excrement of animals is attacked by both the microbes and the fauna. The entire decomposing mass, along with mineral soil particles, may be ingested by earthworms, thereby producing an intimate mixing of organic and mineral matter. The net result is the humification of organic matter with both microorganisms and fauna playing vital roles. In the process the major role of the animals is the fragmentation and mixing, which greatly increase the surface area and prepare the organic matter for the microorganisms. Major credit for the mineralization and recycling of mineral elements goes to the microorganisms. It is now time to consider in more detail the nature and activities of the most important animals involved in organic matter decomposition.

The Most Important Soil Consumers and Decomposers

The soil animals function as consumers and are concentrated in the surface soil layers enriched with organic matter. Thousands of species are present. Space permits consideration of only a few of the most abundant and important. Many are arthropods and are in the phylum arthropoda. The most important soil arthropods that will be discussed include mites, centipedes, millipedes, springtails, insects (ants, termites, beetles), and insect larvae. Important nonarthropod soil animals that will be considered are nematodes and earthworms.

Nematodes—Parasitic Consumers

Nematodes are worms that are mainly microscopic in size and are the most abundant animals in soils. On the basis of their food requirements, three groups are distinguished: (1) those that feed on decaying organic matter; (2) those that feed on earthworms, other nematodes, plant parasites, bacteria, protozoa, and the like; and (3) those that infest the roots of higher plants, passing a part of their life cycle embedded therein. Nematodes are sometimes called eel worms; they are round or spindle-shaped and usually have a pointed posterior.

Under a 10-power hand lens they appear as transparent threadlike worms (see Fig. 6-6). Nematodes lose water readily through their cuticles and live mainly in the water films surrounding soil particles or in plant roots. When soils dry or other conditions become unfavorable, they encyst and become active again when conditions are favorable.

Parasitic nematodes are considered the most important in agriculture. Many plants are attacked, including tomato, peas, carrot, alfalfa, turfgrass, ornamentals, and fruit trees. Parasitic nematodes have a needlelike anterior end (stylet) used to pierce plant cells and suck out the contents. Host plants respond in numerous ways, for example, the development of galls or knots or deformed roots (Fig. 6-7).

Investigations indicate that nematode damage is much more extensive than originally thought. Nematodes are a very serious problem for pineapple production in

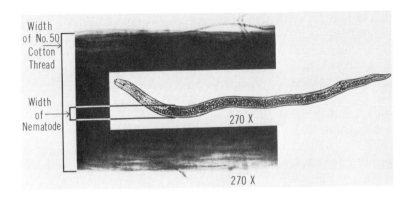

Width of No. 50 Cotton Thread

Width of Nematode

270 X

270 X

Fig. 6-6 A nematode and a piece of ordinary cotton thread photographed at the same magnification (270). The nematode is 1/15 as thick as the thread and is not visible to the naked eye. (Courtesy H. H. Lyon, Plant Pathology Department, Cornell University.)

Hawaii where the soil for new plantings is routinely fumigated to control nematodes. Nematodes may also become serious pests in greenhouse soils unless special care is taken to avoid infestation. Not only do the nematodes injure the plant roots themselves, but, by puncturing the plant, they also prepare an entrance for other parasites. The main role of nematodes in soils appears to be economic, as consumer parasites.

Earthworms—Consumers and Soil Mixers

Perhaps the best-known group of larger animals inhabiting the soil is the common earthworm, of which there are several species. These organisms prefer a moist environment with an abundance of organic matter and a plentiful supply of available calcium. Consequently, earthworms are found most abundantly, as a rule, in fine-textured soils that are high in organic matter and are not strongly acid; they occur only sparingly in acid sandy soils that are low in organic matter.

Obviously, the number and activity of the earthworms vary greatly from one location to another and, as with other soil organisms, figures indicating numbers are merely suggestive. The number of earthworms in the plowed layer of an acre may range from a few hundred or even less to more than a million. It has been estimated that between 200 and 1000 pounds of earthworms are commonly present in an acre of soil (or 200 to 1000 kilograms per hectare).

The common earthworm, *Lumbricus terrestris*, was imported into the United States

Fig. 6-7 Response of roots to nematodes. Root galls or knots on tomatoes on left and excessive branching of carrots on right.

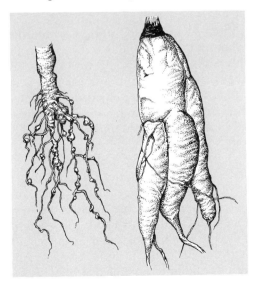

from Europe. *Lumbricus terrestris* makes a shallow burrow and forages on plant material at night. Some of the plant material is dragged into the burrow. Other kinds of earthworms exist by ingesting organic matter that exists in the soil. Excrement or castings are deposited both on and in the soil. The intimate mixing of soil materials, the creation of channels, and the production of castings leaves the soil more open and porous. Channels left open at the soil surface increase water infiltration.

Earthworms normally avoid water-saturated soil. If they emerge during the day when it is raining they are killed by ultraviolet radiation unless they quickly find protection. Casts deposited on the surface of lawns are objectionable. On the other hand, earthworms feed on thatch and help to prevent thatch buildup by turfgrass. Arable lands in the temperate climates are commonly low in earthworms. Research has shown that field soils freeze quickly in the fall and earthworms are killed before they have time to migrate deeper in the soil to avoid the cold. The slow freezing in a forest because of the presence of surface organic layers permits their escape to deep soil layers before freezing. As the soil warms in the spring, earthworms migrate back to the surface. Their direct effect on plant growth is minimal and attempts to increase plant growth by increasing earthworm activities in soils have been disappointing. Earthworms do not convert eroded soils low in organic matter into productive soils.

Arthropod Consumers and Decomposers

A high proportion of soil animals are arthropods. They have an exoskelton and jointed legs. Most have a kind of heart and blood system and usually a well-organized nervous system. The most abundant are springtails and mites. Other arthropods include spiders, insects (including larvae), centipedes, millipedes, wool lice, snails, and slugs. Several of the most important will be discussed.

Springtails—Consumers with a Spring in Their Tail. Springtails are primitive insects less than 1 millimeter (1/25 inch) long. They have a spring like appendage under their posterior end that permits them to "flit to and fro" (see Fig. 6-8). They are very abundant (see Table 6-1) and are distributed worldwide. They live in the macropores of the litter layers and feed largely on dead plant and animal tissue, feces or dung, humus, and fungal mycelia. Water is lost rapidly through the cuticle so they are restricted to moist soil layers. Those in the lower litter layers are the most primitive and are without eyes and pigment. Springtails are in the order *Collembola* and are commonly called Collembola. Their enemies includes mites, small beetles, centipedes, and small spiders.

Mites—Vegetarian and Carnivorous Consumers (Arachnids). My experience with mites has been largely with red spider mites that bother plants and ear mites that bother dogs. Ticks are bloodsucking mites. Mites are the most abundant air-breathing soil animals. They commonly have a saclike body with protruding appendages and are related to spiders (Fig. 6-8).

Most mites feed on dead organic debris of all kinds, such as fungal hyphae and spores. Some mites are predaceous consumers and feed on nematodes, insect eggs, and other small animals such as springtails. Activities of mites include breakup and decomposition of organic materials, move-

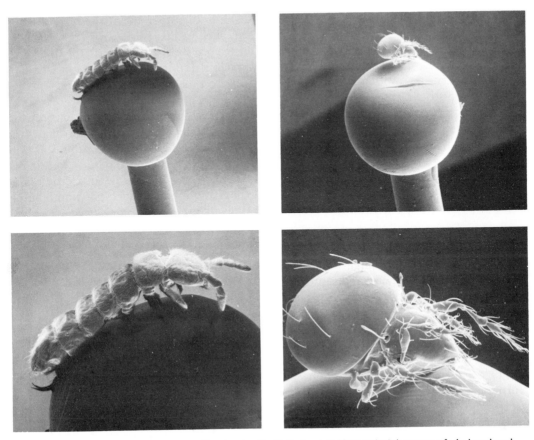

Fig. 6-8 Two of the most common soil animals. (*Left*) Springtail (Collembola) on top of pin head and magnified 50 to 100 times at bottom. (*Right*) Similar photos of a mite. (Courtesy Michigan State University Pesticide Research Electron Microscope Laboratory.)

ment of organic material to deeper soil layers, and maintenance of pore spaces (runways).

Vegetarian Millipedes and Carnivorous Centipedes (Myriapods). Millipedes and centipedes are elongate, fairly large soil animals with many pairs of legs. They are common in forests, and overturning almost any stone or log will send them running for cover. Millipedes have many pairs of legs and are vegetarians or decomposers; they feed mainly on dead organic matter (saproghagous). Some browse on fungal mycelia. Centipedes typically have fewer pairs of legs compared to millipedes and are carnivorous consumers. Centipedes will attack and consume almost any animal of any size they can master. Because they are fairly large compared to mites and springtails, their numbers in soils are small.

White Grubs—Consumers of Roots. In some cases the insect larvae play a more important role in soils than adults, as in the case of the white grub. White grubs are larvae of the familiar brown May beetle or June bug. The grubs are round, white, about 1 inch long, and curl into a C shape when disturbed. The head is black with three pairs of legs just behind the head. They feed mainly on grass roots, causing dead spots in lawns. A wide variety of other plants are also attacked, making white grubs an important agricultural pest. Moles feed on insect larvae and earthworms, resulting in greater likelihood of mole damage in lawns when white grubs are present.

Ants and termites are important soil insects that will be considered later in relation to earth-moving activities.

Nutrient Cycling

Nutrient cycling is the exchange of nutrient elements between the living and nonliving parts of the ecosystem. Two broad processes are involved. *Immobilization* is the uptake of inorganic nutrient ions by organisms. *Mineralization* is the conversion of nutrients in organic matter into inorganic ions principally by microbial decomposers. Nutrient cycling conserves the nutrient supply and results in repeated use of the nutrients. The net effect is greater ecosystem productivity than if nutrient cycling did not occur.

Nutrient Cycling Processes

Figure 6-9 shows that the organic matter that is added to the soil consists of a variety of compounds. These include fats, carbohydrates, proteins, and lignins. Incorporation of these organic compounds into the soil stimulates to the greatest extent the organisms that are benefited the most. As decomposition proceeds, the most easily digested materials disappear first. All groups can effectively break down and utilize carbohydrates and proteins, but the fungi are the most effective in decomposing the lignin.

While digesting the plant residues, the microbes utilize some of the carbon, energy, and other nutrients for their own growth. In time the synthesized tissue dies and becomes the substrate for further decomposition. Figure 6-9 indicates this by the subcycle where constituents in the living organisms are temporarily unavailable or *immobilized*. The immobilization of nutrients refers to the use and incorporation of nutrients into living matter by both microbes and higher plants. The immobilized nutrients are again mineralized when the organisms die. In time even the most resistant materials succumb to the enzymatic attack of the microbes. The net effect is the release of energy as heat, the formation of carbon dioxide and water, and the appearance of nitrogen as ammonium (NH_4^+), sulfur as sulfate ($SO_4^=$), phosphorus as phosphate (PO_4^{-3}), and many other nutrients as simple metallic ions (Ca^{++}, Mg^{++}, K^+). Most of these forms are available to living organisms for another cycle of growth.

As soon as some of the ions or elements released in organic matter decomposition appear, other specialized organisms oxidize some of them. These transformations are beneficial to the extent that the oxidized forms are more readily used by higher plants. This is the case for the oxidation of sulfur to sulfate. The oxidation of iron and manganese makes them less soluble and, therefore, less available to higher plants.

When soils become anaerobic, still other microbes become active. The oxidized forms of plant nutrients are reduced. Sulfate is reduced to hydrogen sulfide and insoluble mineral sulfides. Perhaps one of the most important reactions in anaerobic soil is the reduction of nitrate, which results in the formation of nitrogen gas that escapes from the soil (denitrification). Loss

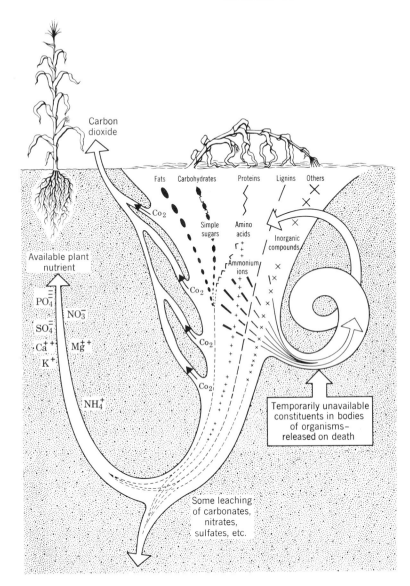

Carbon dioxide

Fats Carbohydrates Proteins Lignins Others

Co₂

Simple sugars

Amino acids

Inorganic compounds

Ammonium ions

Available plant nutrient

$PO_4^=$

NO_3^-

$SO_4^=$

Ca^{++} Mg^{++}

K^+

NH_4^+

Co₂

Co₂

Co₂

Temporarily unavailable constituents in bodies of organisms – released on death

Some leaching of carbonates, nitrates, sulfates, etc.

Fig. 6-9 Schematic diagram of organic matter decomposition and nutrient cycling. (Courtesy Dr. Burns Sabey.)

of nitrogen by dentrification normally involves about 10 percent of the nitrogen turnover per year in well-aerated soils and more than this in poorly aerated soils.

Under the anaerobic conditions of flooded rice paddy soils, the ammonium produced during organic matter decom-position is not oxidized to nitrate, since nitrification occurs only in the presence of O_2. Interestingly, the paddy rice plant shows a preference for ammonium nitrogen over the nitrate form. Not only is the ammonium not oxidized in flooded rice paddy soil, many other compounds are not oxidized as well. Sulfur accumulates as hydrogen sulfide and methane (CH_4) is formed instead of carbon dioxide and water. Hydrogen sulfide and other toxic materials build up in rice paddies, requiring an occasional draining to permit aeration of the soil and the oxidation and elimination of reduced forms of materials that are toxic to higher plants.

A Case Study of Nutrient Cycling

One approach to the study of nutrient cycling is to select a small watershed and measure changes in nutrients over time. Our example involves a 38-acre watershed in the White Mountain National Forest in New Hampshire. For a mature forest the amount of calcium (and other nutrients) taken up from the soil approximates the amount of nutrients returned to the soil in leaves, dead wood, and the like. For calcium (Ca) in New Hampshire this was 49 kilog-

rams per hectare (2.47 acres) or 44 pounds per acre taken up and returned to soil annually (Fig. 6-10). About 9 kilograms of available calcium were added to the system by mineral weathering and 3 kilograms were added in the precipitation per hectare annually. These additions were balanced by leaching losses as measured in the runoff. This means that 80 percent of the calcium in the cycle was recycled or reused by the forest each year. Plants were able to take up 49 kilograms of calcium each year while only 12 kilograms of new calcium was being made available. The researchers concluded that northern hardwood forests have a remarkable ability to hold and circulate nutrients.

The data in Table 6-2 show that 70 to 86 percent of four major nutrients absorbed from the soil were recycled in the vegetation each year. Without such extensive recycling the productivity of the forests would be lower. Nutrient cycling accounts

Table 6-2

Annual Uptake, Retention and Return of Nutrients to the Soil in a Beech Forest

	Nutrients							
	Kilograms per Hectare				Pounds per Acre			
	N	P	K	Ca	N	P	K	Ca
Uptake from soil	50	12	14	96	45	11	13	86
Stored in wood or lost from soil	10	2	4	13	9	2	4	12
Returned to soil in litter	40	10	10	83	36	9	9	74
Percent recycled	80	82	70	86	80	82	70	86

Stoeckler, Joseph H., and Harold F. Arneman, "Fertilizers in Forestry," *Advances in Agronomy* (1960) *12*:127–195. By permission of author and Academic Press.

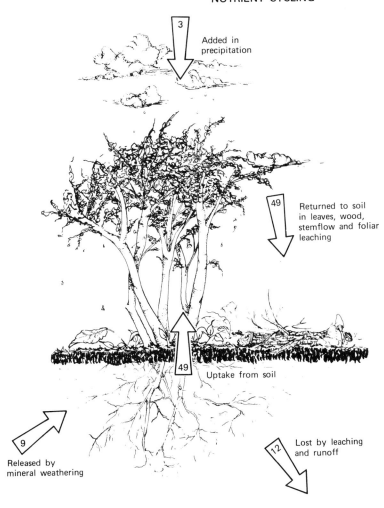

Fig. 6-10 Calcium cycling in hardwood forest in New Hampshire in kilograms (kilos) per hectare per year. (Data from Bormann and Likens, 1970.)

for the existence of enormous forests on some very infertile soils. In this situation most of the available nutrients are in soil organic matter.

Effect of Deforestation on Nutrient Cycling

We have established that the litter layers or surface soil horizons enriched with organic matter are an important nutrient reservoir. Furthermore, vegetation immobilizes the

mineralized nutrients and keeps them in the cycle. The question now is "What is the effect of deforestation on nutrient cycling?"

Researchers in the White Mountain National Forest cut all the vegetation on a watershed and measured the changes in nutrient cycling. The soil was not disturbed to minimize nutrient loss by erosion. Herbicides were applied to prevent nutrient immobilization by plants. Deforestation occurred in early 1966, and increased loss of nutrients in the water leaving the watershed was detected in May. The losses of nitrogen from the watershed became greater than the amount of nitrogen normally immobilized by the vegetation. The concentration of nitrate in the *stream water* leaving the watershed was increased manyfold. Reestablishment of the vegetation resulted in immobilization of nutrients and reduced losses of nutrients from the watershed by mid 1968. In time the nitrogen concentration of the stream waters should equal the concentration that existed before deforestation.

Clearcutting of forests not only removes the trees that immobilize nutrients, but may also expose the soil to increased erosion (Fig. 6-11). Particulate losses of nutrients are greatly increased as the surface soil is eroded away. Streams are enriched with nutrients and soil particles. As in the case of the deforestation experiment, the reestablishment of vegetation quickly reduces nutrient leaching and erosion losses and causes a return to normalcy in several years.

Effect of Crop Harvesting on Nutrient Cycling

A good yield of corn in the corn belt is usually 150 bushels per acre. About 235 pounds (263 kilograms per hectare) of nitrogen are contained in the tops. The grain contains about 135 pounds (151 kilograms per hectare) of nitrogen, which results in a 135-pound loss of nitrogen from the soil if only the grain is harvested. It is not uncommon for farmers to add 150 to 200 pounds of nitrogen per acre (168 to 224 kilograms per hectare) to the soil each year to maintain high corn yields. According to the data in Table 6-2, by contrast, only 9 pounds of nitrogen per acre (10 kilograms per hectare) are needed annually to maintain a beech forest. It is easy to see how the 9 pounds of nitrogen can be added naturally to forests and understand why there has been so little fertilization of forests in the past. There has been little need.

Basically, food production represents nutrient harvesting. The natural nutrient

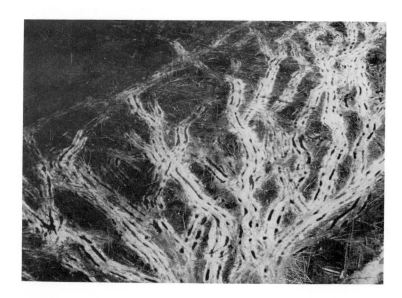

Fig. 6-11 Severe logging damage that has left the land with little vegetation and much soil disturbance conducive to erosion and loss of nutrients from site. (Courtesy Dr. R. G. Campbell, Weyerhaeuser Co.)

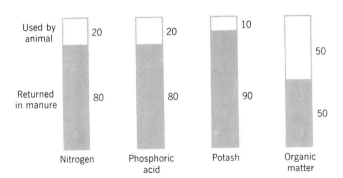

Fig. 6-12 Average proportion of plant nutrients and organic matter in feed consumed by animals and excreted in manure.

cycling process cannot provide enough nutrients to grow the crops with yields high enough to feed the world's people. Yields of grain in many of the world's agricultural systems stabilize at about 8 to 10 bushels per acre without the addition of nutrients in manure or fertilizer.

Role of Manure Use in the Maintenance of Soil Nutrients

The early settlers in America found it difficult to establish a prosperous agriculture on the sandy infertile soils (low in organic matter) along the eastern seaboard. You may recall that the Indian Squanto showed the Pilgrims how to fertilize corn with fish to increase the nutrient supply. Production of plantation crops and their shipment to Europe put a heavy drain on soils for nutrients. Land abandonment and exploitation of new land further west was widespread as early as the 1700s. Subsequently, a permanent and prosperous agriculture was established in the east and in states further west based in part on the judicious use of animal manure.

Nutrients in feed that are not used to build body tissue or milk are excreted in the manure or feces. On the average, farm animals return 75 to 80 percent of the nitrogen, 80 percent of the phosphorus, 85

to 90 percent of the potassium, and about 50 percent of the organic matter in the manure from the feed consumed (Fig. 6-12). Crops can be harvested with only a modest drain on the nutrients in the cycle if the nutrients in the manure are returned to the land. Careful collection and management of the manure and use of legumes to fix nitrogen, along with lime to reduce soil acidity, were the most important practices that resulted in the prosperous agriculture in dairy states like New York and Wisconsin in the nineteenth century.

Nutrient Transference Between Ecosystems

The world does not gain or lose nutrients, practically speaking. Nutrients exist in many forms, are unevenly distributed, and are moved from place to place. The nutrients that leave one ecosystem are added to some other ecosystem. Some nutrients are moved about as dust in the air to create a general movement of nutrients from deserts to more humid regions. Perhaps nutrients lost from the logging site of Fig. 6-11 will enter a nearby lake. Certainly, some of the nutrients will find their way to the mouth of some river. Here, the nutrients will support aquatic plants (primary producers) and serve as food for fish (consumers). Most of the ocean's surface waters, however, are so low in nutrients as to be called a "biological desert." Good fisheries tend to occur at the mouths of rivers that bring in nutrients. Construction of dams on the Nile River has reduced fishing at the mouth and in the adjacent Mediterranean Sea because of reduction in transference of nutrients from the headwaters. Flooding and deposition of silt accounts for the generally high fertility of alluvial soils in the Nile Valley and other alluvial soils of the world. Nutrients are transferred from the headwaters to the floodplains more rapidly than nutrients are removed by leaching, and other methods.

The transference of nutrients from terrestrial ecosystems to aquatic ecosystems is a natural phenomenon. In the larger geologic cycle calcium in soil minerals weathers into available form subject to plant use and removal from soils by leaching. In humid regions the calcium is gradually moved to the oceans. After the calcium reaches the ocean, it may be cycled many times among the primary producers, consumers, and decomposers before calcium is precipitated in the form of limestone. Uplift of the limestone and initiation of a new cycle of soil formation frees calcium to make another trip through the larger geologic cycle.

The creation of cities resulted in movement of nutrients (in food) from farms to cities. In the Orient there has been a considerable transfer of the nutrients in wastes back to the land. The high population density cited in Chapter 1 for China in the early 1900s and today, has been made possible by efficient transference of nutrients from the land to cities or villages and back to the land. In the United States nutrients are moved to the cities, but the nutrients are

mainly disposed of in some water course and are not returned to the land. Fertilizers are used to make up the difference. As nutrients in fertilizers become more expensive there is more interest in returning the nutrients back to the land for reuse.

A key point is that food production depends on nutrients in both terrestrial and aquatic ecosystems. In nature there is considerable conservation and reuse of nutrients. The large shipments of grain from the United States to other countries represent a massive movement of nutrients to foreign countries. Hope for feeding the world's people in the future will likely depend on greater reliance on transference of nutrients back to the land for reuse.

Relation of Soil Organisms to Higher Plants

The role of soil organisms in nutrient cycling has been stressed, and we have seen that some organisms are parasites feeding on plants. In this section we will consider the specialized roles of microorganisms living near or adjacent to plant roots.

The Rhizosphere

Plant roots leak or exude a large number of organic substances. Sloughing of root caps also provides much new organic matter. These substances are food for organisms and cause a zone of intense biological activity near the roots in the area called the *rhizosphere*. Many kinds of organisms inhabit the rhizosphere, the bacteria are benefited most. Bacterial colonies may form a continuous film around the root. Roots supply the microorganisms in the rhizosphere with food.

Fungus Roots or Mycorrhizae

Fungi infect the roots of most plants. Fortunately, most of the fungi form a symbiotic relationship—a relationship that is mutually beneficial. After mycorrhizal spores germinate, hyphae invade rootlets and grow both inside and outside the rootlet (long or major roots are seldom invaded). Fungal hyphae on the exterior of roots serve as an extension of roots for water and nutrient absorption. These fungus roots are called *mycorrhizae*.

There are two types of hyphae: ectotrophic and endotrophic. Ectotrophic hyphae exist between the epidermal root cells, using pectin and other carbohydrates for food. Continued growth of the hyphae outside the root may result in the formation of a sheath or mantle that completely surrounds the root (see Fig. 6-13).

There are literally thousands of species that belong in the group of fungi forming ectomycorrhizae and form mushroom or puffball fruiting bodies (Basidiomycetes). The fungi shown in Fig. 6-4 form mycorrhizae. They are typically associated with trees and thus benefit plants, as illustrated in Fig. 6-14.

Endotrophic mycorrhizae are more abundant than ectomycorrhizae and benefit most field and vegetable crop plants. Hyphae invade roots and ramify within and between cells, usually avoiding invasion of the very center of the root. Coils of hyphae or branched hyphal structures formed within root cells are called *arbuscles*. Swellings that form on the hyphae and contain oil are *vesicles*. These structures form the basis for referring to endomycorrhizae as *vescilar-arbuscular* (VA) mycorrhizae.

The host plant provides the fungus with food. Benefits to the host include:

1. Increasing effective root surface with increased effectiveness in absorption of nutrients (particularly phosphorus) and water.
2. Rootlets functioning longer.
3. Increasing heat and drought tolerance.
4. Rendering soil nutrients more available.
5. Detering infection by disease organisms. This last benefit only occurs in the case of the ectomycorrhizae.

Nitrogen Fixation

We live in a sea of nitrogen because the atmosphere is 79 percent nitrogen. In spite of this, nitrogen is perhaps the most limiting nutrient for plant growth worldwide. Nitrogen in the air exists as an inert gas (N_2) and, as such, is generally unavailable to biological organisms. There are some species of bacteria that can absorb the nitrogen gas from the air and convert it into ammonia, which plants can utilize. This process is *nitrogen fixation*.

Some of the (nitrogen-fixing) bacteria are free living or nonsymbiotic. Most of the important nitrogen fixers are symbiotic and live in association with a host plant. The symbiotic nitrogen-fixing bacteria, called Rhizobia, infect a root, and the plant responds by the formation of a nodule (Fig. 6-15). The host plant supplies the bacteria with food and, in return, the host plant is benefited by the fixed nitrogen. Many plants, both legumes and nonlegumes, are involved in symbiotic nitrogen fixation. Symbiotic nitrogen fixation is the major means by which inert nitrogen in the atmosphere is transferred to soils for use by plant roots and microorganisms.

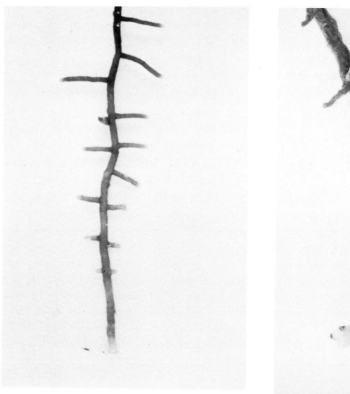 

Fig. 6-13 Ectomycorrhiza. (*Left*) Uninfected pine root. (*Right*) Infected root with fungal mantle. The Hartig net is the hyphae between the root cells. (Courtesy Dr. D. H. Marx, University of Georgia.)

Blue-green algae are nonsymbiotic nitrogen fixers living generally in aquatic ecosystems; they are close relatives of bacteria. Blue-green algae are unique in that they are autrotrophs and can fix nitrogen nonsymbiotically. The limited growth of algae in soils means that they play a very minor role in adding nitrogen to terrestrial ecosystems. Algae fix important amounts of nitrogen in the water of rice paddies and contribute importantly to the yields of rice. Whereas bacteria play the dominant role in transferring nitrogen from the atmosphere to the soil in terrestrial ecosystems, algae play the dominant role in transferring nitrogen from the air into the waters of aquatic ecosystems.

Production of Disease

The soil frequently contains a rather large number of organisms that cause diseases either in plants or animals. Some of these organisms live in the soil only temporarily, and others use it as a permanent habitat.

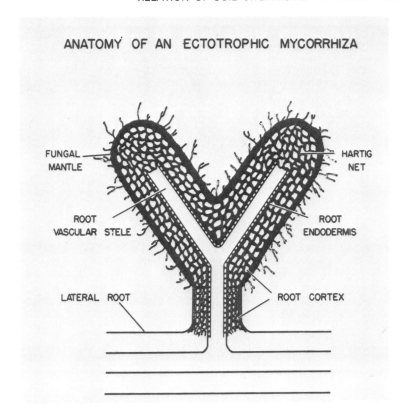

ANATOMY OF AN ECTOTROPHIC MYCORRHIZA

FUNGAL MANTLE

HARTIG NET

ROOT VASCULAR STELE

ROOT ENDODERMIS

LATERAL ROOT

ROOT CORTEX

Fig. 6-13—*cont.*
Mycorrhiza anatomy.

Fig. 6-14 The uneven growth of these red pine seedlings is attributed to nonuniform inoculation of soil with mycorrhizae. This area was a new propagation site and required inoculation for good seedling growth.

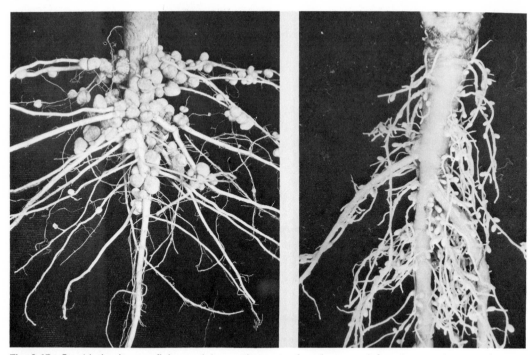

Fig. 6-15 Symbiotic nitrogen fixing nodules on the roots of soybeans on left and sweet clover on the right. (Courtesy Nitragin Co.)

The soil may harbor organisms that cause bacterial diseases such as wilt of tomatoes and potatoes, soft rots of a number of vegetables, leaf spots, and galls. Some of the most destructive parasites are the disease-causing fungi such as those that cause damping-off of seedlings, cabbage yellows, mildews, blights, certain rusts, wilt diseases, scab, dry rot of potatoes, and many others. Certain species of *Actinomyces* may cause diseases like scab in potatoes and sugar beets and pox in sweet potatoes. The catastrophic potato famine in Ireland from 1845 to 1846 was caused by a fungus that produced potato blight.

Soil used for bedding and greenhouse plants is routinely heat treated to kill plant pathogens. Weed seeds are also killed. Field soils are not uncommonly fumigated to control nematodes. Heat treatment does not sterilize the soil, since many bacteria and fungi are not killed.

Competition for Nutrients

The microorganisms or protists have some common nutrient needs with higher plants, absorb nutrients by similar processes, and compete for the same nutrient supply. We have noticed, however, that microorganisms may contribute to a greater available nutrient supply. The major case of competition appears to be for nitrogen. When large quantities of decomposable organic matter are added to soils that are low in available nitrogen, microorganism growth may be stimulated to such an extent that serious competition of nitrogen may occur.

This problem will be discussed in more detail in the next chapter.

Soil Organisms and Environmental Quality

During the billions of years of evolution of living organisms, organisms evolved that could decompose all compounds formed directly or indirectly from photosynthesis. This led to the concept of the "infallibility of soil organisms." This concept is being challenged today because human beings have become important contributors of synthetic compounds to the environment. New questions have been raised, including: "Can the soil organisms destroy any compound that man can synthesize?" The topics of pesticide degradation, disposal of sewage effluent, and contamination of soils with oil spills will be considered as they relate to environmental quality.

Pesticide Degradation in Soils

Pesticides include those substances used to control or to eradicate insects, disease, organisms, and weeds. One of the first and most successful pesticides was DDT, used to kill mosquitoes for malaria control. Today, about 30 years later, there is evidence to indicate that some DDT exists in the cells of "all" living animals. This has dramatized the resistance of DDT to biodegradation and the "fallibility" of the soil microorganisms. Evidence supports the view that the structure of DDT is different from most naturally occurring compounds and, as a consequence, few soil organism have developed enzyme systems that degrade DDT. In general, it is believed that pesticides with structures similar to those found in naturally occurring compounds are degradable and pesticides with "new"

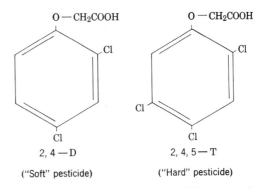

("Soft" pesticide) ("Hard" pesticide)

Fig. 6-16 Structure representing 2,4-D on the left and 2,4,5-T on the right. The structures are very similar except for the additional Cl at the meta position of 2,4,5-T, which is metabolized with great difficulty, if at all, by soil microorganisms.

structures not naturally found are persistent. This is illustrated by comparing 2,4-D(2,4-dichlorophenoxyacetic acid) with 2,4,5-T(2,4,5-trichlorophenoxyacetic acid). The two compounds have very similar structures (Fig. 6-16), except that 2,4,5-T has an extra chlorine at the meta position of the ring. Although 2,4-D is readily decomposed (in fact, some organisms can use 2,4-D as their only source of carbon), 2,4,5-T is very resistant. A chlorine on the meta position is metabolized with difficulty, if at all; consequently, the 2,4,5-T is a very persistent or is a "hard" pesticide.

We have become very dependent on pesticides, and the likelihood of eliminating their use is remote. The challenge for scientists is the development of pesticides that can perform their useful function and disappear from the ecosystem without any undesirable side effects.

Soil as a Living Filter for Sewage Effluent Disposal

A relatively small community of 10,000 people may produce about 1 million gallons of waste water per day from its sewage treatment plant. This is approximately 40 acre inches of sewage effluent a day. The effluent looks much like ordinary tap water; when it is chlorinated, it is safe for drinking. Discharging the effluent into a nearby stream does not create a health hazard. The effluent, however, is enriched with nutrients, and plant growth in streams and lakes may be increased by the addition of nutrients, Increased plant growth in the water increases the consumption of oxygen because more organic matter decomposition will eventually result. As a consequence, oxygen levels in the water may become too low for fish. Weed growth may be stimulated by nutrients, and weeds interfere with boating and swimming. The problem is more acute today because many cities have grown rapidly while the amount of water available for diluting and carrying away the effluent has remained about the same. At the same time, many cities are depleting their underground water supplies, which is a major source of water for use in sewage disposal.

To combat these problems, researchers at Pennsylvania State University set up experiments to use the soil as a living filter for sewage effluent disposal. The researchers expected microorganisms to degrade detergents (similar to the problem of pesticide degradation) and microorganisms and higher plants to immobilize nutrients as the effluent water slowly percolated through the soil. By applying water in excess of the potential evapotranspiration, they expected the "filtered" water to migrate downward and eventually recharge the aquifer for reuse. The water was applied with a sprinkling system (see Fig. 6-17).

The experiment has been a success. For a community of 10,000 people producing 1 million gallons of effluent a day, the

Fig. 6-17 Water application of sewage effluent in a forest with a sprinkler system in winter at Pennsylvania State University. (Photo USDA.)

researchers found that an application of 2 inches per week on 129 acres of land was satisfactory. Under these conditions over 80 percent of the effluent water migrated deep enough to be considered aquifer recharge water. After 3 years the water recovered at a depth of 4 feet showed that over 90 percent of the hard detergents had been removed, and there was almost complete removal of the nitrogen and phosphorus (two major nutrients that have been associated with eutrophication). Furthermore, the trees and crops growing on the land were greatly stimulated.

Oil and Natural Gas Contamination of Soils

Recent construction of the Alaskan Pipeline has created an awareness and concern for oil spills on the land. Many of the soils along the Alaskan Pipeline have pergelic temperature regimes, and there has been uncertainty about the effect of construction on the melting of permafrost. Oil spills and gas leaks near oil wells and pipelines, however, are as old as the petroleum industry. Many studies have been conducted on the effect of petroleum contamination on soils and plant growth, and a summary of the major affects will be presented.

The first result of an oil spill or natural gas leak is the displacement of soil air and the creation of an anaerobic soil. Any vegetation is likely to be killed by a lack of soil oxygen. Over 100 species of microbes can decompose petroleum products. The crude oil and natural gas are a good energy source for microorganisms, and their growth is greatly stimulated. Reducing conditions accompanying the decomposition results in large increases in available iron and manganese. It is suspected that toxic levels of manganese for higher plants are produced in some cases.

In severe cases of oil spills the soil aggregates are broken down and dispersion results. There is usually an increase in micropores and a reduction in bulk density. Cultivation and aeration of the soil hasten the return of the soil to an aerated state and the return of the vegetation. In time the oil or natural gas is decomposed and the soil returns to almost normal conditions. The low soil temperatures in Alaska would slow down dissipation of the oil by the microorganisms. A study of 12 soils exposed to contamination near Oklahoma City showed that the organic matter and nitrogen contents had been increased about 2.5 times. The increased nitrogen supply in soils that have been contaminated probably explains the increased crop yields observed on old oil spills.

Natural gas leaks are not uncommon in cities. The death of grass and trees is due to lack of soil air, which is displaced by the natural gas or the strongly reducing soil conditions. Sometimes the problem can be diagnosed by smelling the loss of natural gas from a crack in a nearby sidewalk or driveway. (Methane is odorless; however, an odor has been added to public gas supplies.)

Use of Soil for Animal Waste Disposal

The concentration of large numbers of animals on specialized livestock farms has created serious animal waste disposal problems. Odors from stored or rotting manure are commonly offensive. Burning animal wastes can pollute the air. Laws exist that prohibit the disposal of animal wastes into water courses. Chemical processing of animal wastes is expensive. As a consequence, animal waste disposal is a major problem for many livestock producers. The soil is a natural medium for waste disposal, and much attention is now being directed toward using the soil for animal waste disposal.

Magnitude of the Animal Waste-Disposal Problem

An interesting comparison between the production of human and animal wastes has been made for Minnesota. In Minnesota there are about 14 million chickens and turkeys, 4 million cattle, $1\frac{1}{2}$ million dairy cows, $2\frac{1}{2}$ million hogs, and $\frac{3}{4}$ million sheep. These animals produce waste equivalent to that produced by over 60 million people. Since the population of Minnesota is about 4 million, the animal wastes produced are about 15 or more times greater than the human wastes.

Some feedlots in the United States carry as many as 10,000 head and produce waste equivalent to a city of 164,000 people (see Fig. 6-18). Some data on animal population and quantity of waste produced in the United States are given in Table 6-3. The data show that about $1\frac{1}{2}$ billion tons of waste are produced annually; about two thirds is solid manure and about one third is liquid. If the waste resulting from dead animal carcasses, paunch manure from abattoirs, and used bedding are included, the total amount of animal waste produced is about 2 billion tons per year.

Capacity of Soil for Waste Disposal

Under normal conditions, the soil microorganisms exist in a near state of starvation. Readily available food or energy supplies are usually limited. The addition of decomposable wastes, such as manure, produce a rapid increase in the numbers and activity of microorganisms. As a result, the soil can dispose of enormous quantities of animal waste. It seems reasonable to expect that a 1 inch layer of manure could be incorporated into the soil every year without difficulty,

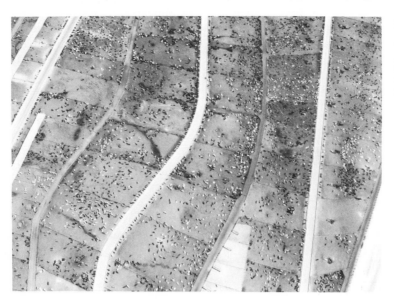

Fig. 6-18 Aerial view at hugh feedlot at Coalinga, Calif. The waste disposal problem for each 10,000 cattle is equal to that of a city of 164,000. (Courtesy EPA.)

Table 6-3
Animal Population and Waste Production in the United States

| Livestock | 1965 Population, millions | Annual Waste Production, million tons | |
		Solid Waste	Liquid Waste
Cattle	107	1004.0	390.0
Horses	3	17.5	4.4
Hogs	53	57.3	33.9
Sheep	26	11.8	7.1
Chickens	375	27.4	—
Turkeys	104	19.0	—
Ducks	11	1.6	—
Total	—	1138.6	435.4

From Wadleigh, 1968.

assuming favorable soil moisture and temperature. Assuming also that the manure is largely water, a 1-inch thick application of manure would be less than 100 tons per acre. If the manure contains about 10 pounds of nitrogen per ton, the nitrogen application would be about 700 to 900 pounds per acre (784 to 1008 kilograms per hectare). On this basis, nitrate pollution of groundwater would likely result in humid regions from 100 tons per acre applications of manure. From the practical standpoint, it appears that considerations other than the decomposing capacity of the soil limit the rate of waste application. Researchers in Canada analyzed the manure on many kinds of farms and concluded that to prevent nitrate contamination of groundwater and not adversely affect corn yields, a minimum $\frac{1}{2}$ acre of land was needed for 1000 broilers, 100 laying hens, 10 hogs (30 to 200 pounds), 2 fedder cattle (40 to 1100 pounds), or 1 dairy cow (1200 pounds). On this basis, a 10,000 head feedlot operation would require 2500 acres (1012 hectares) for the disposal of the manure.

Earth Moving Activities of Soil Animals

All soil animals participate as consumers and aid in the cycling of nutrients and energy. Their harvesting and food storage activities result in transference of nutrients to their nests or burrows. Many of the larger soil animals move soil to such an extent that they affect soil formation. The emphasis in this section is on animals as earth movers.

Earthworm Activities

Earthworms are perhaps the best-known earth movers. Darwin made extensive studies of earthworms and found that they may deposit 10 to 15 tons of castings per acre (4.4 to 6.6 metric tons per hectare) on the soil surface a year, resulting in the buildup of a 1-inch surface layer every 12 years. Other researchers have reported over 100 tons per acre of castings in a 6-month rainy period. This activity produces thicker than normal dark-colored surface layers in some forest soils and buries stones and artifacts that are lying on the top of the soil. The burying of artifacts is important to archeologists.

Ants and Their Activities

Although most persons appear to be more conscious of earthworms and their activities, the activities of ants are perhaps more important. Harvester ants are a pest in many places, including the southwestern part of the United States. Harvester ants denude the area surrounding their nests to distances as great as 10 feet or more. Thorp estimated 20 ant hills per acre with denuded areas ranging from 6 to 20 feet in diameter. Assuming an average denuded area of 13 feet diameter, about 6 percent of the land surface would be denuded. Such harvesting of vegetation by ants can be of economic importance (Fig. 6-19). On range lands the forage for wildlife and cattle is reduced and the bare denuded areas are more subject to erosion. Harvester ants also gather seeds for food; this retards the reseeding of natural grasslands.

Ants transport large quantities of material from within the soil and deposit the material on the surface. Some of the largest ant mounds are a few feet high and more than 10 feet in diameter. The effect of this transport is comparable to that of earthworms in creating thicker dark-colored A horizons and burying objects lying on the surface. A study of ant activity on a prairie in southwestern Wisconsin showed that ants brought material to the surface from depths greater than 5 feet and built mounds about 6 inches high and over 1 foot in diameter (see Fig. 6-20). Furthermore, it was estimated that 1.7 percent of the land was covered with mounds. Assuming that the average life of a mound is 12 years, the entire land surface would be reoccupied every 600 years. The researchers believe that this evidence supports the view that the incorporation of subsoil material (Bt horizon) into the A horizon has helped to produce a thicker dark-colored A horizon with a greater than "normal" clay content. The increase in clay content is supported by the fact that the clay contents of A horizons in a nearby forest are only 10 percent com-

Fig. 6-19 Denuded areas surrounding Red Harvester ant mounds in an alfalfa field. (Photo USDA.)

Fig. 6-20 Ant (*Formica cinera*) in a Prairie soil in southwestern Wisconsin. Upper photo shows ant mounds over 6 inches (15 centimeters) high and over 1 foot (30 centimeters) in diameter. The lower sketch shows soil horizons and location of ant channels; numbers refer to the number of channels observed at the depths indicated. (Courtesy F. D. Hole, Soil Sur. Div., Wisc. Geol. and Nat. History Survey, University of Wisconsin.)

pared to 22 percent clay in the A horizons on the prairie.

Leaf-cutting ants march long distances to cut fragments of plant leaves and stems and bring them to their nests to feed fungi. The fungus is used as food. Organic matter is incorporated into the soil depths and nutrients become concentrated in the nest sites.

Termite Activities

Termites inhabit tropical and subtropical areas. They exhibit great diversity in food and nesting habits. Some feed on wood,

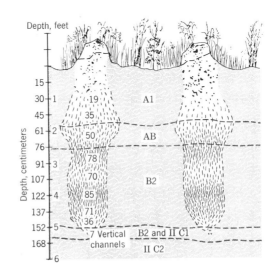

some feed on organic refuse, and others cultivate fungi. Protozoa in the digestive tract of many termites aid in the digestion of woody materials. Some species build huge nests up to 10 feet high and 40 to 50 feet in diameter. Thirty such mounds have been observed per acre, accounting for 10 million pounds of soil per acre (11,400 metric tons per hectare). Most mounds are of a smaller scale, and some termites have nests in the soil and tunnels on the surface to permit foraging for food. Material is brought to the surface from depths as great as 10 feet. Nutrients and soil particles are transferred to the surface. Termites have had an enormous effect on many tropical soils where their earth-moving activities have been going on for hundreds of thousands of years. In summary, ants and termites create channels in soils and transport soil materials that tend to alter or obliterate soil horizons. A concentration of nutrients builds up where the mounds are located because plant materials are stored and fecal material accumulated there. Some farmers in southeast Asia recognize this and make use of the higher fertility of areas occupied by mounds.

Earth-Moving Activities of Rodents

Many rodents, including mice, ground squirrels, marmots, gophers, and prairie dogs, inhabit the soil. A very characteristic microrelief called *mima mounds*, which consists of small mounds of earth, is the work of gophers in Washington and California. The mima mounds occur on shallow soils and appear to be the response of gophers to build nests in dry soil near the top of mounds. Successive generations of gophers at the same site create mounds that range from 1 to 3 feet high and 15 to 100 feet in diameter.

Extensive earth moving by prairie dogs has been documented. An average of 17 mounds per acre were observed near Akron, Colorado; they consisted of 39 tons of soil material. The upper 6 to 10 feet of soil was loess (wind-deposited silt), underlain by sand and gravel. All of the mounds observed contained sand and gravel that had been brought up from depths over 6 feet. Prairie dog activity had changed the surface soil texture from silt loam to loam on one third of the area. Abandoned burrows filled with dark-colored surface soil are common in grassland soils and are called *crotovinas*.

References

Allison, F. E., *Soil Organic Matter and Its Role in Crop Production*, Elsevier, New York, 1973.

Andrews, William A., Editor, *A Guide to the Study of Soil Ecology*, Prentice-Hall, Englewood

Arkley, Rodney J., and Herrick C. Brown, "The Origin of Mima Mound (Hogwallow) Mic-Cliffs, N.J., 1975.
rorelief in the Far Western States," *Soil Sci. Soc. Am. Proc., 18*:195–199, 1954.

Barley, K. P., "The Abundance of Earthworms in Agricultural Land and Their Possible Significance in Agriculture," in *Advances in Agronomy*, Vol. 13, Academic, New York, 1961, pp. 249–268.

Baxter, F. Paul, and F. D. Hole, "Ant (Formica cinerea) Pedoturbation in a Prairie Soil," *Soil Sci. Soc. Am. Proc., 31*:425–428, 1967.

Bormann, F. Herbert, and Gene E. Likens, "The Nutrient Cycles of an Ecosystem," *Sci. Am., 223*:92–101, 1970. Reprint 1202.

Bowen, Ezra, *The High Sierra*, Time, New York, 1972.

Brady, Nyle, C., *The Nature and Properties of Soils*, 8th edition, Macmillan, New York, 1974.

Buol, S. W., F. D. Hole, and R. J. McCracken, *Soil Genesis and Morphology*, Iowa State University, Ames, Iowa, 1973.

Burges, A., and F. Raw, *Soil Biology*, Academic, New York, 1967.

REFERENCES

Clark, Francis E., "Living Organisms in the Soil," in *Soil*, USDA Yearbook, Washington, D.C., pp. 147–165, 1957.

Cole, LaMont C., "The Ecosphere," *Sci. Am.*, April 1958, Reprint 144.

Ellis, Boyd G., "Nutrient Cycling in Agricultural Systems," in *Envir. Quality: Now or Never*, C. L. San Clemente, Editor, Cont. Ed. Service, Michigan State University, 1972.

Ellis, Roscoe, Jr., and Russell S. Adams, Jr., "Contamination of Soils by Petroleum Hydrocarbons," in *Advances in Agronomy*, Vol. 13, Academic, New York, 1961, pp. 197–216.

Kevan, D. Keith McE., *Soil Animals*, Philosophy Library, New York, 1962.

Kardos, L. T., "Waste Water Renovation by the Land—A Living Filter," in *Agriculture and the Quality of our Environment*, AAAS Pub. 85, Washington, D.C., 1967, pp. 241–250.

Martin, W. P., "Soil as an Animal Waste Disposal Medium," *Jour. Soil and Water Conservation*, 25:43–45, 1970.

Marx, D. H., and W. C. Bryan, "The Significance of Mycorrhizae to Forest Trees," in *Forest Soils and Forest Land Management*, Proc. Fourth Nor. Am. Forest Soils Conf., Laval University, Quebec, 1975, pp. 107–117.

Odum, Eugene P., *Ecology*, Holt, Rinehart and Winston, New York, 1963.

Richards, B. N., *Introduction to the Soil Ecosystem*, Longmans, New York, 1974.

Thorp, James, "Effects of Certain Animals That Live in Soils," *Sci. Month.*, 42:180–191, 1949.

Schaller, Friedrich, *Soil Animals*, University of Michigan, Ann Arbor, Mich., 1968.

Stoeckeler, Joseph H., and Harold F. Arneman, "Fertilizers in Forestry," *Adv. in Agronomy*, Vol. 12, Academic, New York, 1960, pp. 127–195.

Walters, E. M. P., *Animal Life in the Tropics*, George Allen and Unwin, London, 1960.

Wadleigh, C. H., "Wastes in Relation to Agriculture and Forestry," *USDA Misc. Pub.*, 1065, 1968.

Webber, L. R., "Animal Wastes," *Dept. Soil Sci. Ann. Prog. Report*, University of Guelph, Ontario, 1967.

Wilson, P. W., *The Biochemistry of Symbiotic Nitrogen Fixation*, University of Wisconsin Press, Madison, Wis., 1940.

7
SOIL ORGANIC MATTER

Almost all of the life in the soil is dependent on organic matter for energy and nutrients. For thousands of years mankind has recognized the importance of organic matter in food production. The story of how the Indian, named Squanto, helped the Pilgrims raise corn by burying a dead fish near each hill is well known. Perhaps the most poetic expression of the effects of organic materials on plant growth was expressed by Omar Khayyam.

I sometimes think that never blows so red
The Rose as where some buried Caesar bled.

Although organic matter in soils is very beneficial, Liebig pointed out over 100 years ago that soils composed entirely of organic matter are naturally very infertile. This chapter clarifies the role and importance of the soil organic matter.

Humus Formation and Characteristics

In the previous chapter we discussed the decomposition of plant residues and the synthesis of many compounds by soil

157

organisms. As a result of these activities, the soil contains an enormous number of organic compounds in various states of decomposition. *Humus* is the word used to refer to the organic matter that has undergone extensive decomposition and is resistant to further alteration.

Humus Formation

The organic residues added to soils are not decomposed as a whole, but the chemical constituents are decomposed independently of one another. In the formation of humus from plant residues there is a rapid reduction of the water-soluble constituents, of the celluloses, and of the hemicelluloses; a relative increase in the percentage of lignin and lignin complexes; and an increase in the protein content. The new protein is believed to be formed for the most part through the synthesizing activities of the microorganisms. The lignin in humus originates mostly from plant residues with perhaps certain chemical modifications. Lignin has a 6-carbon ring structure that resists enzymatic decomposition. Reactions of lignin with amino acids and other substances form very resistant compounds and enhance the accumulation of both lignin

and protein materials in humus (Table 7-1). Fats and waxes have intermediate resistance to decomposition.

Normally, proteins are readily decomposed in soils. Additional mechanisms that have been proposed to account for the accumulation of proteins in humus. Two will be mentioned. First, there is reason to believe that the protein molecules can be adsorbed on the surface of clay minerals and rendered resistant to decomposition. Second, enzymes that decompose proteins may also be adsorbed by clay minerals so that the proteins are less susceptible to decomposition. That the clay plays an important role is supported by the fact that soils high in clay tend to have a high organic matter content. This slow rate of decomposition of humus is obviously of considerable practical importance. It offers a means whereby nitrogen can be stored in the soil and released gradually.

A general schematic summary of the processes leading to humus formation is presented in Fig. 7-1. The material commonly referred to as humus includes the mass of plant residues undergoing decomposition, together with the synthesized cell substance and certain intermediary and end products. It is constantly changing in composi-

Table 7-1

Partial Composition of Mature Plant Tissue and Soil Organic Matter

Component	Percent	
	Plant Tissue	Soil Organic Matter
Cellulose	20–50	2–10
Hemicellulose	10–30	0–2
Lignin	10–30	35–50
Protein	1–15	28–35
Fats, waxes, etc.	1–8	1–8

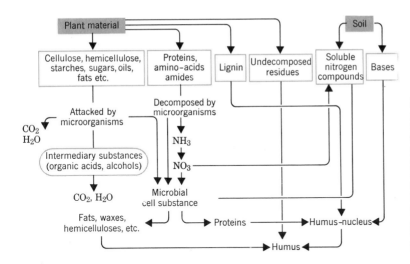

Fig. 7-1 Schematic representation of the mechanism of the formation of humus in the decomposition of plant residues in the soil. (Reproduced from S. A. Waksman, *Humus*, Williams and Wilkins Co. Copyright © 1938. By permission of the Williams and Wilkins Co.)

tion. It is better, therefore, to speak of humus not as a single group of substances but, indeed, as a state of matter, which is different under varying conditions of formation.

Characteristics and Properties of Humus

Humus is practically insoluble in water, although some of it may go into colloidal suspension in pure water. To a large extent it is soluble in dilute alkali, and certain of the humus constituents may dissolve in acid solutions.

One of the most important and characteristic properties of humus is its nitrogen content, which usually varies from 3 to 6 percent, although the nitrogen concentration may be frequently lower or higher than these figures. The carbon content is less variable and is commonly considered to be 58 percent. Assuming 58 percent carbon, the organic matter content can be estimated by multiplying percentage of carbon by 1.724. The carbon–nitrogen ratio (C/N) is of the order of 10 to 12. This ratio varies with the nature of the humus, the stage of

its decomposition, the nature and depth of the soil, and climatic and other environmental conditions under which it is formed.

Humus is also an important reservoir of phosphorus and sulfur. The ratio of C:N:P:S in humus is about 100 to 120:10:1:1.

Another important property of humus is its high cation-exchange capacity. Cation exchange is associated with several chemically active groups in both living and dead organic matter. One of the most important groups is *carboxyl* (—COOH). During humification of organic matter, lignin is altered in such a way that there is a decrease in noncation-exchanging groups such as methoxyl (—OCH_3) and an increase in the cation-exchanging carboxyl groups. As a result, the cation-exchange capacity of humus is many times greater than that of the organic residues originally added to the soil. Cation exchange sites adsorb cations such as Ca, Mg, and K and, in so doing, humus acts similarly to clay in retaining available nutrients against leaching and maintaining the nutrients in a form available to higher plants and microorganisms. The cation-exchange phenomenon of humus (and other kinds of soil organic matter) is illustrated by the following equation.

$$-R-C \overset{O}{\underset{OH}{\diagdown}} + KCl \longrightarrow$$

$$-R-C \overset{O}{\underset{OK}{\diagdown}} + HCl \qquad (1)$$

The equation shows how water-soluble potassium chloride reacts with carboxyl groups of humus. The potassium (K) is exchanged for the H of the carboxyl group. The K is adsorbed with enough energy to retard its loss from the soil by leaching, but the K is still readily available for plant use.

Humus absorbs large quantities of water and exhibits the properties of swelling and shrinking. It does not exhibit as pronounced properties of adhesion and cohesion as the mineral colloids do and is less stable because it is subject to microbial decomposition. It has already been shown that soil humus is an important factor in aggregation (structure formation). Humus possesses other physical and physiochemical properties that make it a highly valuable soil constituent.

The Dual Nature of Humus

Radiocarbon age determinations of soil organic fractions commonly show ages between 500 and 2000 years. This means far less than 1 percent mineralization or decomposition per year. On the other hand, about 2 to 3 percent of the *N in humus* is mineralized each year in well-drained agricultural soils. These rather contradictory facts lead to the concept that soil humus is composed of at least two kinds of organic materials that differ greatly in resistance to decomposition.

When lignin reacts with various soil constituents, it forms compounds of great resistance. Inclusion of organic matter into the clay-mineral matrix of the soil may render organic matter inaccessible for many years. These humus forms that comprise the bulk of humus contribute only a small amount of nutrients to plant growth in a year. Most of the nutrients mineralized in a given year come from a smaller but active humus fraction that is very important in nutrient cycling. This fraction consists of dead plant and animal residues in various states of decomposition and short-lived organisms that soon become the substrate of other organisms.

Amount and Distribution of Organic Matter in Soils

As the rocks and minerals of the earth's crust decomposed, mineral elements were made available to plants; and, as supplies of nitrogen in usable chemical combinations were produced from the store of nitrogen in the air, plants grew, died, and contributed their remains to the soil. Thus, organic matter began to accumulate. As the supply of available plant nutrients in the soil increased, the accumulation of soil organic matter increased accordingly. This condition continued until an equilibrium was reached at which the rate of organic matter accumulation was equal to the rate of decomposition. This section will discuss factors that influence the amount of organic matter in soils including climate, vegetation, drainage conditions, cultivation, and soil texture.

Influence of Climate and Vegetation on Organic Matter Content of Soils

Generally, as the quantity of organic residues added annually to soils is increased, there is an increase in the total organic matter content. One would expect the soils in deserts to contain very little organic matter because the annual additions of organic matter from plant growth are very small. With increasing precipitation and an accompanying increase in the annual production of organic matter, there is an increase in the organic matter content of soils. On the plains in the United States from eastern Colorado to Indiana the annual precipitation increases from about 15 to 35 inches. This is accompanied by a shift from widely spaced bunch and short grasses to tall grass and an increase in the organic matter content of soils from about 80 to 160 tons per acre to a depth of 40 inches (see Fig. 7-2). Similar changes occur in Argentina from the Andes Mountains to Buenos Aires and in the southern part of the Soviet Union from a south to north direction.

The eastern United States is a forested region and the soils have considerably less organic matter than nearby soils developed under tall grass. Furthermore, with increasing average annual temperature in the forested area from north to south, the organic matter content of the soils decreases (Fig. 7-2). A major cause is the increased rate of microbial activity and decomposition of organic matter with increasing temperature (also see Fig. 10-5). In the tropics, however, soils generally have more organic matter than the soils in the southeastern United States. It appears that the absence of a killing frost in the tropics favors more the production of organic matter than it favors decomposition of organic matter. Many tropical soils have high clay content and more amorphorus clays (with high specific surface), resulting in the effective protection of organic matter from decomposition.

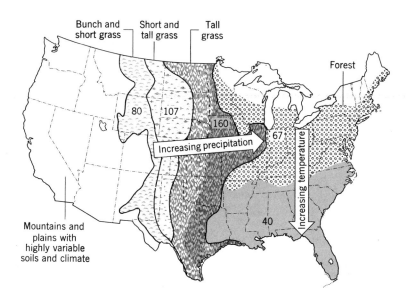

Fig. 7-2 Generalized map showing organic matter content of soils (tons per acre to 40 inches) as related to climate and vegetation. (Adapted from Schreiner and Brown, 1938.)

Organic Matter in Forest Versus Grassland Soils

The settlers that colonized America adapted to farming on soils developed under forest vegetation. By the early 1800s the settlers had spread westward to the tall grass prairie lands of the central United States. The tough sod made it difficult to plow the land and new tools and techniques had to be developed. At first, the prairie lands were avoided but once they were broken, the superiority of grassland soils over forest soils was readily apparent. Now, the great productivity of the American prairies and the Argentine pampa are well known. One reason for the high productivity of these grassland soils is related to the amount and distribution of organic matter. Studies show that grassland soils, as compared to nearby forest soils, have (1) about twice as much organic matter in the soil profile, and (2) a more gradual decrease of organic matter with increasing soil depth (Fig. 7-3).

The explanation for the differences in amount and distribution of organic matter in forest and grassland soils is related to differences in the growth of the plants and how the plant residues become incorporated into the soil. The roots of grasses are short-lived, and each year the decomposition of dead roots contributes to the quantity of humified organic matter. Furthermore, the quantity of roots decreases gradually with increasing soil depth (see Fig. 3-9 for the root system of oats). In the forest, by contrast, the roots are long-lived and the annual addition of plant residues is largely as leaves and dead wood that fall onto the surface. Some of the residues decompose on the surface but small animals transport and mix some of the surface litter with a relatively thin layer of top soil.

In the hardwood forest in southern Wisconsin (where earthworms are active) it was found that 36 tons of organic matter per acre existed in the upper 6 inches of soil (A1 horizon) and only 11 tons per acre in the next deeper 6-inch layer of soil (A2 horizon).

Another interesting fact shown in Fig. 7-3 is that there is a similar amount of total organic matter in each ecosystem but, in the forest, most of the organic matter exists in the *standing trees* while in the prairie ecosystem over 90 percent of the organic matter exists within the *soil*. When settlers cleared forests, they burned or harvested the trees and, in so doing, removed about half of the organic matter. Breaking of the prairie land, by comparison, left virtually all of the organic matter in the soil, even if the grass was burned off before plowing. The differences in amount and distribution of organic matter is one of several explanations for the larger crop yields on the grassland soils. Even today, with good soil management, the average yields of corn on well-drained grassland soils is about 10 to 20 bushels per acre greater than for well-drained forest soils in the central United States. Let us consider next the changes in soil organic matter that were produced when the established equilibrium level of organic matter was disturbed by putting the land into cultivation.

Organic Matter Changes by Cultivation

Even on nonerosive land that is brought under cultivation, rapid losses of organic matter usually occur. It has been observed that the losses are most rapid immediately after farming is started and, thereafter, the rate of disappearance is decreased; ultimately the organic content of the soil will reach a new equilibrium level.

AMOUNT AND DISTRIBUTION OF ORGANIC MATTER IN SOILS

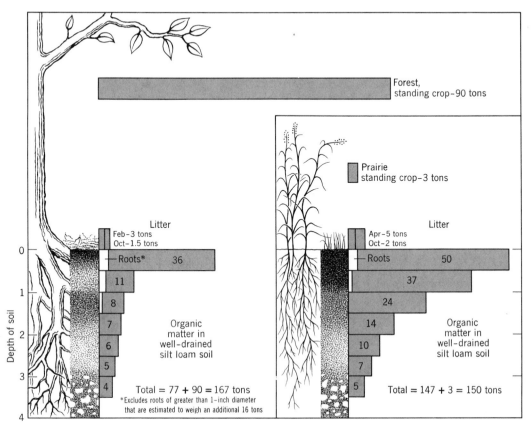

Fig. 7-3 The distribution of organic matter in forest (white oak, black oak) and prairie (big bluestem, Indian grass) ecosystems in south central Wisconsin. (Adapted from Nielsen and Hole, 1963. Courtesy of F. D. Hole, Soil Survey Division, Wisconsin Geological and Natural History Survey, University Extension, University of Wisconsin.)

It has been found at the Missouri Agricultural Experiment Station that, as a result of cultivation over a period of 60 years, soils in a noneroded condition lost over one third of their organic matter, the losses being much greater during the earlier than the later periods. The organic matter losses amounted to about 25 percent the first 20 years, about 10 percent the second 20 years, and only about 7 percent the third 20 years. In other words, a new equilibrium level was almost attained after about 30 years (Fig. 7-4).

Soils in arid regions naturally have very low organic matter contents. Irrigating arid-region land and producing crops result in large increases in the amount of organic matter returned to the soil each year. As a consequence, irrigating arid lands and growing crops results in the establishment of a new equilibrium organic matter level higher than the original level.

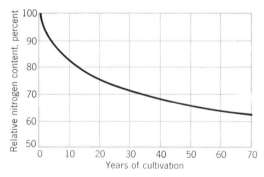

Fig. 7-4 Decline of soil nitrogen (or organic matter) with length of cultivation period under average farming practices in the Midwest. (After Jenny, *Missouri Agr. Exp. Sta. Bull.*, 324, 1933.)

Maintenance and Restoration of Soil Organic Matter

Although there is a rapid depletion of the organic matter in the soil immediately after virgin lands of humid regions are brought under cultivation, there is some consolation in the fact that this high rate does not continue indefinitely. It has been emphasized that after a period of heavy loss, a fairly constant level is attained during a long period of continued cultivation; the level is determined by the environment associated with a particular soil. Once the organic content has reached a low level, restoring the organic matter to its original level would require that the original vegetation be reestablished. In time a new equilibrium level of organic matter content would be achieved that would be the same or similar to the level that existed before the land was cultivated. While the land is being farmed, it is virtually impossible and much too expensive to maintain the organic matter content similar to that which existed in the virgin soil. It is, therefore, usually unwise and uneconomical to maintain the organic matter above a level consistent with good crop yields. Attention should be directed toward the frequent additions of small quantities of fresh organic materials instead of toward practices of maintaining the organic matter content at any particularly high level.

In a consideration of the maintenance of soil organic matter, the amount of crop residues that must be returned to maintain a given organic matter content depends on soil and climatic conditions. It is interesting to note that 3600 pounds per acre of crop residues were needed annually to maintain the organic matter content of blackland soils near Dallas, Texas (Fig. 7-5). Cropping systems that returned more than this amount showed an increase in the organic matter content after 12 years, whereas those returning less than this amount caused a decline in the organic matter content. Only the tops were measured and root residues in addition to top residues were returned to the soil. It would be reasonable to allow 1000 pounds for roots, making the total weight of residues added per year to maintain the organic matter content of these soils equal to about 4600 pounds.

Fig. 7-5 Relation of crop residue returned to the soil to the change in soil organic matter content. (From "Farming Systems for Soil Improvement in the Blacklands," Texas Research Foundation, Bulletin 10, 1961, p. 18. By permission of the Texas Research Foundation, Renner, Texas.)

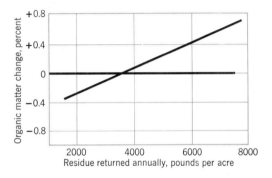

Organic Matter Content versus Soil Texture

Locally, there tends to be a correlation between the clay content of the soil and the content of organic matter. The greater combined supply of water and nutrients favors the production and accumulation of more organic matter in the finer textured soils. Clay also adsorbs decomposing enzymes that become inactivated. Organic molecules adsorbed on clays are partially protected from decomposition by microorganisms. As the content of organic matter in the soil increases, the content of nitrogen and phosphorus increase, since they are important constituents of organic matter. The content of nitrogen and phosphorus in some New York soils is used to illustrate the general relationship between soil texture and organic matter content in Fig. 7-6.

Organic Soils

In the shallow water of lakes and ponds, plant residues accumulate instead of decomposing under anaerobic conditions in the water. Consequently, soils consisting almost entirely of organic matter develop. Plant tissue and pollen grains in peat can be easily identified. The remarkable preservation capacity of some swamp and bog waters can be illustrated by the fact that human bodies up to 3000 years old have been found in peat bogs. One of the best preserved is the 2000-year-old Tollund man found in 1950 in Denmark. The facial expression at death and the bristles of the beard of the Tollund man were well preserved. Excellent finger prints were made and an autopsy revealed that his last meal consisted mainly of seeds, many of them weed seeds. When found, the Tollund man was buried under 7 feet of peat that had formed in the 2000 years after his burial.

In Chapter 3 organic soils were distinguished from mineral soils by having 20 to 30 percent organic matter (depending on the texture of the mineral soil material). A characteristic feature of organic soils is a

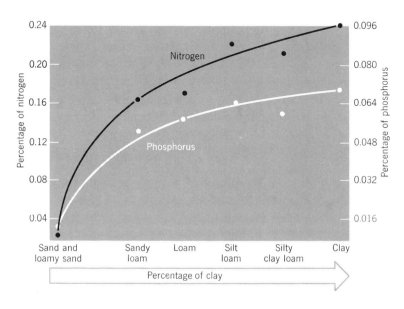

Fig. 7-6 Nitrogen and phosphorus contents of several New York soil classes illustrate the geographically local trend of increasing organic matter content with increasing clay content of the soil.

stratification or layering that represents changes in the kind of plants that produced the organic matter as a result of changes in climate or water level. Such a series of horizons that developed in Sweden is shown in Fig. 7-7. The woody peat layer is indicative of a dry period and was preceded and followed by wet periods when the vegetation was sphagnum moss. The 15 feet of peat accumulated in 9000 years or at the rate of 1 foot each 600 years. About 1 foot every 200 to 800 years is the usual range.

Fig. 7-7 Stratification found in a peat soil in Sweden where changes in climate produced changes in the kind of vegetation that formed the peat. (From Davis and Lucas, 1959.)

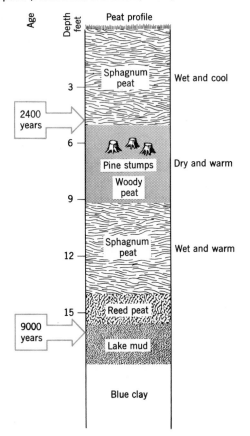

Organic soils must be drained before they can be used for crop production. The upper soil layers then become aerobic and the peat begins to decompose. This converts undecomposed peat into well-decomposed muck. In time the entire organic soil above the underlying mineral soil may disappear as a result of decomposition. This is a serious problem in warm climates, as in southern Florida, where the high annual temperature stimulates rapid decomposition.

Some Organic Matter Management Considerations

Organic matter plays many important roles in soils. Since soil organic matter originates from plant remains, soil organic matter originally contained all nutrients needed for plant growth. Organic matter per se influences soil structure and tends to promote a desirable physical condition. Soil animals depend on organic matter for food and contribute to a desirable physical condition by mixing soil and creating channels. Naturally, there is much interest in managing organic matter to make soils more productive. A discussion of these kinds of considerations will help place the value of organic matter in soils in better perspective.

Organic Matter Decomposition and Mineralization of Nutrients

The "pool" of organic matter in a soil can be compared to a lake. Changes in the level of water in a lake depend on the difference between the amount of water entering and leaving the lake. This idea applied to soil organic matter is illustrated at the top of Fig. 7-8. Soil organic matter decomposes in mineral soils at a rate equal to about 1 to 4 percent per year. Assuming a 2 percent rate and 40,000 pounds of organic matter per

SOME ORGANIC MATTER MANAGEMENT CONSIDERATIONS

General Equation for the Loss or Gain in Soil Organic Matter

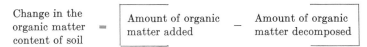

| Change in the organic matter content of soil | = | Amount of organic matter added | − | Amount of organic matter decomposed |

A Case Where the Loss and Gain of Soil Organic Matter Is in Equilibrium

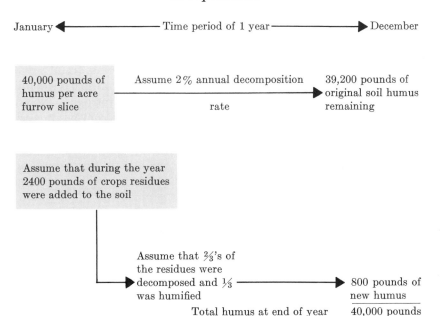

January ◄——————— Time period of 1 year ————————► December

| 40,000 pounds of humus per acre furrow slice | Assume 2% annual decomposition rate ——————► | 39,200 pounds of original soil humus remaining |

Assume that during the year 2400 pounds of crops residues were added to the soil

Assume that ⅔'s of the residues were decomposed and ⅓ was humified ——————► 800 pounds of new humus

Total humus at end of year   40,000 pounds

Fig. 7-8 Schematic illustration of the equilibrium concept of soil organic matter as applied to a representative plow layer (2 million pounds) containing 2 percent organic matter.

acre furrow slice, 800 pounds of soil organic matter would be lost or decomposed each year. On the other hand, if 800 pounds of humus was formed from the residues added to the soil, the organic matter content of the soil would remain the same from one year to the next and the soil would be at the equilibrium level (Fig. 7-8).

Decomposition of 800 pounds of humus would result in the mineralization of about 40 pounds of nitrogen, assuming that soil organic matter is 5 percent nitrogen (800 × 0.05). Assuming that the ratio of N:P:S in soil organic matter is 10:1:1, there would

be 4 pounds of phosphorus and sulfur mineralized. Other nutrients are also mineralized, but the availability of most of the other nutrients is more related to mineral weathering, soil pH, and the like. The more organic matter that is added to soils each year, the more nutrients will be mineralized for plant growth.

Carbon–Nitrogen Ratio and the Decomposition of Organic Residues

The soil microbes are the primary agents for organic matter decay and have certain dietary requirements. Of major concern from a practical standpoint is the amount of carbon relative to nitrogen in the decomposing organic matter. A problem arises when the nitrogen content of decomposing organic matter is small, because the microbes may become deprived of nitrogen and compete with the higher plants for whatever available nitrogen exists in the soil. Since the carbon content of organic materials is relatively constant between about 40 to 50 percent, while the nitrogen content varies manyfold, the carbon–nitrogen ratio is a convenient way to express the relative content of nitrogen. Thus, the carbon–nitrogen ratio of organic materials is an indication of the likelihood of a nitrogen shortage and competition between microbes and higher plants for whatever nitrogen is available in the soil.

The carbon–nitrogen ratios of some organic residues that are frequently added to soils are given in Table 7-2. They range from 10 to 12 for humus and immature sweet clover tissue to 400 for sawdust.

Table 7-2

The Carbon–Nitrogen Ratio of Some Organic Materials

Material	C/N Ratio
Soil humus	10
Sweet clover (young)	12
Barnyard manure (rotted)	20
Clover residues	23
Green rye	36
Cane trash	50
Corn stover	60
Straw	80
Timothy	80
Sawdust	400

Data are taken from several sources. The values are approximate only, and the ratio in any particular material may vary considerably from the values given.

Materials with small or narrow ratios are relatively rich in nitrogen, while those with higher or wider ratios are relatively low in nitrogen.

Mature plant residues that provide the raw material for microbial decomposition contain about 50 percent carbon and 1 percent nitrogen ($C/N = 50$). The carbohydrates are quickly decomposed, and a large increase in microbial activity results. During decomposition, mineralization and immobilization of nutrients occur simultaneously. Of particular concern is whether the immobilization of nitrogen exceeds mineralization. If so, microorganisms will compete with higher plants for any available nitrogen, including that being mineralized from humus decomposition, and higher plant growth will be reduced (Fig. 7-9).

The *nitrogen factor* is a convenient term to express the extent to which a material is deficient in nitrogen for decomposition. It is defined as the number of units of inorganic nitrogen that must be supplied to 100 units of organic material in order to prevent a net immobilization of nitrogen from the environment. A factor of approximately 0.9 represents straw. The nitrogen factor can be calculated for straw as follows. Assume that 100 pounds of straw contains about 40 pounds of carbon and $\frac{1}{2}$ pound of nitrogen. Assuming that 35 percent of the carbon will be assimilated by the microbes and that one tenth as much nitrogen will be assimilated as carbon, the nitrogen factor will be 0.9.

$$40 \times 0.35 = 14 \text{ pounds of carbon assimilated}$$

$$\tfrac{14}{10} = 1.4 \text{ pounds of nitrogen assimilated}$$

$$1.4 - 0.5 = 0.9 \text{ pound of nitrogen deficient}$$

Fig. 7-9 Paper mill sludge reduced the growth of young corn plants grown in the greenhouse when mixed with soil in quantities of one quarter or one half of the total soil volume. The sludge contains about 40 percent paper fiber (cellulose) and little if any nitrogen, so it has a wide or high carbon-nitrogen ratio. Addition of the sludge reduced the supply of available nitrogen for the corn.

The addition of 0.9 pound of inorganic nitrogen to the soil at the time 100 pounds of straw are incorporated should prevent the immobilization of nitrogen from the soil environment and prevent competition for nitrogen between a crop and the microbes. The above calculations assume that 65 percent of the carbon in the straw is converted to carbon dioxide during respiration in the decomposition process.

Immobilization exceeds nitrogen mineralization when carbon–nitrogen ratios are above 30 (Fig. 7-10). In the range 15 to 30, immobilization and mineralization are about equal. Mineralization exceeds immobilization when carbon–nitrogen ratio of the decomposing material is less than 15, as is the case with soil humus.

Several things may be done to prevent competition for nitrogen between the microbes and higher plants. Straw and similar residues may be burned instead of incorporating them into the soil. Such a practice deprives the soil of a source of organic matter, and the organic matter content of the soil will be less than where straw is periodically returned to the soil. Second, nitrogen fertilizer can be added if crops are to be grown immediately after the turning under of wide carbon–nitrogen ratio materials. This nitrogen can be used early in the year for the decomposition of residues of a previous crop and, a month or so later, when the decomposition of residues is largely completed, the nitrogen can be used by the crop.

Many urban people have organic matter residues with wide carbon–nitrogen ratios such as tree leaves, grass clippings, or other plant wastes from a garden. The carbon–nitrogen ratio of these materials can be lowered by *composting*. Composting consists

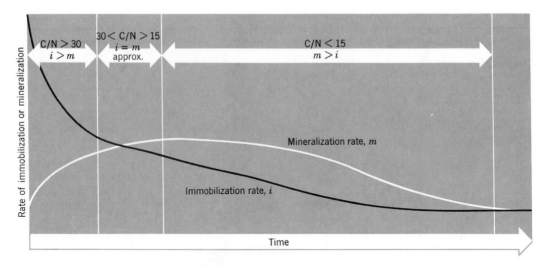

Fig. 7-10 The relationship between carbon-nitrogen ratios and nitrogen availability during decomposition of plant residues. (From Broadbent, 1957.)

of storing the organic materials in a pile while maintaining favorable moisture, aeration, and temperature relationships. As the organic matter is decomposed, much of the carbon, hydrogen, and oxygen are released as carbon dioxide and water. Nutrients, like nitrogen, are continually reused by the microbes and are conserved. Thus, while there is a loss of carbon, the amount of nitrogen remains about constant, resulting in a narrowing of the carbon–nitrogen ratio. There is also general enrichment of all plant nutrients. The rotted material is easily incorporated into the soil or can be used as an organic matter mulch.

The low nitrogen content of the composting materials may greatly retard the rate of decomposition; for this reason most composters add some nitrogen fertilizer. The quality of compost can be further improved by adding other materials. Some recommendations of the United States Department of Agriculture are given in Table 7-3.

Table 7-3

Materials Recommended for Making Compost

	Cups per Tightly Packed Bushel
For general purposes, including acid-loving plants	
Ammonium sulfate	1
Superphosphate (20 percent)	$\frac{1}{2}$
Epsom salt	$\frac{1}{16}$
or:	
10-6-4 fertilizer	$1\frac{1}{2}$
For plants not needing acid soil	
Ammonium sulfate	1
Superphosphate (20 percent)	$\frac{1}{2}$
Dolomitic limestone or wood ashes	$\frac{2}{3}$
or:	
10-6-4 fertilizer	$1\frac{1}{2}$
Dolomitic limestone or wood ashes	$\frac{2}{3}$

From Kellogg, 1957.

Effects of Green-Manuring

One of the oldest agricultural practices is the growing of legumes for soil improvement. The yields of nonleguminous crops are usually greater when grown after legumes, like alfalfa or clover, because of an increased nitrogen supply. In these cases the legume crop is harvested and the benefit to crops grown later is a by-product. In green-manuring a crop is planted just to be plowed under to add some organic matter to the soil. This is particularly beneficial for sandy soils that are very low in organic matter content. In these soils little nitrogen is mineralized from soil organic matter and added nitrogen fertilizer may be leached out of the soil before the crop has utilized it. In such cases the green-manure crop is planted after harvest in the late summer or fall and plowed under just before planting the next crop in the spring. The gradual decomposition of the plowed-under crop provides plant nutrients, particularly nitrogen, for some weeks after planting. The major effect of the green-manure crop in this case is to increase the supply of nitrogen (and other nutrients) instead of providing a significant increase in the organic matter content of the soil. Other benefits commonly cited are protection of soil from erosion and reduced loss of nutrients by leaching. However, it has been difficult to attribute economic benefits of the practice to any other effect than that of increased nitrogen supply. Where green-manure crops deplete soil moisture and contribute to a droughty situation, green-manure crops may cause a reduction in crop yields.

Use of Peats

Peats used for soil amendments are generally classified as moss peat, reed-sedge peat, and peat humus. Moss peat forms from moss vegetation, reed-sedge peat from reeds, sedges, cattails, and other associated plants, while peat humus is any peat that has undergone considerable decomposition. The peats are used largely for mulches and greenhouse-soil mixes.

Some properties of common horticultural peats are given in Table 7-4. The nitrogen content (wide C/N ratio) of sphagnum peat indicates that its incorporation into soil may temporarily lower the available soil nitrogen supply. The low pH of sphagnum peat, however, makes it desirable as a mulch for acid-requiring plants such as azaleas and rhododendrons. Peat moss makes a neat looking surface that "sets off" plants and protects the soil from the disruptive force of rain drips, thereby keeping the soil porous so that water rapidly enters the soil when irrigated.

Nutrient Accumulation Under Shifting Cultivation

When one flies over the jungles of the humid tropics, one can see small clearings where food crops are grown. These people have few animals that produce manure for the fields, chemical fertilizers are not available, and the highly weathered tropical soils are very infertile. How can these people produce the subsistence food crops they need? They use a system of cultivation known as *shifting cultivation.*

We have already noted the relatively efficient recycling of nutrients that can occur in forests (Chapter 6) and the potential for organic matter accumulation in trees. The shifting cultivator utilizes the nutrients in the forest (trees, vines, leaves, etc.) through controlled cutting and burning that kills most of the trees but does not destroy all of the organic matter. Crops are

Table 7-4

Characteristics of Common Horticultural Peats

Type	Range in Nitrogen[a] %	Range of Water-Absorbing Capacity[a] %	Range in Ash Content[a] %	Range in Volume Weights[a] lb/ft³	Range in pH
Sphagnum moss peat	0.6–1.4	1500–3000	1.0– 5.0	4.5– 7.0	3.0–4.0
Hypnum moss peat	2.0–3.5	1200–1800	4.0–10.0	5.0–10.0	5.0–7.0
Reed-sedge peat (low lime)	1.5–3.0	500–1200	5.0–15.0	10.0–15.0	4.0–5.0
Reed-sedge peat (high lime)	2.0–3.5	400–1200	5.0–18.0	10.0–18.0	5.1–7.5
Decomposed peat	2.0–3.5	150–500	10.0–50.0	20.0–40.0	5.0–7.5

From Lucas, et al., *Ext. Bull.*, 516, Michigan State University.
[a] Oven-dry basis.

planted among the few remaining living trees, stumps, and fallen trees. Nutrients are made available to the crops as organic matter decomposes. After about 1 to 5 years, nutrients are depleted and weed and diseases invade the cultivated land, resulting in extremely poor yields. The land is abandoned and the forest quickly reestab-lishes itself. Perhaps 10 to 20 years are needed before enough nutrients have accumulated in the trees to permit another short period of cultivation. Nutrient accumulation by a regenerating forest fallow in the Congo is given in Table 7-5. The 18- to 19-year-old forest vegetation contained about five times more nutrients than

Table 7-5

Nutrient Accumulation (or Immobilization) in Forest Fallow in the Congo

Age of Forest Fallow	N	P	S	K	Ca+Mg
2 years	168	20	33	166	143
5 years	505	29	92	406	375
8 years	516	31	90	748	595
18–19 years	625	96	175	535	732

Nutrients Immobilized in Vegetation, lb/acre

Adapted from "Shifting Cultivation," C. E. Kellogg, *Soil Sci.*, *95*:221–230, 1963. Used by permission of Williams and Watkins Co.

Fig. 7-11 Landscape in Assam showing land use under shifting cultivation. Various stages of forest fallow can be observed as well as a small burning in the right background. (Photo courtesy Charles E. Kellogg.)

were contained in the 2-year-old forest fallow.

A common practice is to grow crops for 2 or 3 years and then use about 15 years for forest fallow. A farmer would need 17 or 18 parcels. Each year a new field would be brought into cultivation, and a field would be abandoned to forest fallow (see Fig. 7-11). A shifting cultivation system may require as much as 50 acres per person.

Grasslands are not such effective nutrient accumulators as forests, since grasses are shallow rooted and do not have as much potential for storage of biomass as standing vegetation. It is an interesting situation that in the humid and subhumid tropics forest soils have greater productivity than grassland or savannah soils under shifting cultivation, while in the humid temperate regions the grassland soils are usually considered the most productive. One cannot help but be impressed by the ingenuity that must have been required to perfect the shifting cultivation system. Today over 200 million people depend on the system for their livelihood.

References

Allison, F. E., *Soil Organic Matter and Its Role in Crop Production*, Elsevier, New York, 1973.

Broadbent, F. E., "Organic Matter," in *Soil*, USDA Yearbook, Washington, D.C., 1957, pp. 151–157.

Davis, J. F., and R. E. Lucas, "Organic Soils," *Michigan Agr. Exp. Sta. Spec. Bull.*, *425*, 1959.

Glob, P. V., "Lifelike Man Preserved 2,000 Years in Peat," *National Geographic Magazine*, *105*:419–430, 1954.

Jenny, Hans, "Causes of the High Nitrogen and Organic Matter Contents of Certain Tropical Forest Soils," *Soil Science*, *69*:63–69, 1950.

Kellogg, Charles E., "Home Gardens and Lawns," in *Soil*, USDA Yearbook, Washington, D.C., 1957, pp. 665–688.

Kellogg, C. E., "Shifting Cultivation," *Soil Science*, *95*:221–230, 1963.

Laws, W. Derby, "Farming Systems for Soil Improvement in the Blacklands," *Tex. Res. Found.*, Bull. *10*, 1961.

Liebig, Justus, *Chemistry in Its Application to Agriculture and Physiology*, L. Playfair, Editor, Wiley, New York, 1952.

Lucas, Robert E., Paul E. Rieke, and Rouse S. Farnham, "Peats for Soil Improvement and Soil Mixes," *Extension Bulletin 516*, Michigan State University, East Lansing, Mich.

Nielson, G. A., and F. D. Hole, "A Study of the Natural Processes of Incorporation of Organic Matter into Soil in the University of Wisconsin Arboretum," *Wis. Acad. Sciences*, *52*:213–227, 1963.

Schreiner, O., and B. E. Brown, "Soil Nitrogen," in *Soils and Men*, USDA Yearbook, pp. 361–376, Washington, D.C., 1938.

Smith, R. M., George Samuels, and C. F. Cernuda, "Organic Matter and Nitrogen Build-ups in Some Puerto Rican Soil Profiles," *Soil Science*, *72*:409–427, 1951.

Waksman, S. A., *Humus*, Williams and Wilkins, Baltimore, 1938.

8

CHEMICAL AND MINERALOGICAL PROPERTIES OF SOILS

As recently as the eighteenth century, the soil was viewed as little more than a mixture of rock and organic matter particles. It was believed that plant roots ingested soil particles and that these particles provided the sustenance of plants. Jethro Tull, an Englishman, believed that the swelling of plant roots caused a pressure that aided the entrance of soil particles into the roots. As a consequence, Tull invented the grain drill to plant grain in rows and a horse hoe (cultivator) to pulverize the soil when the grain was growing. Grain yields were increased and Tull thought this resulted from the effect of cultivation in loosening fine soil particles, which would make the particles more easily ingested. However, it was later learned that the grain grew better when cultivated because weed competition was reduced.

Shortly after 1800, de Saussure of Geneva used improved chemical techniques to discover the basic elements of photosynthesis and respiration. He showed that plants utilized carbon in the daytime and released carbon dioxide at night. The ash of the plant was composed of nutrients obtained from the soil. In spite of the excellent experimental work of de Saussure, many continued to believe the *humus theory*, which held that the plant's source of carbon was humus ingested by the roots. It remained for Liebig to bring a "quick" death to the humus theory through the publication of the book *Chemistry in Agriculture and Physiology* in 1840. Liebig, one of the foremost chemists of the nineteenth century, correctly viewed mineral weathering as a source of nutrients that were absorbed by roots as ions.

Roentgen's discovery of X rays in 1895 led to the development of the use of X-ray diffraction to study the arrangement of atoms in minerals. Using X-ray diffraction and more recently developed methods has resulted in great strides in understanding the role of mineral structure in the weathering and availability of plant nutrients.

Consideration of the characteristics of minerals found in soils and their transformation from one form to another is essential in understanding the nature of the soil's chemical properties and the origin of its fertility. Since the soil develops from material composed of rocks and minerals of the earth's crust, we will direct our attention first to the chemical and mineralogical composition of the earth's crust.

Chemical and Mineralogical Composition of the Earth's Crust

About 92 chemical elements are known to exist in the earth's crust. When one considers the number of possible combinations of such a large number of elements, it is not surprising that some 2000 minerals have been recognized. Relatively few elements and minerals, however, are of real importance in soils.

Chemical Composition of the Earth's Crust

Approximately 98 percent of the crust of the earth is composed of eight chemical elements (Fig. 8-1). In fact, two elements, oxygen and silicon, compose 75 percent of it. Many of the elements important in the growth of plants and animals occur in very small quantities. Needless to say, these elements and their compounds are not evenly distributed throughout the earth's surface. For example, in some places phosphorus compounds are so concentrated that they are mined; in many other areas there is a deficiency of phosphorus for maximum plant growth.

Mineralogical Composition of Rocks

Most of the elements of the earth's crust have combined with one or more other elements to form compounds called *minerals*. The minerals generally exist in mixtures to form the *rocks* of the earth. The

Fig. 8-1 The eight elements in the earth's crust comprising over 1 percent by weight. The remainder of elements make up 1.5 percent. (Data of Clark, 1924.)

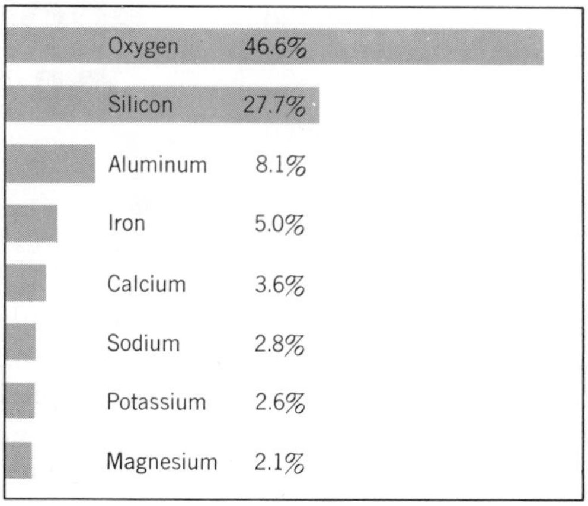

Element	Percent
Oxygen	46.6%
Silicon	27.7%
Aluminum	8.1%
Iron	5.0%
Calcium	3.6%
Sodium	2.8%
Potassium	2.6%
Magnesium	2.1%

mineralogical composition of igneous rocks, shale, and sandstone are given in Table 8-1. Limestone is also an important sedimentary rock and is composed largely of calcium and magnesium carbonates, with varying amounts of other minerals as impurities. The dominant minerals in these rocks are feldspar, amphibole, pyroxene, quartz, mica, clay minerals, limonite (iron oxide), and carbonate minerals.

Weathering and Mineralogical Composition of Soils

Numerous examples of weathering abound and can be observed every day. Rusting of metal, cracking of sidewalks and loss of mortar between bricks are a few examples. Weathering in soils results in the destruction of existing minerals and synthesis of new minerals. Nutrients are made available for plants and clay minerals are formed. In

a real sense, all life on earth is "locked" in the minerals and, through weathering, nutrients essential to life are made available. It is interesting to contemplate the amount of food or lumber that could be produced from the nutrients contained in the rocks of the Rocky Mountains. Even life in the seas awaits nutrients released by weathering on the land and carried to the sea by rivers. For these reasons we need to consider weathering and the mineralogical composition of soils.

Weathering—The Response of Rocks and Minerals to a New Environment

Rocks and minerals that are at or near equilibrium deep in the earth adjust to the greatly reduced pressure and temperature in the soil environment. The resulting adjustments or changes are called *weathering*. The changes are in the direction of a

Table 8-1
Average Mineralogical Composition of Igneous and Sedimentary Rocks

Mineral Constituent	Origin	Igneous Rock, %	Shale, %	Sandstone, %
Feldspars	Primary	59.5	30.0	11.5
Amphiboles and pyroxenes	Primary	16.8	—	a
Quartz	Primary	12.0	22.3	66.8
Micas	Primary	3.8	—	a
Titanium minerals	Primary	1.5	—	a
Apatite	Primary or secondary	0.6	—	a
Clay	Secondary	—	25.0	6.6
Limonite	Secondary	—	5.6	1.8
Carbonates	Secondary	—	5.7	11.1
Other minerals	—	5.8	11.4	2.2

Data from Clarke, 1924.
[a] Present in small amounts.

lower energy state and, to a large extent, are self-generating (exothermic). The response to reduced pressure is seen in the increased volume during *unloading*. Unloading is the removal of thick layers of sediment overlying deeply buried rocks by erosion or uplift. The release of pressure results in an accompanying bit of expansion that produces cracks and fissures (Fig. 8-2). Strains from temperature changes and the pressures of freezing water as well as the erosive action of water, wind, and ice also cause a slow and unceasing breaking up of hard rocks.

The response of minerals to reduced temperature is seen in the exothermic chemical reactions among minerals and the water, oxygen, and carbon dioxide in the soil. The abundance of water, oxygen, and carbon dioxide accounts for the fact that the major chemical weathering reactions are hydration, oxidation, and carbonation. An increase in volume accompanies these reactions and causes a peeling off of rock surfaces, producing exfoliation and spheroidal weathering. Hydration is considered to be very effective in producing spheroidal weathering (see Fig. 8-3).

Crystal Structure—A Clue to Weathering Rate

Particle size, through its effect on specific surfaces, is an important factor influencing the weathering rate of minerals. Ionic bonding within the mineral crystal, however, is ultimately the major factor.

Sodium chloride exists as the natural mineral halite. Sodium and chlorine exist in the crystal as ions that attract each other. At the crystal faces, the sodium and chloride ions, respectively, attract the negative and positive poles of water molecules. Adsorption of water molecules dislodges the sodium and chlorine from the crystal and greatly increases the solubility of the sodium and chlorine. The mineral dissolves easily in water and is readily leached from soils. This accounts for the general absence of halite in the soils of the humid regions.

Fig. 8-2 Unloading is the process whereby cracks and fissures are caused by expansion of rock that accompanies the removal of overlying layers by erosion or uplift. Limestone is the rock in this case.

Fig. 8-3 Hydration of minerals in the outer layer of a rock results in an increase in volume to cause a concentric spalling known as "spheroidal weathering." Weathered rock is basalt in Hawaii.

We have noted that oxygen and silicon make up about 75 percent of the earth's crust on a weight basis, which explains the great abundance of silicate minerals in the crust. From Table 8-2, one can see that the large size of the oxygen atom, coupled with its abundance, causes oxygen to occupy over 90 percent of the volume of the earth's crust. It is productive to think of soil minerals as being composed of ping-pong balls, represented by oxygen atoms, and that marbles, representing the smaller metallic cations, occupy the interstices between the ping-pong balls. The cations are arranged on the basis of their size and valence to produce electrically neutral crystals. The larger the cation, the greater the number of oxygens surrounding it (coordination number) (see Table 8-2).

The silicon atom fits in an interstice formed by four oxygens. The covalent bonding between oxygen and silicon is directional and causes the oxygen and silicon to form a tetrahedron (Fig. 8-4). Silicon-oxygen tetrahedra are the basic units in the silicate minerals. Different silicate minerals exist, depending on the way in which the tetrahedrons are linked together.

Olivine, $MgFeSiO_4$, is composed of individual silicon-oxygen tetrahedra bound together by magnesium and iron. Each magnesium and iron atom is surrounded by six oxygens—three from each of two faces of adjacent tetrahedrons (see Fig. 8-5). The bonds between the oxygen and the iron and magnesium are weak in comparison to the bonds in the tetrahedra between oxygen and silicon. During weathering, exposed magnesium along the crystal face react with hydroxyl ions to form magnesium hydroxide. The ferrous iron along the crystal face readily looses an electron and its size is reduced by about 20 percent (see Table 8-2). The smaller size of the ferric iron atom makes it easier for iron to leave the crystal and likely also to form a hydroxide. For these reasons the weathering of olivine can be visualized, as shown in Fig. 8-5, as consisting of the separation of

Table 8-2

Size, Percent Volume in Earth's Crust, and Coordination Number of the Most Abundant Elements in Soil Minerals

Element	Atomic Radius, A°	Volume, Percentage in Earth's Crust	Coordination Number with Oxygen
O	1.32	93.8	—
Si	0.39	0.9	4
Al	0.57	0.5	6 (and 4)
Fe	0.83 (Fe^{+3} 0.67)	0.4	6
Mg	0.78	0.3	6
Na	0.98	1.3	8
Ca	1.06	1.0	8
K	1.33	1.8	8

silicon-oxygen tetrahedra with the release of iron and magnesium. The magnesium and iron are then available for plant growth. The magnesium compounds formed are sufficiently soluble to be leached from the soil, but the iron compounds formed may be insoluble enough to accumulate. The silicon-oxygen tetrahedra often regroup or polymerize and form some other mineral by combining with other constituents in the soil solution. As with halite, olivine weathers fairly easily, making it impossible for olivine to remain long in the soils of the humid regions.

In most of the silicate minerals the silicon-oxygen tetrahedra are linked together by the common sharing of some oxygen atoms. The number of cations other than silicon needed to neutralize the crystal is reduced to the extent that oxygen sharing

Fig. 8-4 Models showing the tetrahedral (four-sided) arrangement of silicon and oxygen atoms in silicate minerals. The silicon-oxygen tetrahedron on the left has the apical oxygen set off to one side to show the position of the silicon atom.

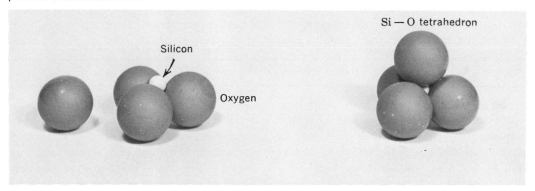

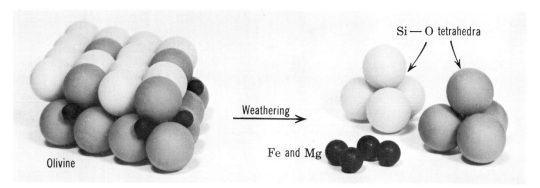

Fig. 8-5 Models representing the weathering of olivine. Olivine is composed of silicon-oxygen tetrahedra held together by iron and magnesium. Every other tetrahedron is "inverted," as shown by the light-colored tetrahedra in the olivine model on the left. During weathering, the silicon-oxygen tetrahedra separate with the release of iron and magnesium.

occurs. Increased oxygen sharing by silicon generally causes greater resistance to weathering because a greater percentage of the crystal bonds are the strong covalentlike bonds that exist between silicon and oxygen. The ratio of the oxygen to silicon in olivine is 4. It can be seen in Fig. 8-6 that increased oxygen sharing in the series olivine to pyroxene to amphibole to mica to quartz is associated with a decrease in the oxygen-silicon ratio from 4 to 2. The decreased ratio is associated with an increase in weathering resistance. In quartz (SiO_2) all of the oxygen atoms are neutralized by silicon, and quartz tends to persist in many soils as the more weatherable minerals disappear.

Mineralogical Composition of Soils versus Weathering Stage

The differences in weathering resistance of minerals were used by Jackson and Sherman to establish 13 weathering stages that relate mineralogical composition of soils to weathering intensity. Representative minerals and typical soils associated with the weathering stages are given in Table 8-3.

Soils of weathering stage 1 may contain some gypsum and halite. Notice that soils containing significant amounts of olivine are representative of stage 3 and biotite mica is representative of stage 4. This sequence is in agreement with our discussion of oxygen-silicon ratio and weathering resistance. Soils with minerals representative of stages 1 to 5 are considered to be in the early weathering stages. Such soils are frequently the youthful soils throughout the world, but are primarily the soils of arid regions where limited water restricts chemical weathering.

Soils of the intermediate weathering stages (stages 6 to 9) include most of the soils of the humid temperate regions. Quartz is often abundant in these soils and is representative of weathering stage 6. Here we notice minerals that we have not mentioned previously such as illite, vermiculite, and montmorillonite (see Table 8-3). These latter three minerals are typically synthesized in soils and increase in abundance as feldspars and micas weather and disappear. Some of the elements released in weathering recrystallize to form illite, vermiculite, and montmorillonite,

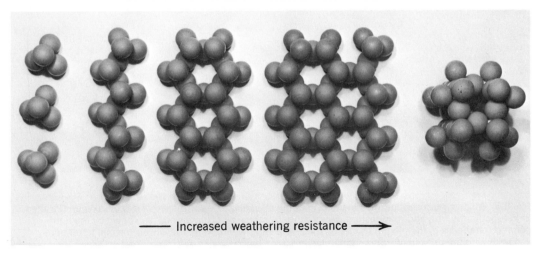

———— Increased weathering resistance ————>

Arrangement of Si-O tetrahedra and representative minerals

Individual	Single chain	Double chain	Sheet	3-dimensional
Olivine	Pyroxene augite	Amphibole hornblende	Biotite (mica)	Quartz
		Oxygen–silicon ratio		
4	3	2.7	2.5	2

Fig. 8-6 Models showing the common arrangements of silicon-oxygen tetrahedra in silicate minerals and their relation to weathering resistance.

which accumulate as fine-sized particles in the clay fraction.

Soils of the advanced weathering stages include the intensely weathered soils of the humid tropics. Soils dominated by minerals of weathering stages 10 to 13 may have lost almost all of the original minerals of the parent material and may consist mainly of stable minerals that have been synthesized during weathering. These minerals are kaolinite, gibbsite, hematite, and anatase (TiO_2). Thus, soils of the advanced weathering stages are usually characterized by extreme infertility and, as pointed out in Chapter 7, where these soils exist most of the nutrients in the ecosystem are circulat-

ing through the vegetation. In a recently developed soil classification system, which we consider in a later chapter, many of the intensely weathered soils of the humid tropics are classified as *Ultisols*, soils that have undergone the "ultimate in weathering and leaching."

Origin, Structure, and Properties of Clay Minerals

The discussion so far has emphasized the degradational aspects of weathering. When something has been taken apart, the pieces remain. When minerals weather, these pieces are the atoms or groups of atoms set

ORIGIN, STRUCTURE, AND PROPERTIES OF CLAY MINERALS

Table 8-3
Representative Minerals and Soils Associated with Weathering Stages

Weathering Stage	Representative Minerals	Typical Soil Groups
	Early weathering stages	
1	Gypsum (also halite, sodium nitrate)	Soils dominated by these minerals in the fine silt and
2	Calcite (also dolomite, apatite)	clay fractions are the youthful
3	Olivine-hornblende (also pyroxenes)	soils all over the world, but mainly soils of the desert regions
4	Biotite (also glauconite, nontronite)	where limited water keeps chemical weathering to a
5	Albite (also anorthite, microcline, orthoclase)	minimum.
	Intermediate weathering stages	
6	Quartz	Soils dominated by these
7	Muscovite (also illite)	minerals in the fine silt and
8	2:1 layer silicates (including vermiculite, expanded hydrous mica)	clay fractions are mainly those of temperate regions developed under grass or trees. Includes
9	Montmorillonite	the major soils of the wheat and corn belts of the world.
	Advanced weathering stages	
10	Kaolinite	Many intensely weathered soils
11	Gibbsite	of the warm and humid
12	Hematite (also goethite, limonite)	equatorial regions have clay fractions dominated by these
13	Anatase (also rutile, zircon)	minerals they are frequently characterized by their infertility.

Based on Jackson and Sherman, 1953.

free from crystal degradation. We know that some of these atoms are used by plants; some are leached from soils, but many of the atoms regroup to form new minerals. Synthesis of minerals begins when a few atoms group together to form the nucleus of a crystal. The crystal grows and the mineral particle increases in size. This synthesis, or building up, accounts for the small particle size and great specific surface of the clay fraction as compared to silt and sand. The clay fraction of soils then tends to be composed of fine-sized secondary minerals formed in the soil called *clay minerals*. An understanding of their origin, structure, and properties is essential to an understanding of how soils retain soluble plant nutrients, the nature and importance of soil pH, and certain physical properties of soils.

Origin of Clay Minerals

There are two major groups of clay minerals: the *silicate clays*, which include *illite*, *montmorillonite*, *vermiculite*, and *kaolinite*, and the *oxide clays*, which include primarily *iron and aluminum oxides*. Soils with clay fractions dominated by silicate clays are representative of weathering stages 7 to 10 and are widely distributed throughout the world. Soils with clay fractions dominated by oxide clays are representative of weathering stages 11 to 13 and are common in the humid tropics.

These secondary clay minerals are formed by alteration of existing minerals or by synthesis. Their origin is related to the kinds of minerals in the soil that can be altered and the amount and kinds of constituents that can recombine. The complexity of the processes makes a clear-cut exposition of their origin impossible, but several general ideas will be presented.

Evidence indicates that micas, feldspars, and ferromagnesian minerals may weather directly to silicate clay minerals like *illite*, *montmorillonite, kaolinite*, or even hydroxide clays like *gibbsite* (hydrated aluminum oxide). The nature of the weathering environment plays an important role in determining when a given mineral will be formed. Illite formation is common in the temperate regions where weathering has not been intense. Alteration of mica minerals by the partial loss of structural potassium and hydration is a common mode of illite formation. Montmorillonite formation requires an abundant supply of magnesium and a neutral or only slightly acid

environment. In the temperate regions illite can weather by alteration into montmorillonite.

Prolonged leaching of a soil in which montmorillonite has formed can lead to the development of high acidity and conditions favorable for the alteration or breakdown of montmorillonite and the formation of kaolinite. Kaolinite is frequently formed directly from primary minerals in soils of the humid tropics. Although kaolinite is very stable, it can weather to form gibbsite, $Al(OH)_3$. The diagram in Fig. 8-7a indicates some possible alteration routes in the formation of secondary minerals.

Kaolinite can also be formed by the resilication of aluminum oxide if the weathering zone is invaded by silica-rich waters. The other silicate clays can also be formed by synthesis so that more than one method is available for their formation. This undoubtedly causes slight differences in their properties, depending on the mode of formation. It means that each kind of clay mineral category represents minerals with much similarity but, within each group, there is a range in properties.

The discussion thus far has emphasized the weathering of crystalline minerals. Noncrystalline or amorphorus minerals exist where volcanoes throw ash and cinders into the air, and they solidify before the atoms become organized into crystalline lattices. In the humid tropics amorphorus minerals weather rapidly into an amorphorus clay called *allophane*. The soils formed lack discrete mineral particles and are like a gel and feel like mucky soil when wet. There is an enormous amount of surface area and great protection against organic matter decomposition. Characteristic features of soils high in allophane include high organic matter content and water-holding capacity.

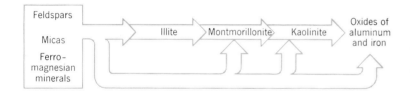

Fig. 8-7a Some possible routes in the formation of silicate and oxide clays from the weathering of crystalline minerals.

Fig. 8-7b Mineral weathering sequence for soils forming from amorphorus parent material in humid areas of Central America.

Slowly, the allophane clays become more organized or crystalline. A mineral sequence from the weathering of volcanic ash is shown in Fig. 8-7b as allophane-halloysite-kaolinite-oxides of aluminum and iron. Soils high in allophane are important in the Andes Mountains, Central America, Oregon, Washington, Alaska, and Hawaii.

Although much more information is needed before the origin of these minerals can be viewed in more certain terms, there is sufficient information about their characteristics to know that they have pronounced effects on plant growth. A discussion of these characteristics follows.

Structure of Silicate Clay Minerals of Soils

The silicate clay minerals consist of two basic components. One component is the silicon-oxygen tetrahedron (silicon in four coordination with oxygen) and the other component is the aluminum octahedron (eight sided with aluminum in six coordination with oxygen and/or hydroxyl). Silicon-oxygen tetrahedra in silicate clays are arranged in a sheet, as shown in Fig. 8-6 (as in biotite). The upper or apical oxygens of the tetrahedral sheet have one bond satisfied by silicon and one bond of the oxygen unsatisfied. The formula for the tetrahedral sheet is $Si_2O_5^{-2}$.

We can visualize that the apical oxygens of the tetrahedral sheet can attract cations that, in turn, are surrounded by enough anions to neutralize all the cation charge. For simplicity, a pair of tetrahedra instead of a sheet are used to illustrate how the tetrahedral and octahedral sheets exist in kaolinite in Fig. 8-8. Items 1 and 2 of Fig. 8-8 show the apical oxygens, which have one free bond, and the position of the aluminum. Items 3 to 6 show the progressive addition of OH to complete the aluminum octahedron. Model 6 is representative of the structure of kaolinite consisting of one tetrahedral and one octahedral sheet. For this reason kaolinite is called a 1:1 clay (ratio of tetrahedral to octahedral sheets). Item 7 of Fig. 8-8 shows that the apical oxygens are common to both the tetrahedral and octahedral sheets. In

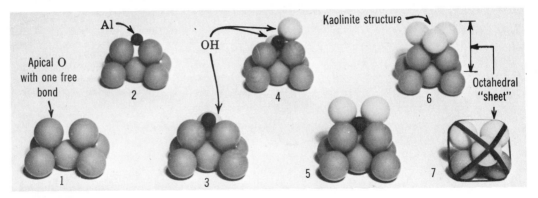

Fig. 8-8 Models illustrating the structure of kaolinite. Item 1 shows that the apical oxygen of the tetrahedral sheet has a free bond that is neutralized by aluminum, as seen in item 2. Items 3 to 6 show the progressive addition of 4 OH to complete the octahedron or to put aluminum in six coordination. Item 6 represents the kaolinite structure, and item 7 shows that the aluminum is surrounded by two oxygen and four hydroxyl. Item 7 is an example of an octahedron—four faces are seen and four faces are hidden.

the octahedron, each Al neutralizes half a valence bond from each of the surrounding six anions, which results in a neutral structure.

The properties of kaolinite are derived from its structure. The tetrahedra and octahedra sheets form layers. Many layers are stacked one on top of the other (Fig. 8-9). A kaolinite particle about 2 microns wide would have the length of about 20,000 oxygens side by side. If the particle was one tenth as high as wide, being 0.2 micron high, it would contain about 50 layers stacked on top of each other. As an analogy, a clay particle is like a stack of sheets of plywood.

The hydroxyls of the aluminum octahedral layer are adjacent to the oxygens of the next layer above in kaolinite. This results in hydrogen bonding that firmly holds successive layers together (see Fig. 8-9). This bonding has two effects. First, it tends to favor the formation of large particles. Many of the kaolinite particles are 2 microns or larger in diameter, and soils with a high content of this clay may have surprisingly high permeability.

Second, water cannot permeate successive layers or units of the particles to cause expansion and contraction with wetting and drying. To carry the analogy of the structure further, it is similar to a stack of plywood sheets glued together. The electron micrograph in Fig. 8-10 shows the hexagonal shape and platy nature of kaolinite particles.

Montmorillonite has a 2:1 lattice. An aluminum octahedral sheet is sandwiched between two silica tetrahedral sheets to form the basic layer (see Fig. 8-11). The stacking of layers on top of each other results in a side-by-side arrangement of oxygen atoms from adjacent tetrahedral sheets so that hydrogen bonding between layers does not exist. In fact, the adjacent planar oxygen surfaces have a natural tendency to repel each other. The layers readily expand and contract with wetting and drying to give an "accordian" effect. Particles also tend to be smaller than those of kaolinite, and soils with a high content of this clay will crack markedly upon drying and tend to be impermeable when wet. It gives the soil highly plastic characteristics.

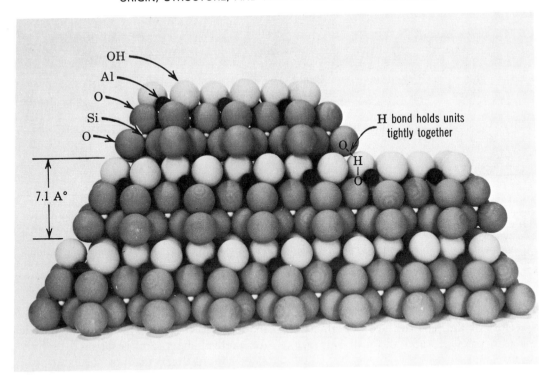

OH

Al

O

Si

O

H bond holds units
tightly together

O
|
H
|
O

7.1 A°

Fig. 8-9 Model of the 1:1 clay mineral kaolinite showing three units stacked one on top of the other and held together tightly by hydrogen bonding (attraction of hydrogen of the hydroxyl for the oxygen of the next adjacent unit). Each unit is 7.1 angstroms thick.

Fig. 8-10 An electron micrograph of kaolinite particles magnified 35,821 times. Note the flaky and hexagonal shape of the particles. (Courtesy Mineral Industries Experiment Station, Pennsylvania State University.)

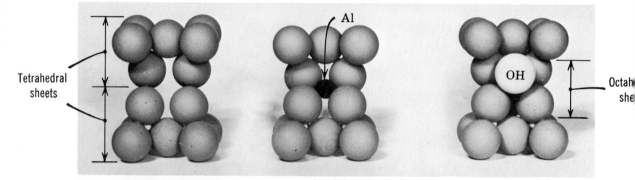

Fig. 8-11 Models illustrating a 2:1 structure consisting of two tetrahedral sheets and one octahedral sheet. On the left are two silicon-oxygen tetrahedral "sheets" facing each other. The center model shows the addition of aluminum that is shared by four apical oxygens. The right model shows the addition of two hydroxyls (one in rear is invisible) to place the aluminum in six coordination in an octahedral arrangement. The aluminum shares half of a valence bond from each of the six surrounding anions (four oxygen and two hydroxyl).

It also means that the interiors as well as the exteriors of the particles are available for the adsorption of water and nutrients (Fig. 8-12).

During the formation of montmorillonite, some of the aluminum atoms in the interstices of the octahedral layer are replaced by magnesium. It has been mentioned that an abundance of magnesium in the weathering environment is a prerequisite for montmorillonite formation. The substitution of aluminum by magnesium can occur because the two atoms are sufficiently similar in size so that the replacement of only one sixth of the aluminum atoms does not cause excessive strain in the lattice. This substitution is called *isomorphous substitution* (see Fig. 8-12).

Since the valence of the original lattice aluminum is three and the valence of the substituting magnesium is two, each substitution leaves the lattice with one unsatisfied negative charge or valency bond. In montmorillonite about one sixth of the aluminum ions has been replaced and the result is a highly negatively charged lattice or layer. This negative charge is permanent; it originates within the lattice (octahedral sheet) and is satisfied by hydrated cations that remain on the exterior of the particle or in close proximity of the surface. Thus a most important function of the clays is evident—that of holding hundreds or thousands of pounds of nutrient ions per acre furrow slice. These ions are adsorbed strongly enough to retard their movement out of the soil by leaching but weakly enough to be readily used by plants. In fact, one theory holds that plant roots can exchange hydrogen ions directly with cations on the colloidal surfaces.

Illite also has a 2:1 lattice. Some of the tetrahedral Si^{+4} has been replaced by Al^{+3}, as indicated in Fig. 8-13. As in montmorillonite, a negatively charged lattice results. In illite, however, most of the lattice charge is neutralized by potassium ions that sit in the "hexagonal holes" of the planar surface of adjacent units. The potassium forms a bonding mechanism called a "potassium bridge" (O—K—O), which holds adjacent layers together and produces a nonexpand-

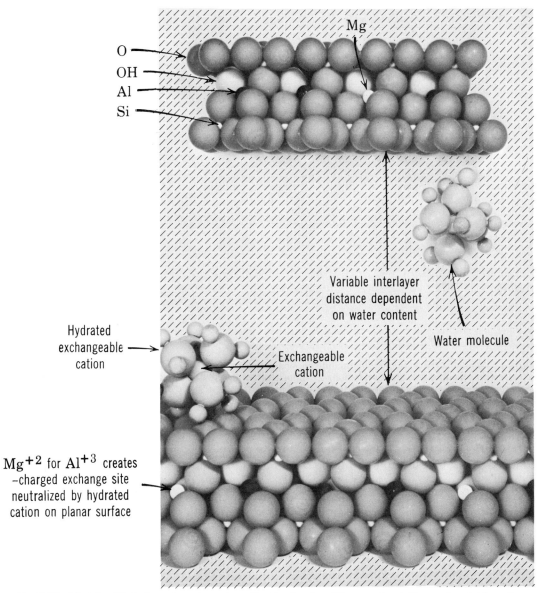

O

OH

Al

Si

Mg

Variable interlayer
distance dependent
on water content

Water molecule

Hydrated
exchangeable
cation

Exchangeable
cation

Mg^{+2} for Al^{+3} creates
–charged exchange site
neutralized by hydrated
cation on planar surface

Fig. 8-12 Models of 2:1 clay mineral montmorillonite showing (1) expanding lattice and variable water content, (2) cation-exchange site originating from isomorphous substitution of Mg^{+2} for Al^{+3}, and (3) adsorption of two hydrated cations on the planar surface cation-exchange sites. In a moist soil the interlayer water is believed to be all or primarily water associated with exchangeable cations.

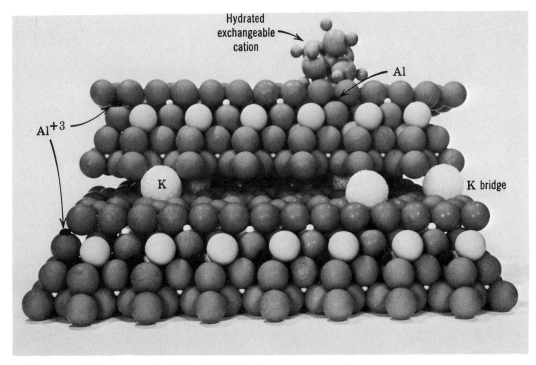

Fig. 8-13 Model of the 2 : 1 clay mineral illite showing (1) some tetrahedral Al^{+3} substituted for Si^{+4}, which creates a negatively charged lattice, (2) "potassium bridges" that neutralize most of the lattice charge and hold the layers together to form a nonexpanding lattice, and (3) adsorption of a hydrated cation on an exchange site created by aluminum substitution.

ing lattice (Fig. 8-13). Some of the tetrahedral Al around the edges of the particles are not involved in forming potassium bridges and serve as sites for cation exchange. The lattice charge for cation exchange is about one third as great as it is for montmorillonite. During weathering, if all or most of the potassium is removed from the interlayer positions, the clay becomes an expanding lattice type of clay called *vermiculite*. The cation-exchange capacity of the clay is thereby greatly increased, since none of the lattice charge is neutralized by K bridges.

In all clays cation-exchange sites originate from the dissociation of H from exposed OH along the edges of particles. Cation-exchange sites also exist where exposed oxygens are bonded to silicon and have one "free" bond. Isomorphous substitution in montmorillonite accounts for about 80 percent of the cation-exchange sites, and exposed edge bonds account for about 20 percent whereas, in kaolinite, most or all of the sites result from exposed edge bonds. A summary of cation-exchange capacity and other properties of layered-silicate clays is given in Table 8-4.

Some expanding clays, like montmorillonite, can hold water many times the volume of the clay itself. The water films between the layers and between the clay particles

Table 8-4
Some Generalized Properties of Clay Minerals

Clay Mineral	Ratio of Tetrahedra to Octahedra Sheets	Relative Particle Size	Cation-Exchange Capacity, me/100 g	Isomorphous Substitution	Structural Formula
Kaolinite	1:1	Large	8	Little or none	$(OH)_8Si_4Al_4O_{10}$
Montmorillonite	2:1	Small	100	Mg for Al in octahedral layer	$(OH)_4Si_3(Al_{3.34}Mg_{.66})O_{20}$
Illite	2:1	Intermediate	30	Al for Si in tetrahedral layer	$(OH)_4K_2(Si_6Al_2)Al_4O_{20}$

give soil its shrink and swell characteristics, its plasticity, and its cohesion (Fig. 8-14). Expanding clays can push in basement walls. Some scientists at Iowa State University are experimenting with the use of expanding clays as jacks. For example, dry clay is placed under road pavements that settle and, when the clay is wetted, the expansion raises the pavement. The leaning of the Tower of Pisa and the settling of buildings in Mexico City are associated with changes in the water content of clays.

Fig. 8-14 Large cracks formed by drying of montmorillonite clay. The montmorillonite originated from weathering of basalt in the hills and was moved to the valley floor by erosion. The clay formed under an ustic moisture regime where bases remain in the soil.

Soil Water and the Exchangeable Cations

The water existing in the immediate vicinity of clay particles is considered to be basically water associated with exchangeable cations. This is explained on the basis that exchangeable cations attract water molecules more strongly than the exposed oxygen atoms held in clay particles. In the case of sand and silt particles, with essentially none or very low cation-exchange capacity, the water in the immediate vicinity of the particle surface is mostly hydrogen bonded, as shown in Fig. 8-15. These innermost water molecules probably exist in a "quasi" crystalline state.

The innermost adsorbed water of both clay and silt and sand particles is unavailable to plants and exists in soils in the air dry state. Further out from the soil particle surfaces, cations are much less abundant than near the clay surfaces, and water exists in a "free" state. This water contains hydrated cations in fairly low concentration and can be considered equivalent to the soil solution.

Oxide Clays

All or most soils contain at least a small amount of colloidal-sized particles composed of oxides of iron and aluminum. Representative oxide clays include gibbsite, $Al(OH)_3$, of weathering stage 11, and hematite, Fe_2O_3, of weathering stage 12. Soils with properties dominated by oxide clays and kaolinite (weathering stage 10) are in the advanced weathering stages and are usually found in the humid tropics.

Soils dominated by oxide and kaolinite clays are characterized by very stable soil aggregates, and they exhibit a low degree of plasticity. Large amounts of oxide and

Fig. 8-15 Model to show hydrogen bonding of water molecules to exposed oxygen atoms of a mineral surface where the oxygen are coordinated with silicon in tetrahedrons.

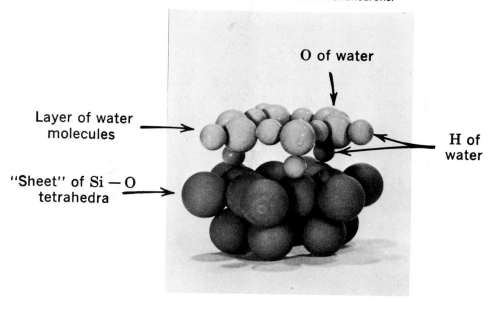

kaolinite clays in a soil contribute to the formation of extremely stable soil aggregates because the clays tend to neutralize each other. Kaolinite, as we have seen, has a net negative charge, while the oxide clays have a net positive charge (in acid soils). This promotes flocculation. Oxide gels also probably contribute to the hardness of aggregates that resist crushing when moistened and rubbed in the palm of the hand. Such soils act as "sands," even though they contain 100 percent clay, as is common for soils in the Hawaiian Islands that have weathered from basalt. Infiltration of water

release of one or more ions held by the micelle in termed *cation exchange*. For example, assume that the micelle has one half of its capacity satisfied with Ca ions, one quarter with K ions, and one quarter with H ions. The situation would be as shown in the diagram below. Now suppose that the colloidal material is treated with a solution of strong KCL. In time, the K ions from the KCL will replace virtually all the cations on the micelle, creating a micelle that is entirely potassium saturated, and the adsorbed calcium and hydrogen will exist in solution as chlorides.

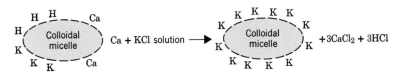

is rapid and the soil resists erosion. Plasticity is low when wet and soils are easily tilled. Adsorption and release of water to plants occurs largely at low soil moisture tensions. Soil textures by feel are frequently silt loam and silty clay loam or, in some cases, sandy clay loam.

Cation and Anion Exchange in Soils

Clay and humus are of utmost importance in soils. Because they are in a colloidal state, they expose a relatively large amount of surface area for adsorption of water and ions. Nutrients set free in solution during weathering tend to be adsorbed on the humus and clay surfaces. In this section we emphasize the adsorption and storage of nutrients in exchangeable form on colloidal surfaces.

The Nature of Cation Exchange in Soils

The adsorption of a cation by a colloidal nucleus or micelle and the accompanying

The efficiency with which ions will replace each other is determined by factors such as (*a*) relative concentration or numbers of the ions, (*b*) the number of charges on the ions, and (*c*) the speed of movement or activity of the different ions. The first factor is an application of the well-known chemical law of mass action. The greater the number of charges carried by the ion, the greater is its efficiency, other factors being the same. The speed or activity of an ion is primarily a function of its size, but the degree of hydration must also be considered. If we consider the ions Li, Na, K, and Rb, we find that they are listed in Table 8-5 in the ascending order of size, and hence we would expect their efficiency of replacement to be Li Na K Rb. During hydration, however, the Li ion associates itself with so many water molecules that its speed is much reduced. Because of its large hydrated radius, Li can not get as close to the micelle as other ions. Likewise, the Na ion is more highly hydrated than the K ion. The result is that the order of replacement

Table 8-5
Ionic Radii, Hydration, and Exchange Efficiency of
Several Monovalent Ions

Ion	Radii of Ions Angstroms, $(10^{-8}$ cm)		Order of Cation-Exchange Efficiency
	Dehydrated	Hydrated	
Li	0.78	10.03	4th
Na	0.98	7.90	3rd
K	1.33	5.32	2nd
Rb	1.49	5.09	1st

is reversed: Rb K Na Li (see Table 8-5).

Considering some of the most common cations in soils, the replaceability series is usually $Al > Ca > Mg > K > Na$. Exchangeable H is difficult to put in the series because of the uncertainty of its hydration properties. It is usually considered quite easily replaced, as are the other monovalent ions.

Cation Exchange Capacity of Soils

The colloidal fraction carries a positive as well as a negative charge. The negative charge, however, is of much greater magnitude and of greater significance for plant growth in most soils. The cation exchange capacity (CEC) is an expression of the *number* of cation adsorption sites per unit weight of soil. It is defined as the sum total of exchangeable cations adsorbed, expressed in *milliequivalents per 100 grams of oven-dry soil*. An equivalent weight is that quantity that is chemically equal to 1 gram of hydrogen. The number of hydrogen atoms in an equivalent weight is Avagadro's number (6.02×10^{23}). A milliequivalent weight is equal to 0.001 gram of hydrogen. If there was 1 milliequivalent weight of cation exchange capacity in a teaspoon of

soil, it would contain 6.02×10^{20} negatively charged adsorption sites. Such large numbers are difficult to comprehend but will be better understood when these numbers are converted into weight of adsorbed cations later in this chapter. It suffices at this time to mention that on an acre furrow slice basis, this may amount to several thousands of pounds of exchangeable cations.

The total cation exchange capacity of the soil is the total number of exchange sites of both the organic and mineral colloids. For our purposes it can be assumed that the organic matter of a mineral soil will have a cation-exchange capacity of 200 milliequivalents per 100 grams. Each constituent in the clay fraction of the soil has an average cation-exchange capacity that is rather characteristic. The approximate cation-exchange capacities of some of the common clay minerals are: montmorillonite, 100; illite, 30; and kaolinite, 8 (Table 8-4). The exchange capacity of the hydrous oxides of iron and aluminum is much less than that of the other clay minerals. For many soils of the temperate zone, the average cation-exchange capacity of the clay fraction as a whole might be in the vicinity of 50 milliequivalents per 100 grams because the clay fraction is a mixture of clay minerals. Under these conditions, the cation-exchange capacity of 100 grams of soil is 2 milliequivalents for each percent of organic matter and $\frac{1}{2}$ milliequivalent for each percent of clay. A soil with 3 percent organic matter and 24 percent clay would have a cation-exchange capacity of approximately 18 milliequivalents per 100 grams $[2(3) + \frac{1}{2}(24) = 18]$.

The grassland soils of western Canada contain clay and organic matter that averages 57 and 250 milliequivalents per 100 grams, respectively. Cation exchange capacity for these soils can be approximated with the following equation.

$$CEC = \text{percent organic matter} \times 2.5 + \text{percent clay} \times 0.57$$

The cation-exchange capacity of the sand and silt fractions is omitted in the approximation of exchange capacity, since it is so small. An accurate determination can be made by saturating all exchange positions with a single cation, such as ammonium, and then determining the total amount of ammonium adsorbed.

The cation-exchange capacity ranges from less than 5 for soils containing very little clay or organic matter to about 200 for a muck soil. The cation-exchange capacity of the horizons of a typical prairie (Mollisol) soil are given in Table 8-6.

Kinds and Amounts of Exchangeable Cations

The amounts of the exchangeable cations for the horizons of a typical prairie soil are also given in Table 8-6. For practical purposes, the sum of the cations shown in the table is considered synonymous with cation-exchange capacity, recognizing that very small but important amounts of exchangeable iron, copper, manganese, and other cations are present.

The upper 6 inches of the prairie (Mollisol) soil cited in Table 8-6 has 13.9 milliequivalents of exchangeable calcium per 100 grams of soil. To calculate the pounds of exchangeable calcium per acre furrow slice requires that the milliequivalence of calcium be converted to a weight unit. The equivalent weight of calcium is 20, making

Table 8-6

Exchangeable Cations, Cation Exchange Capacity, Percent Base Saturation, and pH of Horizons of Tama Silty Clay Loam

Depth, in.	Horizon	Exchangeable Cations, Me/100 g					Cation-Exchange Capacity	Percent Base Saturation	pH
		Ca	Mg	K	Na	H			
0–6	Ap	13.9	3.4	0.5	0.1	9.3	27.2	66	5.7
6–11	Al2	13.8	4.2	0.4	0.1	11.4	29.9	62	5.8
11–20	AB	14.5	6.1	0.4	0.1	9.0	30.1	70	5.8
20–35	B2t	14.7	6.7	0.3	0.1	7.5	29.3	74	5.7
35–51	B3	14.8	5.7	0.3	0.1	5.6	26.5	79	6.0
51–61	C	16.1	5.6	0.4	0.2	4.1	26.4	84	6.5

Adapted from "Soil Survey Investigations Report," No. 3, Iowa, SCS, USDA, 1966.

the milliequivalent weight equal to 0.02 gram. Thus the A1 horizon contains 0.278 gram (13.9×0.02) of exchangeable calcium per 100 grams of soil. Conversion to pounds can be made by simply stating that there would be 0.278 pound of exchangeable calcium per 100 pounds of soil. Assume this 6-inch layer weighs 2 million pounds or 20,000 times more than 100 pounds. Multiplying 0.278 by 20,000 produces 5560, which is the pounds of exchangeable calcium. The formula for calculating the weight of any cation per acre furrow slice is:

Pounds per acre furrow slice

$= (20,000)$ (milliequivalents/100 grams)

(milliequivalent weight in grams)

The total quantity of exchangeable calcium in the soil ramified by the root system of plants is many times the amount in this particular soil. The importance and magnitude of the nutrients available to plants as exchangeable cations in soils is readily apparent. Taking into consideration the differences in the milliequivalent weight of the various cations, there would be 816 pounds of exchangeable magnesium and 390 pounds of exchangeable potassium in the plow layer. Knowledge such as this is useful in predicting the response, if any, of crops to these cations applied as fertilizers.

Percent Base and Hydrogen Saturation

The exchangeable bases include Ca, Mg, K, and Na.[1] Note that they are found in this

[1] Technically a base is a proton acceptor like the OH ion while an acid is a proton donor like the H ion. However, the exchangeable cations Ca, Mg, K, and Na are all associated with compounds in the soil like $CaCO_3$, $MgCO_3$, K_2CO_3, and Na_2CO_3, which are more basic than acidic in reaction. For this reason Ca, Mg, Na, and K are commonly referred to as *exchangeable bases*, while H is commonly called an *exchangeable acid*.

order of decreasing abundance in Table 8-6. As exchangeable cations they act in the formation of hydroxyls as follows.

$$\boxed{\text{micelle}} - K + H_2O$$

$$\rightleftharpoons \boxed{\text{micelle}} - H + K^+ + OH^-$$
(in solution)

The percentage base saturation is the percentage of the cation-exchange capacity saturated by these cations. Thus, in Table 8-6, it can be seen that all the horizons have well over 50 percent base saturation, a common characteristic of prairie soils. The AB horizon has a base saturation of 70 percent, calculated as follows.

Percent base saturation

$$= \frac{\text{milliequivalents exchangeable bases}}{\text{cation exchange capacity}}$$
$$\times 100$$
$$= \frac{21.1}{30.1} \times 100 = 70$$

The percentage of hydrogen saturation is calculated by substituting milliequivalent hydrogen in the place of the milliequivalence of bases in the formula for determining base saturation. Its importance will be brought out in Chapter 9 when the relationship of percentage of hydrogen saturation and pH will be discussed.

Cation Exchange and Plant Growth Relationships

There are two important points about cation exchange and plant growth. One point concerns the total amount of nutrients available to plants as exchangeable cations; the other point concerns the degree to which the exchange is saturated with bases as contrasted to hydrogen. The exchangeable hydrogen contributes to soil acidity, so it produces a special effect, which we will consider in detail in the next chapter. For our purposes, two specific cases will be cited to show the importance of the kinds and amounts of cations in the exchange.

Over a period of some years researchers had collected soil samples under good productive stands of many tree species. Wilde reasoned that the soil properties found where trees naturally grow well in the forest would be desirable properties to have in nursery soils for growing the seedlings. On this basis he formulated the nursery standards given in Table 8-7. The most demanding tree species require both a greater total amount of exchangeable Ca, Mg, and K, but also a higher percentage base saturation (and pH) than the less demanding species that do well on soils of moderate or low fertility.

Horticulturists have studied highbush blueberry plantations to determine the soil conditions best suited for blueberries. Soil samples were collected and analyzed from 55 locations where the growth of the blueberries had been judged to be either good or poor. Soils were analyzed for the major exchangeable cations, cation exchange capacity, and pH. The only factor that consistently correlated with the growth of the blueberries was exchangeable calcium. The results plotted in Fig. 8-16 show that the best growth of blueberries occurred on soils with less than 10 percent exchangeable calcium. This is consistent with the very low base saturation and highly acid soil conditions required.

Most garden and agricultural crops grow best when the base saturation is 80 percent or more and the pH is 6 or above. These examples illustrate the great importance of cation exchange relationships for plant growth. The entire next chapter on pH will use these ideas to explain the nature, importance, and control of pH.

Anion Exchange in Soils

Although cation exchange in soils has been studied extensively and its importance in

Table 8-7

Standards for Raising Nursery Stock

Fertility Level	Species	CEC, me/100 g	Exchangeable Cations me/100 g			Percent Base Saturation	pH Range
			Ca	Mg	K		
High	White oak, pecan, hard maple, basswood, white cedar	10	5.0	2.0	.3	73	5.5–7.3
Moderate	White spruce, white pine, Douglas fir, yellow birch	7	2.5	1.0	.2	53	5.0–6.0
Low	Jack pine, Scotch pine, Virginia scrub pine	4	1.5	.5	.1	52	4.8–5.5

Adapted from Wilde, 1958.

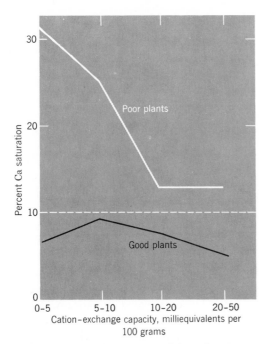

Fig. 8-16 Effect of percentage of the soil cation-exchange capacity occupied by calcium and the general vigor of blueberry plants. Good vigor occurred when exchangeable calcium was less than 10 percent over a wide range of cation-exchange capacity. (Adapted from Ballinger, et al., 1958.)

supplying nutrients to plants is recognized, the phenomenon of the exchange of anions in the colloidal complex has received less attention. It has been established that such exchange does occur, although apparently to a much lesser extent in most soils than that of cations. Anions may replace the OH groups in the clay minerals and, as these groups are much more plentiful in the kaolinite minerals than in the montmorillonite minerals, the kaolinite minerals are considered the seat of most anion exchange in temperate or arid region soils.

Although the areas are not extensive, some of the most weathered soils in the humid tropics have little or no kaolinite, and the clays are almost entirely oxides of iron and aluminum (to a lesser extent also manganese and titanium). Such soils may have a *net positive* charge where the charge due to the clay over balances the negative charge of the organic matter. Imagine what this means. Instead of having calcium adsorbed and nitrate very mobile and subject to leaching, the reverse occurs. We need to consider the reasons for such an important reversal in soil properties.

Gibbsite is an example of an important oxide clay. Gibbsite consists of aluminum in six coordination (aluminum in octahedral position) surrounded by six hydroxyls. In the normally acid environment of highly weathered tropical soils, the hydroxyls take on hydrogen atoms (protons) or are protonated as follows.

$$
\begin{array}{ccc}
\mid & & \mid \\
Al & - & Al & - \\
\mid \diagdown & & \mid \diagdown \\
O \quad OH + H^+ & \xrightarrow{\text{protonated}} & O \quad OH_2^+ \\
\mid \diagup & & \mid \diagup \\
AL & & Al \\
\mid & & \mid
\end{array}
$$

neutral   proton   protonated and
oxide   positively charged
surface   oxide surface

Few soil horizons are positively charged. When they have a net positive charge, however, the importance of the phenomenon can hardly be overestimated, since most important soil-plant nutrition relationships are reversed. In summary, soils with net positively charged colloids: (1) adsorb anions like nitrate and chloride ions. (2) Cations like calcium, magnesium, and potassium are repelled and remain in the soil solution very susceptible to leaching. In addition the term percentage base saturation is meaningless. (3) Phosphate and sulfate ions are tightly fixed by replacement of hydroxyls and the soils have very high phosphorus fixing capacity and naturally a very low level of available phosphorus. One reason for some failures in the use of temperate region practices to crop production on some tropical soils has been because of the lack of understanding concerning positively charged soil colloids. Use your imagination to think of some specific reasons for such failures.

pH-Dependent Charge

The silicate clays carry a permanent negative charge that is unaffected by pH of the soil environment. A good example is the lattice charge of montmorillonite resulting from the isomorphous substitution of aluminum by magnesium. We noted in Chapter 7 that cation-exchange capacity or

Fig. 8-17 Relation of cation exchange capacity to pH. Kaolinite and montmorillonite have some pH-dependent charge above pH 6. All of the CEC of humus is pH dependent and increases linearly with pH.

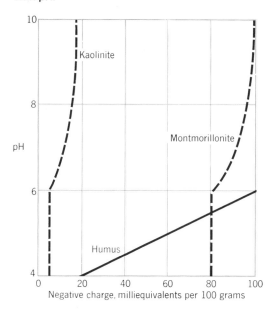

negatively charged sites arise in humus by the dissociation of hydrogen ions from hydroxyls of carboxyl groups. A reduced concentration of hydrogen ions in solution, as occurs when the pH increases, encourages more dissociation of hydrogen from hydroxyls, and thus produces more negatively charged sites or cation exchange capacity. To a limited extent, the same occurs for the exposed hydroxyls of the silicate clay minerals (Fig. 8-17).

The charge of oxide clays is entirely pH dependent. For example, consider what happens when an acid oxide clay soil is limed (pH increased).

$$
\begin{array}{c}
| \\
\text{Al} \\
| \quad \diagdown \\
\text{O} \quad \text{OH}_2^+ + \text{OH}^- \xrightarrow{\text{deprotonation}} \\
| \quad \diagup \\
\text{Al} \\
|
\end{array}
\qquad
\begin{array}{c}
| \\
\text{Al} \\
| \quad \diagdown \\
\text{O} \quad \text{OH} + \text{H}_2\text{O} \\
| \quad \diagup \\
\text{Al} \\
|
\end{array}
$$

positive   OH⁻ from lime   neutral   water
charged   oxide
oxide   surface
surface   (isoelectric
                               point)

If more liming occurs, further deprotonation occurs.

$$
\begin{array}{c}
| \\
\text{Al} \\
| \quad \diagdown \\
\text{O} \quad \text{OH} + \text{OH}^- \xrightarrow{\text{deprotonation}} \\
| \quad \diagup \\
\text{Al} \\
|
\end{array}
\qquad
\begin{array}{c}
| \\
\text{Al} \\
| \quad \diagdown \\
\text{O} \quad \text{O}^- + \text{H}_2\text{O} \\
| \quad \diagup \\
\text{Al} \\
|
\end{array}
$$

neutral   OH⁻ from lime   negative   water
oxide   charged
surface   oxide
                              surface

We have just noted two important things: (1) the charge on soil colloids is pH dependent, and (2) the exchange sites are both negative and positive. Positive sites are important in tropical soils high in oxide clays. In some cases subsoil horizons low in organic matter have a net positive charge at low pH. Increasing the pH increases the number of cation-exchange sites, and a point will be reached where the number of negatively and positively charged sites are equal. This point is called *zero point of charge* (ZPC).

This effect of pH on the charge of soil colloids calls attention to the importance of soil pH—the topic of the next chapter.

References

Ballinger, W. E., A. L. Kenworthy, H. K. Bell, E. J. Benne, and S. T. Bass, "Production in Michigan Blueberry Plantations in Relation to Nutrient-Element Content of Fruiting-Shoot Leaves and Soil," *Mich. Agr. Exp. Sta. Quart. Bull.*, *40*:896–914, 1958.

Clarke, Frank Wigglesworth, "Data of Geochemistry," *U.S. Geol. Sur. Bull.* 770, Washington, D.C., 1924.

Coleman, N. T., and A. Mehlich, "The Chemistry of Soil pH," in *Soil*, USDA Yearbook, Washington, D.C., 1957, pp. 72–79.

Forsythe, Warren M., "Soil-water Relations in Soils Derived from Volcanic Ash of Central America," in *Soil Management in Tropical America*, North Carolina State University, Raleigh, S.C., 1975, pp. 155–167.

Jackson, M. L., and G. D. Sherman, "Chemical Weathering of Minerals in Soils," *Adv. Agronomy*, *5*:219–318, 1953.

Keng, J. C. W., and G. Uehara, "Chemistry, Mineralogy, and Taxonomy of Oxisols and Ultisols," *Soil and Crop Sci. Soc. of Florida*, *33*:119–126, 1973.

Liebig, Justus, *Chemistry in its Application to Agriculture and Physiology*, Lyon Playfair, Editor, Wiley, New York, 1852.

Schofield, R. K., "The Effect of pH on Electric Charges Carried by Clay Particles," *Jour. Soil Sci.*, *1*:1–8, 1949.

Smith, G. D., W. H. Allaway, and F. F. Riecken, "Prairie Soils of the Upper Mississippi Valley," *Adv. Agronomy*, *2*:157–205, 1950.

St. Arnaud, R. J., and G. A. Septon, "Contribution of Clay and Organic Matter to Cation-Exchange Capacity of Chernozemic Soils," *Can. Journ. Soil Sci.*, *52*:124–126, 1972.

Wilde, S. A., *Forest Soils*, Ronald, New York, 1958.

9

SOIL pH—CAUSES, SIGNIFICANCE, AND ALTERATION

The particular pH measured in a soil is caused by a particular set of chemical conditions. Therefore, a determination of soil pH is one of the most important tests that can be made to diagnose plant growth problems. For example, suppose some diseased plants have light green leaves that could be caused by several factors. If the pH of the soil is as low as 5.5 or less, the disease is probably not an iron deficiency, since iron compounds are soluble under acid conditions. If the soil pH is 8, one should seriously consider the possibility of iron deficiency because iron compounds are very insoluble in soils with a pH of 8. As an analogy, the pH of the soil is like the temperature of an animal. Both tests are easily made and provide basic information useful in diagnosing what is likely to be the disease or problem. Our attention in this chapter will be focused on the causes, significance, and alteration of soil pH.

Definition and Causes of Soil pH

Usually, soils of humid regions are acid and soils of arid regions are alkaline. In acid soils the soil solution contains more hydrogen ions (H^+) than hydroxyl (OH^-); in alkaline soils the soil solution contains more OH^- than H^+. Our purpose is to examine the causes for various concentrations of H^+ and OH^- that are responsible for the range of pH from about 4 to 10 which is normally found in soils.

pH Defined

Water is neutral because the concentration of H^+ and OH^- are equal. At neutrality the pH is 7. For our purposes it is sufficient to consider water as composed of water molecules, hydrogen ions, and hydroxyl ions. Water dissociates or ionizes as follows.

$$HOH \rightleftharpoons H^+ + OH^- \qquad (1)$$

At equilibrium the reaction is strongly to the left, producing the following composition of water expressed as moles per liter: (at 25°C a liter of water weighs 997 grams and 1 mole of water weighs 18 grams, resulting in 55.4 moles of water per liter).

HOH   55.399,999,8 moles per liter
H^+ 0.000,000,1 moles per liter
OH^- 0.000,000,1 moles per liter

Two points should be noted. First, only one water molecule in 554 million is ionized. Second, the number, concentration, or moles per liter of H^+ is equal to OH^-.

The pH scale has been devised for conveniently expressing the extremely small concentrations of H^+ found in water and in many important biological systems. The pH is defined as:

$$pH = \log \frac{1}{[H^+]}$$

(where $[H^+]$ equals moles of H^+ per liter)
$$(2)$$

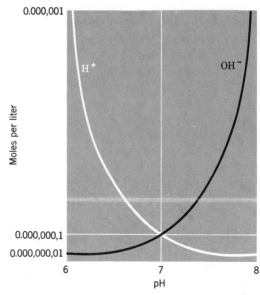

Fig. 9-1 Changes in concentration of H^+ and OH^- with changes in pH of solutions.

The pH of pure water is calculated as follows.

$$pH = \log \frac{1}{0.000,000,1}$$

$$= \log 10,000,000 = 7 \qquad (3)$$

Each unit change in pH is associated with a tenfold change in the concentration of H^+ and OH^- (Fig. 9-1).

Exchangeable Cations as Sources of H^+ and OH^-

From the previous chapter we know that the exchangeable cations are adsorbed with sufficient energy to retard their leaching from the soil, but a significant number dissociate from the cation exchange surfaces and exist in the solution where they are readily utilized by plants. On dissociation the exchangeable bases cause hydrolysis and OH ions are produced (Fig. 9-2).

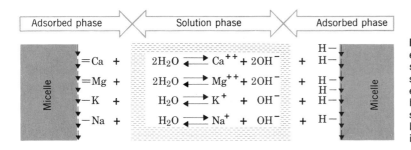

Fig. 9-2 Hydrolysis of exchangeable bases as a source of OH⁻ for the soil solution. (The exchangeable bases are hydrated, but have been shown to be not hydrated for simplicity of illustration.)

Exchangeable H dissociates and contributes H^+ to the soil solution according to Equation 4.

$$\text{micelle-H} \rightleftharpoons H^+ \qquad (4)$$

(adsorbed)   (in solution)

Exchangeable H is the principal source of H^+ until the pH of the soil goes below 6, when Al in the Al octahedral sheet of clays becomes unstable and is adsorbed as exchangeable Al. The exchangeable Al is a source of H^+ according to Equation 5.

$$\text{micelle-Al} + 3H_2O \rightleftharpoons$$

$$Al(OH)_3 + \text{micelle} \underset{\text{(insoluble)}}{\overset{H}{\underset{H}{-H}}} \rightleftharpoons H^+ \qquad (5)$$

The net effect of the hydrolysis by exchangeable Al is an increase in the H^+ concentration of the soil solution resulting from the dissociation of the exchangeable (micelle) H that is produced (same as in Equation 4). For all practical purposes, exchangeable H and Al are both sources of H^+ and, for simplicity, the term exchangeable H will be used with no attempt to indicate whether Al is involved.

Relationship Between Soil pH and Percent Base and Hydrogen Saturation

One of the most logical questions at this point is, "What conditions produce equal concentrations of H^+ and OH^- and a pH of 7?" In every soil there is a relationship between the percentage of base (or H) saturation and pH. For broad geographic areas where soils have similar mineralogy, a general relationship exists. In southern Michigan the data from thousands of soil samples were analyzed and, for mineral soils, the relationship shown in Equation 6 was found.

$$pH \times 24 =$$

$$187 - 0.3 \text{ (CEC)} - \text{percent H} \\ \text{saturation} \qquad (6)$$

Using Equation 6, Fig. 9-3 was constructed. From the diagram it can be seen that a pH of 7 was associated with a 15 percent H saturation and 85 percent base saturation (assuming a CEC of 13). Furthermore, for the soils studied, the minimum and maximum pH that could be developed by the dissociation of cations from the exchange was 3.5 and 7.6, respectively.

At 50 percent H and base saturation the pH was 5.5. This is because the exchangeable H is less tightly adsorbed to the micelles than are the exchangeable bases, which are predominately divalent calcium and magnesium. In kaolinitic soils, where the exchange sites are mainly from dissociation of H from OH on the exposed edges of the clay particles, a pH between 6 and 7 may be associated with 50 percent base saturation

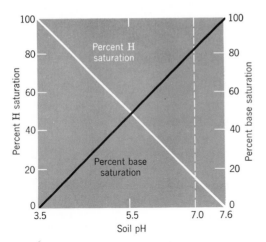

Fig. 9-3 Relationship between soil pH and percentage base and hydrogen saturation for loamy textured soils of southern Michigan using equation 6 and assuming a cation-exchange capacity of 13. (Courtesy Dr. John Shickluna.)

because OH bound H is much more tightly adsorbed than is H adsorbed on an exchange site, resulting from isomorphous substitution.

Earlier it was mentioned that the pH of some soils was as high as 10. The exchangeable cations in a leached humid region soil can account for a pH of 7 and slightly more but, to account for pH values of as high as 9 or 10, one must examine some other soil conditions. The first to be examined will be the conditions found in calcareous soils.

pH of Calcareous Soils

A *calcareous* soil contains calcium carbonate ($CaCO_3$) and, when treated with hydrochloric acid (HCl), a bubbling can be observed representing the evolution of carbon dioxide. Calcium carbonate is relatively insoluble, but when present in soils it creates a constant pressure to saturate the exchange with calcium as follows.

$$CaCO_3 + H_2\text{-micelle}$$
$$\rightarrow Ca\text{-micelle} + H_2O + CO_2 \qquad (7)$$

For this reason calcareous soils are 100 percent base saturated and the pH is mainly controlled by the hydrolysis of calcium carbonate as follows.

$$CaCO_3 + 2H_2O \rightarrow Ca(OH)_2 + H_2CO_3 \qquad (8)$$

The greater dissociation of the calcium hydroxide and production of OH^-, as compared to the production of H^+ from the weak carbonic acid, creates an alkaline effect. As a result the pH of calcareous soils usually ranges from 7^+ to a maximum of 8.3.

Effect of Sodium on Soil pH

Sodium is released from the weathering of minerals. In the humid regions leaching readily removes the sodium because of the weak attraction of sodium for cation-exchange sites. In the arid regions the sodium may accumulate as sodium carbonate and the sodium will tend to occupy some of the exchange positions. The hydrolysis of sodium carbonate and exchangeable sodium produces a very strong base, NaOH. When the soil is 15 percent or more sodium saturated or a significant amount of sodium carbonate exists in the soil, the pH value may be between 8.5 and 10.

Effect of Soluble Salts on Soil pH

Saline soils have sufficient soluble salts to impair plant growth, mainly by increasing the osmotic pressure of the soil solution and restricting water uptake. Soluble salts may accumulate naturally in soils in arid regions or as a result of the addition of irrigation water. Excessive use of soluble fertilizer salts produces saline soils and is a particular problem in managing greenhouses. The major potassium fertilizer is KCl. Its hydrolysis produces both a strong base, KOH, and a strong acid, HCl, which are about equal in their ability to produce H^+ and OH^-. Saline soils tend to have a pH at or near 7 because of the hydrolysis by soluble salt.

Other Factors Affecting Soil pH

Carbon dioxide released from respiration dissolves in the soil solution and forms weak carbonic acid. This influence has existed in calcareous and other alkaline soils for thousands of years, indicating that the formation of carbonic acid in soils plays a very minor role in causing a particular soil pH. Carbonic acid in soil does facilitate weathering and plays a role in leaching (see Fig. 9-4).

A few other factors influencing soil pH are worthy of mention. Sulfur is a by-product in industrial gases and is sometimes responsible for soil acidity in nearby soils as a result of the formation of sulfuric acid. Some unusual soils that contain a significant amount of iron sulfide (FeS) also contain significant amounts of acid. Plants are killed when the sulfur is oxidized and converted to sulfuric acid and the soil pH becomes very low. A small amount of nitric acid is a natural component of rain, but its effect appears to be insignificant.

About 20 years ago a marked decrease occurred in the pH of rainfall in the northeastern United States. Rain with pH as low as 2.1 was collected. There has been an increase in the amount of fossil fuel burned, and the taller stacks distribute the SO_2 over a wider area. The resulting increased acidity of rain is affecting lakes, vegetation, soil acidity, and weathering of buildings. The effect of acid rain on plants and soil acidity appears to have reduced the growth of forests in the northeastern United States and Sweden. Much of the SO_2 causing acid rain in Sweden likely comes from industrial areas of England and the Ruhr.

A survey of rain acidity was conducted by 16,000 high school students in March 1973. Many low readings of 3.5 were reported for Chicago, New York, Cleveland, Boston,

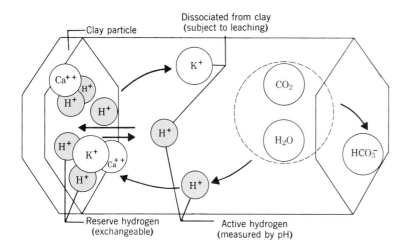

Fig. 9-4 Schematic diagram showing the relationship between reserve and active acidity. Carbon dioxide and water form carbonic acid, which provides hydrogen ions that replace exchangeable bases. In this way carbonic acid contributes to the development of soil acidity. (From *Laboratory Manual for Introductory Soil Science*, 3rd ed., W. C. Brown and Co., Dubuque, Iowa. Used with permission.)

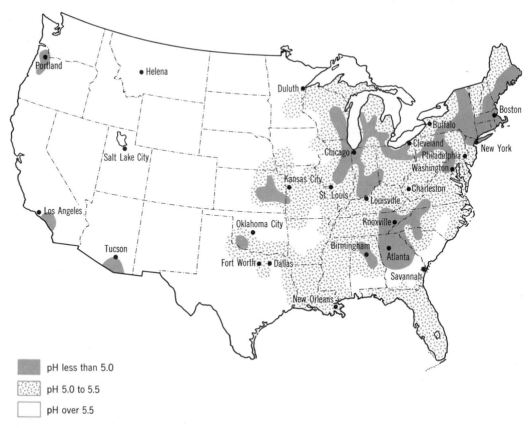

Fig. 9-5 The pH of rain in the United States. (From A. E. Klein, "Acid Rain in U.S.," *Science Teacher*, pp. 36–38, May 1974.)

and Los Angeles. Normal rainfall has a pH of about 5.7. A map based on the survey showing the distribution of rain acidity in the United States in given in Fig. 9-5.

Significance of Soil pH

The major effects of soil pH are biological. Some organisms have rather small tolerances to variations in pH, but other organisms can tolerate a wide pH range. Studies have shown that the actual concentrations of H^+ or OH^- are not very important except under the most extreme cir-

cumstances. It is the associated conditions of a certain pH value that is most important.

Nutrient Availability and pH Relationships

Perhaps the greatest general influence of pH on plant growth is the effect of pH on the availability of nutrients. In Fig. 9-3 the pH was seen to be related to percentage base saturation. When the base saturation is less than 100 percent, an increase in pH is associated with an increase in the amount of calcium and magnesium in the soil solution, since calcium and magnesium are usually the dominate exchangeable bases. Many studies have been conducted that relate increases in plant growth with increases in the percentage of calcium in plants and with increasing pH or percentage base saturation. The general relationship between pH and availability of calcium and magnesium are shown in Fig. 9-6.

Another nutrient whose availability is increased as the pH is increased on the low end of the pH range in soils is molybdenum. At low pH molybdenum forms insoluble compounds with iron and is rendered unavailable. Under these conditions plants susceptible to molybdenum deficiency, such as cauliflower, clover, and citrus, will show a growth response to an increase in pH. The fact that increased availability of molybdenum occurs as the pH increases is substantiated by the data in Fig. 9-7, which show that molybdenum concentration in cauliflower leaves increase with increasing pH.

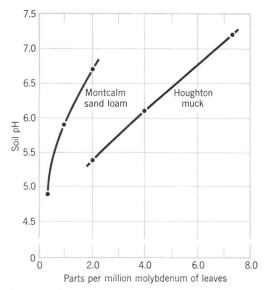

Fig. 9-7 Correlation between molybdenum content of cauliflower leaves and soil pH in a greenhouse experiment. (From Turner and McCall, 1957.)

Fig. 9-6 General relationship between soil pH and availability of plant nutrients: the wider the bar, the more available the nutrient.

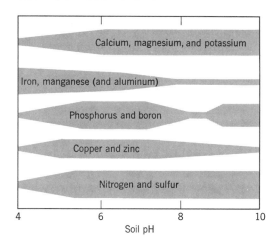

Potassium availability is usually good in alkaline soils that reflect the limited leaching of exchangeable potassium (Fig. 9-6).

The availability or solubility of some plant nutrients decreases with an increase in pH. Iron and manganese are two good examples (see Fig. 9-8). Iron and manganese are commonly deficient in calcareous soils. Phosphorus and boron also tend to be unavailable in calcareous soils resulting from reactions with calcium. Phosphorus and boron also tend to be unavailable in very acid soils. Copper and zinc have reduced availability in both highly acid and alkaline soils. For the plant nutrients as a whole, good overall nutrient availability is found near pH 6.5.

Effect of pH on Soil Organisms

The greater capacity of fungi over bacteria to thrive in highly acid soils was mentioned

Fig. 9-8 Pin oak is a popular ornamental tree that is susceptible to iron deficiency when grown on neutral or alkaline soil. Iron-deficient leaves have dark green veins and yellow intervein areas. This tree is seriously deficient in iron, as indicated by the death and defoliation of many leaves on the upper branches.

9-6). High soil acidity has also been shown to inhibit earthworms in soils. Peter Farb relates an interesting case where soil pH influenced both earthworm and mole activity.

On a certain tennis court in England, each spring after the rains and snow had washed away the court marking, the marking lines could still be located because moles had unerringly followed them. This intriguing mystery was investigated, and it produced the following results. The tennis court had been built on acid soil and chalk (lime) was used to mark the lines. As a result, the pH in the soil below the marking lines was increased. Earthworms then inhabited the soil under the markings, and the remainder of the soil on the tennis court remained uninhabited. Since earthworms are a primary food of moles, the moles restricted their activity to the soil only under the marking lines.

Aluminum, Iron, and Manganese Toxicity in Acid Soils

The data in Fig. 9-6 show that iron (aluminum) and manganese are most available in highly acid soils. The relationship between manganese in corn leaves and soil pH is shown in Fig. 13-15. Manganese toxicity occurs when the pH is about 4.5 or less. Evidence indicates that the high exchangeable aluminum in many acid soils of the southeastern United States restricts root growth in many subsoils. Aluminum toxicity was probably a factor in an unsuccessful attempt to grow Gaines wheat in Maryland. Gaines is a short-strawed variety that was developed on high base-saturated soils of the state of Washington. In Washington yields as high as 209 bushels per acre were attained. When Gaines wheat

in Chapter 6. The pH requirements of some disease organisms can be used by soil management practices as a means to control diseases. One of the best-known cases is that of the maintenance of acid soils to control potato scab. Damping-off disease in nurseries is controlled by maintaining the pH at 5.5 or less. Nitrifying organisms also become inhibited when the pH is less than 5.5. The availability of nitrogen in soils is related mainly to the effect of pH on decomposition of organic matter (see Fig.

was grown in Maryland, it yielded only 17 bushels per acre, although locally adapted wheat varieties yielded 60 bushels per acre. Plants, even varieties of the same species, exhibit differences in tolerance to high levels of aluminum, iron, or manganese as well as other soil conditions associated with soil pH. This gives rise to pH preferences of plants.

pH Preferences of Plants

Blueberries are well known for their highly acid soil requirement. In the previous chapter it was shown in one study that poor blueberry growth occurred when the calcium saturation exceeded 10 percent. Normally this would mean that the soil had a very low base saturation and pH. In Table 9–1 the optimum range for highbush blueberries is given as 4.0 to 5.0. Other plants preferring a pH of 5 or less include azaleas, orchids, sphagnum moss, jack pine, black spruce, and cranberry. The plants listed in Table 9–1 that prefer a pH of 6 or less are italicized. Most field and vegetable crops prefer a pH of about 6 or higher.

The fact that plants have specific soil pH requirements gives rise to the need to alter soil pH for successful growth of many plants. We examine this topic next.

Alteration of Soil pH

There are two approaches to assure that plants will grow without serious inhibition from unfavorable soil pH: (1) plants can be selected that will grow well with the existing soil pH, or (2) the pH of the soil can be altered to suit the preference of the plants. Considerations for altering soil pH will be examined first.

Theory of Changing Soil pH

In leached soils where there is no significant amount of salts, such as calcium or sodium carbonate, the soil pH is determined mainly by the base or H saturation (Fig. 9-3). In theory, if a soil represented by the data plotted in Fig. 9–3 had a pH of 5.5, the base saturation would be 50 percent. If the base saturation of such a soil was increased to 85 percent, the soil pH would be 7. Conversely, if the base saturation was lowered from 85 to 50 percent, the pH would be lowered from 7 to 5.5. Thus, alteration of soil pH under these conditions is dependent on changes in base (or H) saturation. Increases in base saturation produce increases in pH, and decreases in base saturation produce decreases in pH.

The case depicted in Fig. 9-3 can be used for illustration. The cation exchange capacity of the soil was 13 milliequivalents per 100 grams and, as we have already indicated, at 50 percent base saturation the pH is 5.5. To increase the soil pH to 7 would require increasing the base saturation 35 percent (from 50 to 85 percent). The quantity of bases needed is calculated as follows.

13 milliequivalents × 35 percent = 4.55 milliequivalent bases needed per 100 grams of soil

The amount of lime required is usually expressed on the basis of some convenient volume of soil such as an acre furrow slice, commonly considered to weigh 2 million pounds. In our example, if $CaCO_3$ were used as source of base (Ca), the 4.55 milliequivalents of base needed per 100 grams of soil would require 0.2275 gram of $CaCO_3$ per 100 grams of soil calculated as follows.

Table 9-1
Optimum pH Ranges of Selected Plants

Field Crops		Alyssum	6.0–7.5	Oak, Pin.	5.0–6.5
Alfalfa	6.2–7.8	*Azalea*	*4.5–5.0*	Oak, White	5.0–6.5
Barley	6.5–7.8	Barberry, Japanese	6.0–7.5	*Pine, Jack*	*4.5–5.0*
Bean, field	6.0–7.5	Begonia	5.5–7.0	*Pine, Loblolly*	*5.0–6.0*
Beets, sugar	6.5–8.0	Burning bush	5.5–7.5	*Pine, Red*	*5.0–6.0*
Bluegrass, Ky.	5.5–7.5	Calendula	5.5–7.0	*Pine, White*	*4.5–6.0*
Clover, red	6.0–7.5	Carnation	6.0–7.5	*Spruce, Black*	*4.0–5.0*
Clover, sweet	6.5–7.5	Chrysanthemum	6.0–7.5	Spruce, Colorado	6.0–7.0
Clover, white	5.6–7.0	*Gardenia*	*5.0–6.0*	*Spruce, White*	*5.0–6.0*
Corn	5.5–7.5	Geranium	6.0–8.0	*Sycamore*	*6.0–7.5*
Flax	5.0–7.0	*Holly, American*	*5.0–6.0*	Tamarack	5.0–6.5
Oats	5.0–7.5	Ivy, Boston	6.0–8.0	Walnut, Black	6.0–8.0
Pea, field	6.0–7.5	Lilac	6.0–7.5	Yew, Japanese	6.0–7.0
Peanut	5.3–6.6	Lily, Easter	6.0–7.0		
Rice	5.0–6.5	*Magnolia*	*5.0–6.0*	Weeds	
Rye	5.0–7.0	*Orchid*	*4.0–5.0*	Dandelion	5.5–7.0
Sorghum	5.5–7.5	*Phlox*	*5.0–6.0*	Dodder	5.5–7.0
Soybean	6.0–7.0	Poinsettia	6.0–7.0	Foxtail	6.0–7.5
Sugar Cane	6.0–8.0	Quince, flowering	6.0–7.0	Goldenrod	5.0–7.5
Tobacco	5.5–7.5	*Rhododendron*	*4.5–6.0*	Grass, Crab	6.0–7.0
Wheat	5.5–7.5	Rose, hybrid tea	5.5–7.0	Grass, Quack	5.5–6.5
		Snapdragon	6.0–7.5	*Horse Tail*	*4.5–6.0*
Vegetable Crops		Snowball	6.5–7.5	*Milkweed*	*4.0–5.0*
Asparagus	6.0–8.0	Sweet William	6.0–7.5	Mustard, Wild	6.0–8.0
Beets, table	6.0–7.5	Zinnia	5.5–7.5	Thistle, Canada	5.0–7.5
Broccoli	6.0–7.0				
Cabbage	6.0–7.5	Forest Plants		Fruits	
Carrot	5.5–7.0	Ash, White	6.0–7.5	Apple	5.0–6.5
Cauliflower	5.5–7.5	*Aspen, American*	*3.8–5.5*	Apricot	6.0–7.0
Celery	5.8–7.0	Beech	5.0–6.7	Arbor Vitae	6.0–7.5
Cucumber	5.5–7.0	*Birch, European*		Blueberry, High	
Lettuce	6.0–7.0	*(white)*	*4.5–6.0*	Bush	4.0–5.0
Muskmelon	6.0–7.0	*Cedar, White*	*4.5–5.0*	Cherry, sour	6.0–7.0
Onion	5.8–7.0	*Club Moss*	*4.5–5.0*	Cherry, sweet	6.0–7.5
Potato	4.8–6.5	*Fir, balsam*	*5.0–6.0*	Crab apple	6.0–7.5
Rhubarb	5.5–7.0	Fir, Douglas	6.0–7.0	*Cranberry, large*	*4.2–5.0*
Spinach	6.0–7.5	*Heather*	*4.5–6.0*	Peach	6.0–7.5
Tomato	5.5–7.5	*Hemlock*	*5.0–6.0*	*Pineapple*	*5.0–6.0*
		Larch, European	5.0–6.5	Raspberry, Red	5.5–7.0
Flowers and Shrubs		Maple, Sugar	6.0–7.5	Strawberry	5.0–6.5
African violet	6.0–7.0	*Moss, Sphagnum*	*3.5–5.0*		
Almond, flowering	6.0–7.0	Oak, Black	6.0–7.0		

Data from Spurway, 1941. Plants with an optimum pH range of 6.0 and below are italicized.

molecular weight of CaCO$_3$
= 100 grams = 2 equivalents
= 2000 milliequivalents

$$\frac{100 \text{ grams CaCO}_3}{2000 \text{ milliequivalents}}$$

= 0.05 gram of CaCO$_3$ per milliequivalent

4.55 milliequivalents × 0.05 gram
= 0.2275 grams of CaCO$_3$ per 100 gram soil

We need not convert grams to pounds, but we can just state that 0.2275 *gram* of CaCO$_3$ per 100 *grams* of soil is equal to 0.2275 *pound* of CaCO$_3$ per 100 *pounds* of soil. To put the theoretical lime requirement on an acre furrow slice basis, we use a simple proportion.

$$\frac{0.2275 \text{ pound CaCO}_3}{100 \text{ pounds soil}}$$

$$= \frac{x \text{ pounds CaCO}_3}{2 \text{ million pounds soil}}$$

= 4550 pounds CaCO$_3$

For each milliequivalent of base needed per 100 grams of soil, the theoretical lime requirement is equal to 1000 pounds of pure calcium carbonate per acre furrow slice.

Turning the situation around, suppose the pH were 7 and a pH of 5.5 was desired. Now, 4.55 milliequivalents of acid-producing material per 100 grams of soil would be required to furnish the H needed to increase H saturation from 15 to 50 percent. Sulfur is commonly used to increase soil acidity and has a milliequivalent weight of 0.16 gram (32 ÷ 2000) because sulfur is oxidized in the soil to sulfuric acid and each sulfur atom is the source of two H$^+$. Substituting 0.0728 for 0.2275 in the above equation results in a theoretical sulfur requirement of 1456 pounds of sulfur per acre furrow slice. For each milliequivalent of exchangeable acidity needed per 100 grams of soil, the theoretical sulfur requirement is 320 pounds per acre furrow slice.

Role of Cation-Exchange Capacity in Altering Soil pH

Suppose the soil just used to illustrate the theoretical lime requirement had a CEC twice as large, 26 instead of 13. Using Equation 6, one finds that at pH 5.5 the soil with twice as much CEC has a 53 percent base saturation compared to 50 for the soil with a CEC of 13. Notice this in Table 9-2. This means that the same concentration of H$^+$ in the soil solution (same pH) is produced with only a slight difference in percentage H or base saturation. The soil with a CEC of 26 has nearly twice the exchangeable H as the soil with a CEC of 13 when both have a pH of 5.5. The data in Table 9-2 also show that similar percentage base saturation increases are needed to increase pH from 5.5 to 7. This results in the need for almost two times more lime for the soil with twice the CEC (Table 9-2), 4.55 versus 9.36 milliequivalents per 100 grams of soil. The greater capacity of soils with greater exchange capacities to adsorb bases for a given pH change is illustrated in Fig. 9-9.

Soil acidity has two components: (1) the active or solution H, and (2) the exchangeable or reserve H. These two forms tend toward equilibrium so that a change in one produces a change in the other. When a base is added to an acid soil, the solution H is neutralized and some exchangeable H ionizes to reestablish the equilibrium. The amount of exchangeable H is slowly decreased, the solution H is decreased, and the pH slowly increases. The resistance of

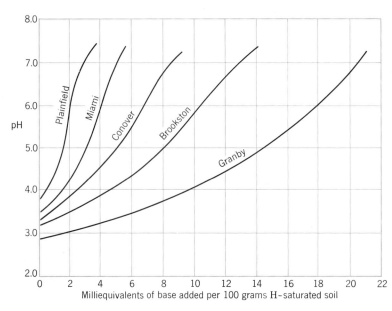

Fig. 9-9 Titration curves showing the different amounts of base needed to increase the pH of soils having increasing cation-exchange capacities from left to right (different degrees of buffering). All soils were hydrogen-saturated when titration was started.

the soil pH to change gives rise to the buffering phenomenon in soils. Soils with the largest CEC capacities offer the greatest resistance to change in pH and are the most strongly buffered.

Liming produces little pH change if the CEC is mainly pH dependent. This results from increased CEC as lime neutralizes soil acidity. There is an increase in milliequivalents of exchangeable bases but little change in percentage base saturation. For this reason liming of many tropical soils rich in oxide clays (Oxisols) is based on reduction of toxic effects of low pH, such as Al toxicity, instead of increasing pH to some predetermined goal as 6.5 (see Fig. 9-10).

Table 9-2

Comparisons of Base Saturation Relationships of Soils with Varying Cation Exchange Capacity as Calculated Using Equation 6

	Cation Exchange Capacity of Soil, me/100 g	
	13	26
Base saturation at pH 5.5	50%	53%
Exchangeable H at pH 5.5	6.5 me	12.2 me
Base saturation at pH 7	85%	89%
Base saturation change needed to increase pH from 5.5 to 7	35%	36%
Base needed per 100 g soil	4.55 me	9.36 me

Fig. 9-10 Response of soybeans to lime on an Oxisol in Brazil where the major benefit of lime is reduction in exchangeable aluminum.

Methods Used to Determine the Lime Requirement

In practice, the first decision in arriving at a recommendation of lime is deciding on the pH value desired. The desired pH depends in part on the type of soil and the kind of plant to be grown. Identification of the factors limiting the growth of alfalfa on 13 highly acid soils of the southeastern United States showed that manganese toxicity was the most limiting in eight soils, aluminum toxicity in one soil, and calcium deficiency in one soil. The results of the study are presented in Table 9-3.

It is common practice in soil-testing laboratories to make a direct measurement of the lime requirement by using a buffer solution. One such test was developed at Ohio State University and is widely used. A buffer solution with pH of 7.5 is mixed with a known quantity of soil. The exchange acidity is replaced from the exchange sites and the depression in pH from 7.5 (buffer solution pH) to the pH of the mixture of buffer solution and soil is a measure of the total acidity. The lime requirement is obtained from tables that have been developed that relate depression of buffer pH to tons of limestone required to raise the pH of the soil to a desired value.

A less accurate but useful method, especially for the homeowner, is to determine the pH of the soil with inexpensive indicator dyes (see Fig. 9-11) and to obtain the lime requirement from a table that relates soil pH and texture (used as an indication of cation-exchange capacity) to lime requirement. Such a table used for liming a garden is presented in Table 9-4. The lower lime requirement values for the soils of the southern states is a reflection of the lower cation-exchange capacity resulting from kaolinitic clays. Caution must be observed to prevent overliming because some of the micronutrients may become unavailable, resulting in nutrient deficiencies.

Forms of Lime

Chemically, lime is CaO, but an extension of the meaning of the word lime is now used

Table 9-3
Identification of Factors Limiting the Growth of Alfalfa on Acid Surface Soils

Soil	pH	Region	Sample Site	Growth Limiting Factor
Zanesville	4.9	Ozarks	Fayetteville, Ark.	Mn toxicity
Johnsburg	5.1	Ozarks	Fayetteville, Ark.	Mn toxicity
Taloka-complex	5.2	Ozarks	Fayetteville, Ark.	Mn toxicity
Waynesboro	5.2	Ozarks	Fayetteville, Ark.	Mn toxicity
Richland	5.1	Loessial Terraces	Marianna, Ark.	Mn toxicity
Loring	5.3	Loessial Hills	Colt, Ark.	Mn toxicity
Dundee	4.8	Mississippi Bottoms	Earle, Ark.	Mn toxicity
Cecil	5.2	Piedmont	Watkinsville, Ga.	Mn toxicity
Bladen	4.8	Lower Coastal Plains	Fleming, Ga.	Al toxicity
Leon	4.2	Lower Coastal Plains	Fleming, Ga.	Ca deficiency
Lakeland	5.1	Lower Coastal Plains	Live Oak, Fla.	Complex of factors not separated
Tifton	5.6	Middle Coastal Plains	Tifton, Ga.	Complex of factors not separated
Rains	4.7	Middle Coastal Plains	Tifton, Ga.	Complex of factors not separated

Reprinted from *Plant Food Review*, Vol. 10, publication of the Fertilizer Institute, 1964.

to include all limestone products used to neutralize soil acidity. Limestone deposits are widely distributed and constitute the most important source of lime (Fig. 9-12). Limestone is a carbonate form of lime with $CaCO_3$ and $MgCO_3$ as the major components. The oxide form, CaO, is produced by heating calcium carbonate and driving off carbon dioxide. Hydroxide forms of lime are produced by "slaking" or adding water to the oxide forms. In the soil all forms act similarly in that the exchangeable H is replaced and converted to water and the base saturation of the soil is correspondingly increased. The reaction of the carbonate and hydroxide forms with exchangeable hydrogen is as follows.

$$CaCO_3 \quad + \quad \genfrac{}{}{0pt}{}{H}{H} - \text{micelle} \rightarrow$$

$$Ca - \text{micelle} + H_2O + CO_2 \qquad (9)$$

$$Ca(OH)_2 \quad + \quad \genfrac{}{}{0pt}{}{H}{H} - \text{micelle} \rightarrow$$

$$Ca - \text{micelle} + H_2O \qquad (10)$$

The oxide form of lime reacts with water in the soil to form calcium hydroxide and then neutralizes soil acidity according to Equa-

Fig. 9-11 Determination of soil pH using pH indicator dye solution and pH color chart.

tion 10. Neutralization or the removal of H^+ by the formation of water drives the reactions to the right. The greater solubility of the oxide and hydroxide forms as compared to $CaCO_3$ results in more rapid reaction with exchangeable hydrogen.

On a weight basis the three forms of lime have different neutralization capacity. The neutralizing value of liming materials is calculated on the basis of pure calcium carbonate as 100 percent. It has been pointed out that when pure calcium carbonate is burned in a kiln, 100 pounds of the dry material give off 44 pounds of carbon dioxide gas, leaving 56 pounds of calcium oxide. When the 56 pounds of calcium oxide are moistened, they react with 18 pounds of water to form 74 pounds of hydrated lime. It is obvious that 100 pounds of pure calcium carbonate, 74 pounds of pure calcium hydroxide, and 56 pounds of pure calcium oxide all contain the same amount of calcium and all have the same power to neutralize soil acidity.

Table 9-4

Suggested Applications of Finely Ground Limestone to Raise the pH of a 7-in. Layer of Several Textural Classes of Acid Soils, in Pounds per 1000 ft²

	pH 4.5 to 5.5		pH 5.5 to 6.5	
Textural Class	Northern and Central States	Southern Coastal States	Northern and Central States	Southern Coastal States
Sands and loamy sands	25	15	30	20
Sandy loams	45	25	55	35
Loams	60	40	85	50
Silt loams	80	60	105	75
Clay loams	100	80	120	100
Muck	200	175	225	200

From Kellogg, 1957.

Fig. 9-12 Many areas have limestone near the surface that can readily be mined, ground, and applied to acid soils as an inexpensive corrective for excess soil acidity. (Monroe County, Wisc.)

The relative ability of different liming materials to neutralize acidity is frequently expressed on a percentage basis. Thus, the neutralizing power of the different forms of lime in the pure state is determined by their molecular weights. The molecular weight of calcium carbonate is 100; of calcium hydroxide, 74; and of calcium oxide, 56. By dividing 100 by 74, the figure 1.35 is obtained, which means that 1 pound of calcium hydroxide supplies the same amount of calcium as 1.35 pounds of calcium carbonate. In other words, if expressed on a percentage basis (1.35 × 100), pure calcium hydroxide has a neutralizing value of 135 percent relative to calcium carbonate. Likewise, calcium oxide (100/56 × 100 = 178) has a neutralizing power of 178 percent. Pure magnesium carbonate with a molecular weight of 84 has a neutralizing value of 119 percent (100/84 × 100 = 119). A limestone containing 80 percent calcium carbonate and 20 percent magnesium carbonate would have a neutralizing value of 103.8 percent (80 + 20 × 1.19 = 103.8). Frequently magnesium limestones have a neut-

ralizing power of 107 or 108 percent. The molecular weights, neutralizing values, and calcium carbonate equivalents for the common chemical forms of liming materials in a pure state are given in Table 9-5. Locally, many other materials are used for liming. These include marl, by-product lime from manufacturing, and oyster shells.

Importance of Limestone Particle Size

With all factors being equal, the finer a limestone is ground, the more rapidly it will dissolve and the more thoroughly it can be mixed with the soil. The effectiveness of a ground substance of a given neutralizing power is determined not only by its rate of solubility, but also by its contact with the colloidal particles. However, the finer the stone is ground, the greater its cost will be and the less its lasting qualities will be. Furthermore, a very finely ground limestone is difficult to handle and unpleasant to distribute. Therefore, it is generally recommended that a ground limestone of medium fineness be purchased. Such a

Table 9-5
Relative Neutralizing Power of Different Forms of Lime

Form of Lime	Molecular Weight	Neutralizing Value, %	Pounds Equivalent to 1 Ton of Pure CaCO$_3$
Calcium carbonate	100	100	2000
Magnesium carbonate	84	119	1680
Calcium hydroxide	74	135	1480
Magnesium hydroxide	58	172	1160
Calcium oxide	56	178	1120
Magnesium oxide	40	250	800

grade can be ground rather cheaply, and it contains a sufficient quantity of fine material to give immediate effects and a sufficient amount of coarse material to give it lasting qualities. Such a ground limestone would be one that would pass an 8-mesh screen, and 25 to 50 percent would pass a 100-mesh screen. It can be seen from Fig. 9-13 that the fractions 8 to 20 mesh and larger are not very effective in increasing the pH of acid soils.

Method and Time of Applying Lime

The principal requirement of any method of applying lime is that it should be distributed evenly and, except when applied to grasslands, it should be thoroughly mixed with the soil. Lime, even as it dissolves, moves to no appreciable extent horizontally and only to a limited extent vertically. Movement is not sufficient to distribute the lime evenly over the field or to mix it thoroughly with the soil. Since soil acidity is due largely to the colloidal clay acids, it is essential that lime come in contact with all the soil particles so far as possible. This requires a thorough and even mixing of the lime with the soil. An even distribution may

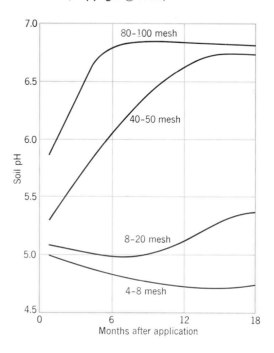

Fig. 9-13 Effect of dolomitic lime of different particle size on the pH of the soil at various times after application. (Adapted from T. A. Meyer and G. W. Volk, "Effect of Particle Size of Limestone on Soil Reaction," *Soil Science, 73*: 37–52, 1952. Used by permission of author and the Williams and Wilkins Co., copyright © 1952.)

be accomplished by the use of most of the standard lime spreaders on the market. The only way of mixing the lime thoroughly with the soil is by tillage operations.

Generally, lime may be applied any time during the year when it is most convenient. Often, however, the type of the rotation, the system of farming, and the form of lime used will be the deciding factors.

It is advisable, of course, to apply lime where it can be used to the best advantage in the rotation, for example, preceding the legume crop or in connection with a green-manure crop. It is usually best to apply lime considerably in advance of seeding legumes, so that the lime will have time to correct the acid condition of the soil. To insure best results, lime should be in the soil at least 6 months prior to seeding legumes, although successful legume seedings sometimes are obtained by applying lime immediately preceding or with the legume seeding.

The data from lysimeter studies conducted in Ohio and cited in Table 4-2 showed that the annual loss of calcium in the drainage water could be equal to or greater than the calcium removed in some harvested crops. This shows that in the humid regions leaching may be as important as crop production in the removal of exchangeable bases and development of soil acidity. Factors that affect leaching, such as soil permeability and slope, will influence the frequency of liming. In general it can be stated that several hundred pounds of lime per acre will be needed annually in the humid regions. Weathering can be expected to supply part of the need, but it is obvious that the maintenance of a pH near 6.5 will require liming once every 5 to 10 years.

Acidulation of Soils

The addition of acid sphagnum peat to soil may have some acidifying effect. However, significant and dependable increases in soil acidity are perhaps best achieved through the use of sulfur. The sulfur is slowly converted to sulfuric acid so the change in soil pH is gradually decreased over several months or a year. Earlier it was noted that 320 pounds of sulfur per acre furrow slice would theoretically increase exchangeable acidity or exchangeable hydrogen equal to 1 milliequivalent per 100 grams of soil. In a forest nursery near Orono, Canada, the use of sulfur resulted in significant increases in soil acidity and the growth of red pine seedlings. About 900 pounds of sulfur per acre lowered soil pH one unit in the experiment.

As with lime, the amount of sulfur required varies with the cation-exchange capacity of the soil. Recommendations for using sulfur on small areas are given in Table 9-6. The effect of nitrogen fertilizer on soil acidity is discussed in Chapter 13.

Managing the pH of Calcareous Soils

Many millions of acres of soil are calcareous in arid regions, on flood plains, and on recently drained lake plains. Some of these soils are excellent for agriculture, such as those located in the Palouse Country in Washington, the flood plains of the Red, Platte and Mississippi rivers, and on the lake plains around the Great Lakes. Plants growing on calcareous soils are sometimes deficient in iron, manganese, zinc, copper, or boron. To lower the pH of calcareous soils appreciably, the calcium carbonate must be leached out, and this is impractical. As a result, crops are fertilized with appropriate nutrients that are deficient or

NATURE AND MANAGEMENT OF SALINE AND SODIC SOILS

Table 9-6

Suggested Applications of Ordinary Powdered Sulfur to Reduce the pH of 100 ft² of an 8-in. Layer of Sand or Loam Soil

	Pints of Powdered Sulfur, for Desired pH									
	4.5		5.0		5.5		6.0		6.0	
Original pH	Sand	Loam	Sand	Loam	Sand	Loam	Sand	Loam	Sand	Loam
5.0	$\frac{2}{3}$	2								
5.5	$1\frac{1}{3}$	4	$\frac{2}{3}$	2						
6.0	2	$5\frac{1}{2}$	$1\frac{1}{3}$	4	$\frac{2}{3}$	2				
6.5	$2\frac{1}{2}$	8	2	$5\frac{1}{2}$	1	4	$\frac{2}{3}$	2		
7.0	3	10	$2\frac{1}{2}$	8	2	$5\frac{1}{2}$	1	4	$\frac{2}{3}$	2

Taken from Kellogg, 1957.

crops are selected that are adapted to the alkaline soils.

Plants show considerable differences in their ability to utilize certain nutrients in calcareous soils. Sorghum may be iron deficient, but alfalfa will not be iron deficient when grown on the same calcareous soil. Varieties of the same species may exhibit remarkable differences to tolerate toxicity concentrations of some nutrients or to remove certain nutrients from the soil. Since the pH of calcareous soils cannot be practically altered, crop selection and fertilization with deficient nutrients are commonly the only practical solutions for growing many crops on calcareous soils.

Nature and Management of Saline and Sodic Soils

In arid regions some soils develop under conditions of poor drainage in spots and there is more water evaporating than there is water coming into the area as precipitation. Under these conditions soluble salts and exchangeable sodium may accumulate in sufficient amounts to impair plant growth and to alter soil properties. The addition of irrigation water, which contains varying amounts of soluble salts, for crop production in arid regions also creates the potential for the accumulation of soluble salts and exchangeable sodium. *Saline* soils contain sufficient soluble salt to impair plant growth, and *sodic* soils contain sufficient exchangeable sodium to impair plant growth and alter soil properties. The quantity, proportion, and nature of salts present may vary in saline and sodic soils. This gives rise to three kinds of soils: saline, saline-sodic, and sodic soils. A consideration of the development, properties, and management of these soils follows.

Saline Soils

Saline soils contain a relatively high concentration of soluble salts, made up largely of chlorides, sulfates, and sometimes nitrates. Small quantities of bicarbonates may occur, but soluble carbonates are usually absent. Frequently relatively insoluble salts, such as calcium sulfate and calcium and magnesium carbonates are also present. The chief cations present are calcium, magnesium, and sodium, but sodium seldom makes up more than one half of the soluble cations and is not adsorbed to an appreciable extent on the colloidal fraction of the soil.

The pH value of these soils is 8.5 or less, and the exchangeable sodium percentage is less than 15. White crusts frequently accumulate on the soil surface, and streaks of salt are sometimes found within the soil (see Fig. 9-14). Saline soils have a favorable structure because the colloids are highly flocculated. Soils in this group are similar to those designated as "white alkali" soils.

Sodic (Alkali) Soils

These soils do not have as high a concentration of soluble salts as saline soils and the exchangeable sodium percentage exceeds 15. The pH values usually range between 8.5 and 10. The sodium dissociates from the colloids, and small amounts of sodium carbonate may form. Organic matter in the soil is highly dispersed and is distributed over the surface of the particles, giving a dark color, hence the term "black alkali," which was formerly used to designate such soils. Sodic soils frequently occur in small irregular areas in regions of low rainfall and are referred to as "slick spots."

Sodic soils may develop as a result of irrigation. Because of the dispersed state of the colloids, the soils are difficult to till and are slowly permeable to water. After a long period of time the dispersed clay may migrate downward, forming a very dense layer with a prismatic or columnar structure. When this phenomenon occurs, a few inches of relatively coarse-textured soil may be left on the surface.

The soil solution of sodic soils contains only small amounts of calcium and magnesium, but larger quantities of sodium. The anions include sulfate, chloride, bicarbonate, and usually small quantities of normal carbonate. In some areas an appreciable amount of potassium salts is also present.

Fig. 9-14 Saline or "white alkali" soils dominated by salt-tolerant greasewood and of little agricultural value unless reclaimed by leaching out salts.

Saline-Sodic (Alkali) Soils

Soils in this group are characterized by a high concentration of soluble salts, but they differ from saline soils in that the exchange sodium percentage is greater than 15. As long as the large quantity of soluble salts remains in the soil, the high sodium content in the colloids may not cause trouble, and the soil pH seldom exceeds 8.5. If, however, the soluble salts are temporarily leached downward, the pH goes above 8.5, the sodium causes the colloids to disperse, and a structure unfavorable for tillage, entry of water, and root development develops. The movement of the soluble salts upward into the surface soil may lower the pH and restore the colloids to a flocculated condition. The management of this group of soils is a problem until the excess soluble salts and the exchangeable sodium are removed from the zone of root growth. Unless calcium sulfate or another source of soluble calcium is present, the drainage and leaching of these soils convert them into non-saline sodic soils.

Detrimental Effects of Saline and Sodic Soils on Plants

High concentrations of neutral salts such as sodium chloride and sodium sulfate may interfere with the absorption of water by plants through the development of a higher osmotic pressure in the soil solution than exists in the root cells. Furthermore, the wilting coefficient of soils is raised by salt accumulations, and hence the quantity of water that a soil will supply to plants may be reduced through the presence of salts. Detriment to plants may result also from soluble salts when the concentration is not sufficient to influence absorption of water. The entrance of nutrient ions into root hairs is influenced by the nature and concentration of other ions present. The salts may therefore result in nutritional difficulties in crops because of their inability to absorb needed nutrients from the soil.

The highly alkaline reaction caused by the presence of sodium carbonate and the large quantity of adsorbed sodium represses the availability of several nutrients, especially iron, manganese, zinc, and phosphorus. Also, the alkaline soil solution has a corrosive action on the bark of roots and stems.

The exchangeable sodium in sodic soils results in a deflocculation of the colloids and hence in a breaking down of the soil structural units. This puddled condition renders the soil more or less impervious, retards entrance of irrigation and rain water, and impedes drainage. In fine-textured soils the penetration of roots may be restricted by the density of the deflocculated zone. Aeration is also much reduced, setting up anaerobic conditions and resulting in the formation of reduced compounds that are toxic to plants.

Reclamation of Saline Soils

Good drainage is necessary for the reclamation of saline soils. It is essential in the reclamation process to remove the excess salts from the root zone, and this can only be done by the application of sufficient water to wash them into the lower soil depths. Unless there is ample drainage, the addition of so much water will raise the water table and hence lead to increased accumulations of salt in the surface soil instead of to a correction of the saline condition. Sufficient drainage should be provided to reduce the ground-water level well below the zone of root penetration. Preferably the ground water should never be less than 8 to 10 feet below the surface, and every reasonable effort should be made to prevent its rising nearer than 5 to 6 feet from the surface even for brief periods."

With ample drainage provided, one may proceed to the leaching out of the salts. In fine-textured soils the reclamation process will be slow, and doubly so if the soil is underlain by dense clay subsoil. In fact, the presence of a dense clay layer makes difficult the removal of salts from even medium- or coarse-textured soils. It is questionable if reclamation of soils with very deep clay subsoils is feasible from an economic standpoint.

Experiments have shown that leaching is all that is needed to reclaim saline soils that have adequate internal drainage. The addition of chemicals, plowing under of manure, or green-manuring crops are unnecessary. No specific directions can be given regarding the frequency of irrigation or the quantity of water to apply at each irrigation. The main points to observe are (1) that the soil be kept moist so that the soil solution will not become sufficiently concentrated to damage the growing crop, (2) that sufficient water be applied at each irrigation to result in some leaching of salts into the drainage water, and (3) that the soil of each irrigation check be carefully leveled so that the water will enter the soil uniformly.

Reclamation of Sodic (Nonsaline Alkali) Soils

All that has been said concerning the need for drainage and the application of sufficient irrigation water to cause leaching is of as much, if not more, importance in the reclamation of sodic soils as in the treatment of saline soil. Although it has been

Fig. 9.15 Illustration of the removal of exchangeable sodium by the addition of calcium sulfate. When irrigation water and calcium sulfate are applied to a sodic soil having dispersed particles and small pores, the calcium ions from the calcium sulfate replace the exchangeable sodium ions, which form sodium sulfate and leach out of the soil. Soil pH is lowered, colloids flocculate, larger pores develop, and soil permeability is increased. (Adapted from "Chemical Amendments for Improving Sodium Soils," *Agr. Inf. Bull.*, 195, USDA, 1959.)

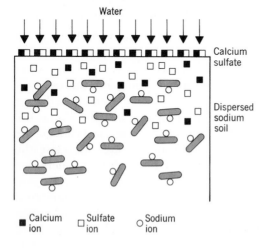
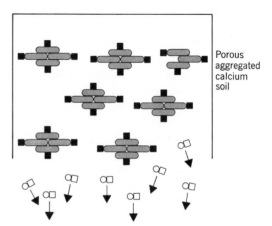

shown that the application of ample irrigation water, coupled with good farming practices, will ultimately result in the removal of exchangeable sodium as well as soluble salt, the reclamation process may be materially hastened through the application of various chemicals. The basis of the treatments is the replacement of exchangeable sodium in the colloidal fraction by calcium and the conversion of the replaced sodium and any occurring as the carbonate into neutral sodium sulfate (see Fig. 9-15).

The desired changes may be brought about by applications of considerable quantities of finely ground calcium sulfate (gypsum). Ground sulfur, however, will accomplish the same results somewhat more slowly. The sulfur must first be oxidized in the soil and then combines with water to make sulfuric acid. Other soluble sulfates such as iron or aluminum have also proved effective. A supply of soluble calcium is needed to complete the reactions. Acid resulting from sulfur additions dissolves $CaCO_3$ that might be present in the soil to supply soluble calcium. Harmful exchangeable sodium can then be replaced by calcium with a resulting improvement in the physical condition of the soil.

References

Allaway, W. H., "pH, Soil Acidity, and Plant Growth," in *Soil*, USDA Yearbook, Washington, D.C., 1957, pp. 67–71.

Coleman, N. T., and A. Mehlich, "The Chemistry of Soil pH," in *Soil*, USDA Yearbook, Washington, D.C., 1957, pp. 72–79.

Farb, Peter, *The Living Earth*, Harper, New York, 1959.

Foy, C. D., and G. B. Burns, "Toxic Factors in Acid Soils," *Plant Food Review, 10,* 1964.

Foth, H. D., et al., *Laboratory Manual For Introductory Soil Science*, Brown, Dubuque, 4th edition, 1976.

Kellogg, Charles E., "Home Gardens and Lawns," in *Soil*, USDA Yearbook, Washington, D.C., 1957, pp. 665–688.

Klamprath, E. J., "Potential Detrimental Effects From Liming Highly Weathered Soils to Neutrality," *Soil and Crop Soc. Florida, 31*:200–203, 1971.

Klein, Asron E., "Acid Rain in the United States," *The Science Teacher, 41*:36–38, May 1974.

Likens, Gene E., and F. Herbert Bormann, "Acid Rain: A Serious Regional Environmental Problem," *Science, 184:* 1176–1197, 1974.

Meyer, T. A., and G. W. Volk, "Effect of Particle Size of Limestone on Soil Reaction," *Soil Science,73*:37–52, 1952.

Mohr, E. C. J., and F. A. van Baren, *Tropical Soils*, Interscience, New York, 1954.

Mullin, R. E., "Soil Acidulation with Sulfur in a Forest Tree Nursery," *Sulfur Jour., 5*:2–3, 1969.

Pearson, R. W., "Soil Acidity and Liming in the Humid Tropics," *Cornell Int. Agr. Bull., 30,* Ithaca, 1975.

Soil Survey Staff, *Soil Survey Manual, USDA Agr. Handbook, 18*, Washington, D.C., 1951.

Spurway, C. H., "Soil Reaction (pH) Preferences of Plants," *Mich. Agr. Exp. Sta. Spec. Bull., 306,* 1941.

Turner, F., and W. W. McCall, "Studies on Crop Response to Molybdenum and Lime in Michigan," *Mich. Agr. Exp. Sta. Quart. Bull., 40*:268–281, 1957.

United States Salinity Laboratory Staff, "Diagnosis and Improvement of Saline and Alkali Soils," *USDA Agr. Handbook, 60,* Washington, D.C., 1969.

10
SOIL GENESIS AND THE SOIL SURVEY

About 10,500 soil series have been described and mapped in the United States. On a particular farm there may be more than half a dozen kinds of soil series. Even though we have many different soils, a few basic processes occur in the development of all soils. These processes proceed at different rates and in different ways to produce the various types. The five groups of factors responsible for the kind, rate, and extent of soil development are: *climate, organisms, parent material, topography,* and *time*. We need to know how these factors influence soil development in order to understand why soils differ, why soils vary in their productivity and, ultimately, how they may be properly used. This understanding is also necessary for the efficient mapping of soils showing their geographic distribution. In studying the influence of variations in each factor on soil properties, we will need to assume that the other factors are constant in the situations cited.

Horizon Differentiation and Soil Genesis

Soil as defined in Chapter 1 included mixtures of organic and mineral matter that were used for growing plants in the greenhouse as well as the soil body on the landscape, which has a profile. The discussion in this chapter is restricted to soils that have experienced some degree of genetic evolution and, consequently, have horizons. The presence of horizons in all genetically developed soils suggests that certain processes are common to the development of all soils and, therefore, each kind of soil is not the product of a distinctly different set of processes. We will next consider the major kinds of horizons found in soils, and we will follow this with a discussion of the processes responsible for horizon differentiation.

Master Horizons or Layers

Broadly speaking, all soil profiles contain two or more master horizons. Short descriptions of these horizons follow.

O—Organic horizons of mineral soils. Horizons: (1) formed or forming in the upper part of mineral soils above the mineral part; (2) dominated by fresh or partly decomposed organic material; and (3) containing more than 30 percent organic matter if the mineral fraction is more than 50 percent clay, or more than 20 percent organic matter if the mineral fraction has no clay. Intermediate clay content requires proportional organic matter content.

A—Mineral horizons consisting of: (1) horizons where organic matter is accumulating or forming at or adjacent to the surface; (2) horizons that have lost clay, iron, or aluminum with resultant concentration of quartz or other resistant minerals of sand or silt size; or (3) horizons dominated by 1 or 2 above but transitional to an underlying B or C.

B—Horizons in which the dominant feature or features is one or more of the following: (1) an illuvial concentration (moved in from some other horizon such as A horizon) of silicate clay, iron, aluminum, or humus, alone or in combination; (2) a residual concentration of sesquioxides or silicate clays, alone or mixed, that has formed by means other than solution and removal of carbonates or more soluble salts; (3) coatings of sesquioxides adequate to give conspicuously darker, stronger, or redder colors than overlying and underlying horizons; or (4) an alteration of material from its original condition that obliterates original rock structure, that forms silicate clays, liberates oxides, or both, and that forms granular, blocky, or prismatic structure if textures are such that volume changes accompany changes in moisture.

C—A mineral horizon or layer, excluding bedrock, that is either like or unlike the material from which the solum is presumed to have formed, only slightly affected by pedogenic processes, and lacking properties diagnostic of A or B.

R—Underlying consolidated bedrock, such as granite, sandstone, or limestone.

Master Horizon Subdivisions

The master horizons usually consist of one or more subdivisions that are indicated by numbers. For example, the O1 refers to the undecomposed litter of twigs and plant leaves. In the O2 only partially decomposed organic matter is found (Fig. 10-1).

That layer in which mineral particles predominate and the dark color or organic residues persists is designated the A1 horizon. The A2 layer, on the contrary, is comparatively light in color and shows the maximum effects of leaching or *eluviation* (washing out) (see Fig. 10-2a).

The horizon of accumulation or *illuviation* (washing in) is technically designated the B horizon. It is often subdivided into the B1, B2, and B3 sections, depending on the degree of accumulation in evidence. These terms and their relationships are presented in Fig. 10-1. Notice that horizons A and B together constitute the *solum*.

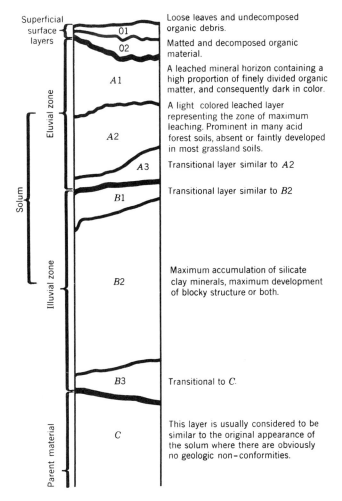

Fig. 10-1 A hypothetical soil profile having all the principal horizons. The thickness of the horizons varies as indicated.

$A_{00} A_0$
A_1
A_2

B

C

A_0

A_1

B

C

A_0
A_1

C

(a)   (b)   (c)

Fig. 10-2 Compare the thin A1 horizon and the strongly developed A2 horizon of the Spodosol profile on the left with the deep A1 horizon and the absence of an A2 horizon in the Mollisol profile in the center. Profile C shows a young soil developing from limestone. Note the absence of a B horizon. (Profiles B and C by courtesy of the late Dr. Harper of the Oklahoma Experiment Station.)

As previously stated, all horizons are not present in every soil. Figure 10-2b shows a profile in which the light-colored A2 horizon is not evident. Limited leaching of the soil because of limited rainfall and grass vegetation are two factors that contribute to the development of this type of profile.

Sometimes neither the A2 nor the B horizon is discernible, and a profile similar to the one in Fig. 10-2c is produced. A high water table that has limited the activity of weathering agencies, limited rainfall, and a comparatively short period of activity of soil-developing processes are some of the conditions that give rise to this type of profile.

Additional features of the horizons are indicated with the use of lowercase letters. For example, cultivated fields may no longer have the original upper layer intact, and possibly the O horizons as well as the

A1 and part of the A2 have been mixed together to form a plow layer. This layer is designated Ap to indicate disturbance by cultivation or pasturing. An accumulation zone in the B horizon may be high in clay or iron oxide. The zone of maximum clay in a soil is commonly labeled a B2t, the t indicating illuvial clay. Spodosols commonly have a Bir horizon, which indicates illuvial iron. Where both iron oxide and humus have accumulated in the same horizon, as in some Spodosols, it is called a Bhir horizon. Illuvial humus is indicated by the "symbol" h. Other characteristics are indicated as follows.

b—Buried soil horizon.

ca—An accumulation of carbonates of alkaline earths, commonly of calcium.

cs—An accumulation of calcium sulfate.

cn—An accumulation of concretions.

f—Frozen soil.

g—Strong gleying.

m—Strong cementation, induration.

sa—An accumulation of salts more soluble than calcium sulfate.

si—Cementation by siliceous material, soluble in alkali.

x—Fragipan character (firmness, brittleness, and high density layer).

The O1 horizon in forest soils is often referred to as the L layer or litter layer and the O2 horizon as the F layer if the struc-

ture of the organic matter is evident, or the H layer if the organic matter is amorphous.

Processes of Horizon Differentiation

Soil genesis or horizon differentiation includes processes that can be viewed as *additions, losses, transformations,* or *translocations.* Plants and animals find a habitat in all soils and become a part of the organic fraction. Carbon in organic matter is lost from the soil as carbon dioxide resulting from microbial decomposition. Nitrogen is transformed from the organic to inorganic

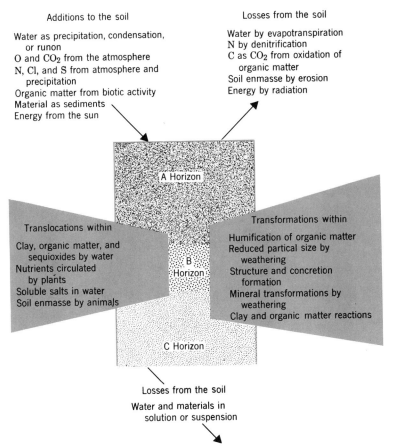

Additions to the soil

Water as precipitation, condensation, or runon
O and CO_2 from the atmosphere
N, Cl, and S from atmosphere and precipitation
Organic matter from biotic activity
Material as sediments
Energy from the sun

Losses from the soil

Water by evapotranspiration
N by denitrification
C as CO_2 from oxidation of organic matter
Soil enmasse by erosion
Energy by radiation

A Horizon

Translocations within

Clay, organic matter, and sequioxides by water
Nutrients circulated by plants
Soluble salts in water
Soil enmasse by animals

B Horizon

Transformations within

Humification of organic matter
Reduced partical size by weathering
Structure and concretion formation
Mineral transformations by weathering
Clay and organic matter reactions

C Horizon

Losses from the soil

Water and materials in solution or suspension

Fig. 10-3 Diagrammatic presentation of additions, losses, translocations, and transformations involved in horizon differentiation.

forms. Furthermore, organic matter is subject to translocation from place to place in the soil by means of water and animal activity.

Mineral constituents undergo changes that can be similarly considered. In all soils minerals weather with the simultaneous formation of secondary minerals and other compounds of varying solubility that may be moved from one horizon to another. A summary of these processes is presented in Fig. 10-3. The great diversity of soils in the world results not from the operation of many distinctly different processes but, instead, from variations in the intensity and length of time the processes have operated.

Soil Development in Relation to Time

Soils are constantly undergoing change. The changes take place slowly, and many people hastily conclude that none occur. In Chapter 1 reference was made to the life cycle of soils. The life cycle includes the stages of parent material, immature soil, mature soil, and old soil. A discussion of these stages and the amount of time required for soil development follows.

Major Stages in Soil Development

The parent material may be transformed into an *immature* or young soil in a relatively short period of time if conditions are favorable. This stage is characterized by organic matter accumulation in the surface soil and by little weathering, leaching, or translocation of colloids. Only the A and C horizons are present and soil properties to a large extent have been inherited from the parent material. The mature stage is attained with the development of the B horizon. Eventu-

ally, if sufficient time has elapsed, the mature soil may become highly differentiated so that large differences exist in the properties of the A and B horizons. This is the *old-age stage*. Many clay-pan soils are characteristic of those in the old-age group, and they have low fertility and productivity. Highest natural productivity is found in the mature and immature soils.

A summary of the stages in the development of soils in the central United States in unconsolidated, medium-texture material and under the influence of prairie vegetation is shown in Fig. 10-4. Under these conditions, development proceeds from parent material, to Entisol (immature), to Mollisol (mature), to Alfisol (old age).

Mohr and van Baren have recognized five stages in the development of tropical soils.

1. Initial stage—the unweathered parent material.
2. Juvenile stage—weathering has started, but much of the original material is still unweathered.
3. Virile stage—easily weatherable minerals have largely decomposed; clay content has increased and a certain mellowness is discernible.
4. Senile stage—decomposition arrives at a final stage, and only the most resistant minerals have survived.
5. Final stage—soil development has been completed and the soil is weathered out under the prevailing conditions.

The names used to refer to the stages are very descriptive, for instance, virile, referring to the stage at which the capacity of the soil to support vegetation is at a maximum.

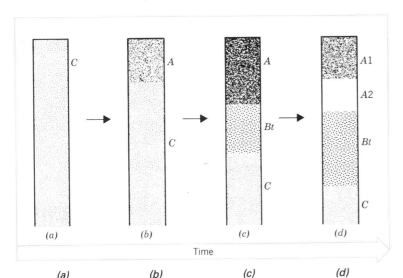

(a) Parent Material	(b) Young Soil Entisol	(c) Mature Soil Mollisol	(d) Old Soil Alfisol
Original material before soil development begins.	Thin solum, organic matter accumulation in A horizon, from which carbonates have been leached. Minimal weathering and eluviation.	Organic matter content is at a maximum. Has moderate clay accumulation in the B horizon and the solum is acid. Stage of maximum productivity for corn.	Very acid in reaction, severely weathered, and has less organic matter than mature stage. Clay accumulation in B horizon has formed a clay pan. An A2 horizon exists.

Fig. 10-4 A summary of the stages in the development of soils in the central United States under tall-grass vegetation.

Amount of Time Required for Soil Development

A question that has many interesting aspects is, "How much time is required to form an inch of soil or for a soil to develop?" For development from hard rock, the time may be very great. On the other hand, development can proceed rapidly in permeable, unconsolidated material in a warm and humid climate. Plant growth can occur on freshly exposed parent material, so soil development need not precede plant growth. This is readily seen in areas where a plant cover is established on freshly exposed road cuts along the highway. The answer to the question, therefore, lies partly in knowledge of the nature of the material from which the soil develops.

From loessial, glacial, volcanic, and other unconsolidated deposits, Entisols can develop in less than 100 years. Mature Spodosols that are about 1000 years old have been found in Alaska. On the Kamenetz fortress in the Ukraine, the modern soil is 4 to 16 inches thick and has developed from limestone slabs since the fortress was abandoned in 1699. On the late

Wisconsin glacial materials, which are about 10,000 years old, most of the soils are in the mature stage and old clay-pan soils are rarely found. The development of argillic horizons versus time for some eastern California soils is shown in Fig. 3-6.

Aridity and the rapid removal of soil by erosion on steep slopes can delay or prevent the development of mature soils. It becomes clear then that the rate of development varies greatly from one soil to another. A given period of time may produce much change in one soil and little in another. For this reason the maturity of the soil is expressed in the degree of horizon development instead of the number of years. Conditions that hasten the rate of soil development are: warm, humid climate; forest vegetation; permeable, unconsolidated material low in lime content; and flat or depressional topography with good drainage. Factors that tend to retard development are cold, dry climate; grass vegetation; impermeable, consolidated material high in lime; and steeply sloping topography.

Rate of Soil Development

As is typical of many processes in nature, the rate of soil development as a whole varies over time, as do many of the individual processes. First, it can be mentioned that the characteristics of a soil change most rapidly when the soil is young and that detectable changes occur more slowly with age. As stated earlier, Spodosols are mature soils and were observed on material deposited about 1000 years ago in Alaska. These soils are very similar to Spodosols in other parts of the world that are many times older in terms of years but are not older in terms of degree of development.

Second, the individual processes vary in intensity over time. Changes in the organic matter content of a soil can be separated into three phases. In young soils, the organic matter content is increasing rapidly because the rate of addition exceeds the rate of decomposition. Maturity is characterized by a constant organic matter content as additions are counterbalanced by losses. Old age is characterized by a declining organic matter content, indicating that the rate of addition is waning as the soil becomes more weathered. The fertility declines and the reduced rate of organic matter production allows decomposition to exceed the rate of addition.

Another illustration of the rate of soil development is silicate clay formation. A youthful soil that has a low clay content and a high content of primary minerals might be characterized by a high rate of clay formation. In a mature or old soil in which most of the primary minerals have already been weathered, silicate clay formation will necessarily be low. The high clay content, however, encourages a relatively high rate of clay decomposition. Thus, it is seen that some of the processes are more operative in youthful soils whereas others are more operative in old soils.

Soil Development in Relation to Climate

Important climatic influences that affect soil development are precipitation and temperature. The climate also influences soil development indirectly in determining the natural vegetation. It is not surprising that there are many parallels in the distribution of climate, vegetation, and soil on the earth's surface.

Climate a Factor in the Organic Matter Content of Soils

The quantity of organic matter in a soil represents the balance between addition and decomposition. Accordingly, climatic factors that affect the quantity of organic material developed in or returned to the soil and the activity of decay organisms have a bearing on the amount accumulated. Studies have shown that when average annual temperatures increase and the moisture and other relations remain constant, the quantity of organic matter decreases in temperate region soils of similar characteristics and covered by the same type of vegetation. The decrease is somewhat greater in grassland soils than in forest soils (Fig. 10-5). Although this relationship exists in the continental United States, it cannot be validly extrapolated to the equatorial regions. Many soils in the humid

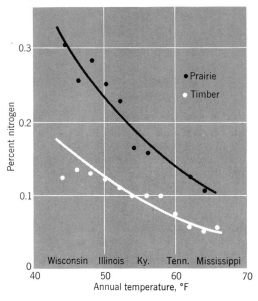

Fig. 10-5 Nitrogen-temperature relation in humid prairie (upper curve) and humid timber soils (lower curve) for silt loams. (From *Missouri Res. Bull.*, 152.)

tropics contain abundant quantities of organic matter.

On the contrary, an increase in moisture supply with temperatures remaining constant results in an increase in organic content in soils of similar characteristics and vegetative cover. Again the change is more pronounced in grassland than in timbered soils (Fig. 10-6).

In view of the long growing season and high rainfall of the southeastern states, a combination that results in a large amount of plant growth, one might expect the organic content of the soils to be correspondingly high. Such is not the case, because the long, warm, moist seasons are also favorable for decomposition, with the result that corresponding soils in the North are higher in organic matter than in the South (see also Fig. 7-2). Humid tropical soils, however, are known for their high organic matter content. As explained in Chapter 7, it appears that the lack of a killing frost in the tropics is more favorable for organic matter production than organic matter decomposition.

Indirectly, precipitation has another influence—if through leaching it causes the pH of the surface layers to become about 4.5 or less. Under these conditions, microbial decomposition of organic matter may become so restricted that the litter added to the surface of the soil accumulates instead of decomposes. This accounts in large part for the existence of prominent O horizons in some forest soils.

Weathering and Clay Formation as Influenced by Climate

Mineral weathering occurs through physical and chemical reactions whose rates are influenced by temperature. All other things being equal, an increase in temperature

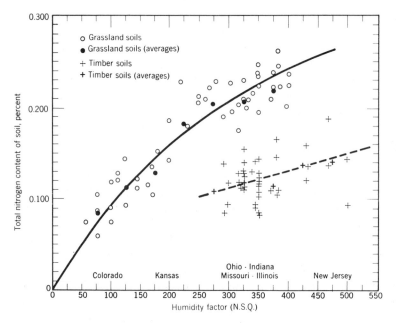

Fig. 10-6 Soil nitrogen-humidity factor relationship along the annual isotherm of 11°C. (From *Missouri Res. Bull.,* 152.) (N.S.Q. is the ratio of precipitation to the absolute saturation deficit of the air. These values are used instead of precipitation values as such, since rainfall alone is not a satisfactory index of soil-moisture conditions because of the great variations in evaporation. The N.S.Q.s include, therefore, the effect not only of temperature but also of air humidity on evaporation.)

causes an increased rate of weathering and clay formation. The rate of weathering is also related to precipitation, as the presence of water enhances the weathering reaction. High average temperature and precipitation tend to encourage rapid weathering and clay formation. The conditions that result in a minimum degree of weathering are found where the climate is warm and dry, cold and dry, or cold and moist.

Chemical Properties

It has been pointed out that an increase in precipitation is commonly associated with an increase in organic matter and clay content. Since the cation-exchange capacity is directly related to the amount of these two fractions, it also increases with precipitation.

Where the annual precipitation is small and evaporation is rapid, insufficient water moves through the soil to leach away the exchangeable bases. This is the general case in the central United States, where the annual precipitation is less than 26 inches and the soils are about 100 percent base saturated. In the more humid areas, the downward movement of water leaches away exchangeable bases. The bases are replaced by hydrogen from the water, and this results in a decrease in the percentage of base saturation. In the humid regions, soils in their natural state tend to be acid in reaction, and the extent of acidity is related to the amount of effective precipitation. The relationships between annual precipitation and cation-exchange capacity, exchangeable bases, and exchangeable hydrogen are shown in Fig. 10-7).

Type of Clay Minerals

Several generalizations concerning the type of clay in soils and climate can be made. In the soils of the northern part of the United

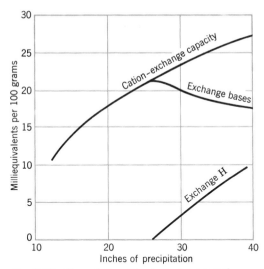

Fig. 10-7 The relationships between annual precipitation and cation-exchange capacity, exchangeable bases and hydrogen in soils of the central United States. (From "Functional Relationships between Soil Properties and Rainfall," Hans Jenny and C. D. Leonard, *Soil Sci.,* 38: 363–381, 1934. Used by permission of the author and the Williams and Wilkins Co., copyright © 1934.)

States, the clay fraction is generally dominated by 2:1 silicate clay minerals (illite, montmorillonite). Kaolinite or 1:1 lattice silicate clays and oxides of iron or aluminum are more common in the soils of the southeastern United States and in many tropical areas. In the humid tropics, intense weathering can result in an almost complete loss of silica, and the clay fraction of the soil will then be high in iron and aluminum oxides. There are many exceptions to these generalizations, since they only apply to the clay formed in the soil.

Soil Development in Relation to Organisms

Soil organisms include both plants and animals. The role of animals as earth movers and their influence on soil development

was discussed in Chapter 6. Here the discussion will emphasize the effect of vegetation on soil development.

Natural vegetation may be divided, very broadly, into the two general classes of trees and grass, and the soils supporting them are termed forest soils and grassland soils, respectively. There are several characteristics in soils developed in association with grass that are of considerable agricultural significance. The different effect of each kind of vegetation on the soil supporting it is brought out in the following discussion.

Amount and Distribution of Organic Matter in Soil

In Chapter 7 the differences in the amount and distribution of organic matter in grassland and forest ecosystems were compared (see Fig. 7-3). Considering the discussion in Chapter 7, it is sufficient now to restate that *under comparable environmental conditions* grassland soil profiles contain more organic matter that is more uniformly distributed with depth than forest soil profiles do (Fig. 10-8).

Differences in Nutrient Cycle

Plants absorb nutrients from the soil and transport the nutrients to the tops of the plants. When the tops die and fall onto the soil surface, decomposition of the organic matter releases the nutrients in a self-fertilizing "do-it-yourself manner." Bases returned to the soil surface in this manner retard the loss of exchangeable bases by leaching and retard the development of soil acidity. Wide differences in the uptake of ions and consequently in the chemical composition of plant tissues have been well substantiated. Even between tree species there are large differences, and this plays a role in

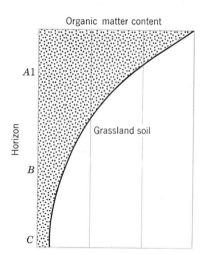
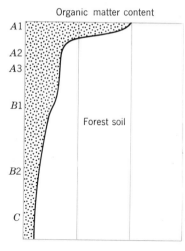

Fig. 10-8 Grassland soil profiles contain about twice as much organic matter that is more uniformly distributed through the profile than forest soils under similar environmental conditions.

soil development. Species that normally absorb large quantities of alkaline earths and alkali metals will delay the development of soil acidity because of the large amount of bases returned to the surface of the soil in vegetative residues. The data in Table 10-1 confirm the fact that hardwoods maintain a higher pH and percentage base saturation than spruce when grown on parent material with the same mineralogical composition.

Rate of Eluviation and Leaching

Under the same climatic conditions, where both forest and grasslands exist side by side and have comparable parent material and slope, the forest soils will show evidence of greater eluviation and leaching. Three possible causes for this difference have been offered. First, the forest vegetation returns fewer alkaline earths and alkali metals to the surface in vegetation each year. Second, water is intercepted for transpiration at a greater depth by trees so the water is more effective in leaching before it is absorbed by roots. Third, water entering the soil is more

Table 10-1
Effect of Tree Species on Soil pH and Base Saturation

Forest Type	Horizons	pH	Percentage Base Saturation
Spruce	O2	3.45	13
	A2	4.60	20
	B1	4.75	27
	B2	4.95	27
	C1	5.05	23
Hardwood	O2	5.56	72
	A1	5.05	47
	B1	5.14	36
	B2	5.24	34
	C1	5.32	34

Adapted from R. Muckenhirn, 1949.

acid. Hydrogen ions dissolved from the organic acids in the O horizon, which is more prominently developed under trees, cause greater replacement and leaching of exchangeable bases.

Closely associated with the leaching of bases is the translocation of clay. That clay eluviates downward is evident from the higher clay content of the B horizon and from the occurrence of more pronounced clay coatings on the peds of the B horizon. Greater movement of clay in the forested soil is based on the higher clay content of the B horizon and the lower clay content of the A horizon of the forested soil as compared to the grassland soil. Thus, the permeability and other physical properties of the subsoil also exhibit a degree of difference.

Two important points stand out in summarizing the differences between forest and grassland soils. The forest soil has about half as much organic matter in the solum as grassland soil does, and it is less uniformly distributed vertically. The forest soil shows evidence of greater age or development. The horizons of the solum are more acid and have a lower pecentage base saturation. Relatively more clay has been translocated from the A to the B horizon. It can be seen that the differences are one of degree and not kind. This supports the view that the same basic processes have been operative in both. Eventually both kinds of soils can evolve into clay-pan soils and these modest but agriculturally important differences become less important. The soils may become strikingly similar in old age regardless of the vegetative cover under which they evolved.

Soil Development in Relation to Parent Material

The nature of the parent material will have a decisive effect on the properties of young soils and may exert an influence on even the oldest soils. Where parent material is derived from consolidated rock, the formation of parent material and the soil may occur simultaneously. Properties of the parent material that exert a profound influence on soil development include texture, mineralogical composition, and degree of stratification.

Consolidated Rock as a Source for Parent Material

Consolidated rock is not strictly parent material, but serves as a source for parent material. Soil formation may begin immediately after the deposition of volcanic ash, but must await the physical disintegration of hard rock where granite is exposed. During the early stages of soil formation, rock disintegration may limit the rate and depth of soil development. Where the rate of rock disintegration exceeds the rate of removal of material by erosion, productive soils with thick solums may develop from bedrock. This is the case in the Blue Grass region of Kentucky where soils developed from limestone (in area 10 of Fig. 10-9).

Water-Deposited Sediments

Alluvial deposits are scattered in narrow, irrgular strips bordering streams and rivers. A common characteristic of this material is its stratification, layers of different-sized particles overlying each other. Mineralogically, alluvium is related to the soils that served as a source of the material.

Most alluvium is carried and deposited during floods because it is at this period that erosion is most active and the carrying capacity of streams is at a maximum. When a flooding stream overflows its banks, its carrying power is suddenly reduced as the flow area increases and velocity decreases. This causes the coarse sands and gravels to settle along the bank, where they sometimes form conspicuous ridges called

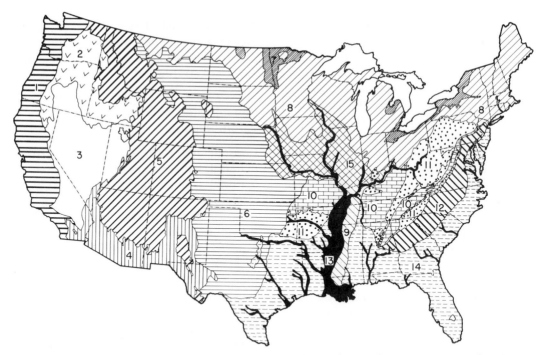

Fig. 10-9 A generalized physiographic map of the United States. (Drawn from a map prepared by the Division of Soil Survey and presented in *USDA Bull.,* 96.) Legend of areas: 1. Pacific Coast region. 2. Northwest intermountain region. 3. Great Basin region. 4. Southwest arid region. 5. Rocky Mountain region. 6. Great Plains region. 7. Glacial lake and river terraces. 8. Glacial region. 9. Loessial deposits. 10. Limestone Valleys and Uplands. 11. Appalachian Mountains and Plateaus. 12. Piedmont Plateaus. 13. River flood plains. 14. Atlantic and Gulf coastal plains. 15. Loessial deposits over glacial material.

natural levees. As the water reaches the *flood plains* of the valley, the rate of flow is slow enough to permit the silt to settle. Finally the water is left in quiet pools, from which it seeps away or evaporates, leaving the fine clay. Levees are characterized by good internal drainage during periods of low water, whereas flood plains exhibit poor internal drainage.

Terraces are developed from flood plains as streams cut deeper channels because of lowered outlets. Several terraces may be found along a stream or a lake that has undergone repeated changes in level. In glaciated areas extensive terraces were formed as the glaciers receded and their

outwash plains were no longer covered with water. Terraces usually are quite well drained and may be droughty. They exhibit stratification.

Streams flowing from hills or mountains into dry valleys or basins drop their sediments in a fanlike deposit as the water spreads out. These *alluvial fans* are usually coarsely textured, being composed of sands and gravels, and are well drained.

Sediments not deposited as flood plains are carried to the lake, gulf, or other body of water into which a stream empties. The decrease in velocity at the stream's mouth together with the coagulating effect of the salt content of the receiving water body results in the deposition of much of the suspended material, thus producing a delta. These deposits are poorly drained but, where drainage is provided, they constitute important crop-producing areas, as is evidenced by the deltas of the Nile, Po, Tigris, Euphrates, and the Mississippi.

Flood plains as well as deltas are in general rich in plant nutrients and comparatively high in organic matter content. Terraces and alluvial fans, on the other hand, are more likely to be less fertile. Special crops such as vegetables and fruits frequently are grown on the latter formations because the soil warms up quickly and their good drainage and coarse texture permit free root development.

It is common to find *marine* deposits along the coastlines. This material was derived from sediments carried by streams and deposited in the ocean and gulf through decreased current velocity and chemical coagulation. Much of it is sandy but is interspersed with beds of silt and clay that were deposited in estuaries or other sheltered bodies of water or farther out in the ocean. When raised above water level, these deposits were subjected to soil-forming processes.

Glacial Materials

Ice was the transporting agent for much of the mantle of northern Europe, Asia, and North America. In the United States the Ohio and Missouri rivers form a general southern boundary for ice-carried material. As the great continental ice sheets moved southward from their accumulation centers, they first followed and filled the great drainage valleys and then gradually spread out over the intervening upland and divides. As the ponderous ice mass moved forward, it pushed before it and gathered within itself a large part of the unconsolidated surface layer. It also scooped up great rock fragments, which scraped at the rock floor over which they passed. Sharp corners and edges of even the hardest rocks were ground smooth by this abrasive action to form the rounded rocks and boulders that are characteristics of glaciated landscapes. Large quantities of weathered and unweathered rocks, varying in size from the fine rock powder to massive boulders, were thus incorporated into the ice and carried along in the glacier.

The movements of this continental ice sheet depended on the changes in climatic conditions that took place during the glacial age. During mild periods the ice melted rapidly. In cold seasons melting ceased and the ice front would creep southward. Sometimes during extremely mild periods the ice would melt faster than it was pushed forward. This would lead to a rapid recession of the ice front, and all debris carried in the ice was, of course, dropped. Generally, after this type of recession, the land surface appeared as a rolling plain, called a *till plain* or *ground moraine*.

At certain times climatic conditions allowed the glacier to melt back just as fast as its rate of advance, and this process resulted in the front of the ice remaining at one place for some time. All debris carried by the ice was brought to the line of the stationary ice front and there dumped as melting proceeded. This process resulted in the formation of ridges or a series of hills, called *terminal* and *recessional moraines*. Lateral moraines formed along the sides of the ice sheets (Fig. 10-10). Moraines are usually composed of an unassorted, heterogeneous mass of boulders, rocks, sand, silt, and clay, briefly called *till* but, in places, a water sorting also occurred. As would be expected, the proportions of

Fig. 10-10 The Bierstadt moraine is a lateral moraine in the Rocky Mountain National Park. Valley glaciers from the mountains to the left deposited the debris along their margins as the ice melted.

these materials vary greatly; hence some moraines are relatively high in sand content whereas others contain a large proportion of fine particles. Not only does the material vary in texture, but the shape of the surface is also variable.

As the ice melted, giving rise to moraines, great volumes of water rushed away. These waters carried quantities of sediment, the coarser of which was deposited as the current diminished. These coarse-textured, comparatively level deposits are known as

outwash plains (Fig. 10-11). Most of the finer silt and clay were carried into slowly moving water or lake basins, where they settled out to form lake bed or *lacustrine* plains. In Michigan there is evidence of as many lakes that have become extinct as there are lakes in existence, whereas along the Great Lakes extensive areas have been exposed through the disappearance of ice barriers, lowering of the outlet at Niagara Falls, and tilting of the land surface. The disappearance of Lake Agassiz has laid bare a great land

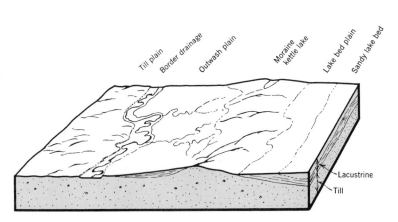

Fig. 10-11 A generalized view of physiographic features in a glaciated area. On the left is a till plain separated from the moraine by an outwash plain and a border-drainage way developed while the ice front was at the moraine line and melting as rapidly as it advanced to deposit the morainic material. To the right a glacial lake has receded to leave a plain of fine-textured sediment partially covered with lacustrine sands (deposit nearer the water's edge).

surface in Minnesota and the Dakotas, and the basin of old Lake Bonneville occupies an immense area in Utah.

The glaciers also produced *kames* and *eskers*, but these formations are only of local significance. Minor readvances of the ice sheet resulted in the modification of these glacial formations in many places. Sometimes till was relaid over a lake plain, or sandy and gravelly outwash was pushed up to form gravelly moraines. Major readvances "streamlined" glacial material in some parts of the country to form *drumlins*. It is customary to designate all the material deposited by glaciers and their melt waters as *glacial drift*. In general, the effect of glaciation was to scrape off and smooth the tops of high land forms while filling valleys and depressions. Thus continental glaciation decreased the local relief intervals and flattened topography.

Wind-Transported or Aeolian Material

There are three classes of wind-moved soil material: (1) sand of variable fineness which may be collected into low swells or steep ridges, like the dunes occurring on the leeward side of large bodies of water and on sandy deserts; (2) volcanic ash; and (3) silt-like material, called *loess*, which occupies large areas in the United States, Europe, and Asia.

Dune sand is of little agricultural value, although crops are produced on it to a limited extent in humid regions. At times dunes are a hazard to agriculture; in their movement they sometimes cover good land.

Loess was deposited in the central United States after the recession of the ice sheet. This material was derived in part from sediments deposited by huge rivers that were fed by the melting continental glaciers

in a broad belt, even beyond the southern limits of glaciation. A period of aridity after the recession of the glaciers with strong westerly winds set the stage for the transportation of this wind-blown material to its present resting place (Fig. 10-12). The great thickness of the deposits of loess on the east and northeast banks of the Mississippi and Missouri rivers is one of the facts that have led to this explanation of the accumulation of the material. Glaciers from the Rockies probably supplied the sediments making up the western part of the loessial deposits. Deserts are also sources of loess. Extensive deposits of loess are also found along the Rhine and its tributaries and over a large part of the immense valley of Hwang Ho. Other deposits occur in southern Russia, several Balkan countries, and northern France, Belgium, Poland, and the Argentine pampa.

Loess is composed largely of silt and is commonly grayish yellow or buff. This wind-blown material stands in almost vertical walls so that gullies and streams cutting through it have very steep banks. Its content of mineral plant nutrients was originally high, as was the quantity of calcium compounds. Loess is one of the most uniform of soil parent materials, but even it varies considerably in texture and mineralogical composition. Soils developed from these deposits are frequently referred to as fertile; however, no single material always gives rise to productive soil, as the parent material is only one of the factors involved in soil formation. Major deposits in the central United States are shown in Fig. 10-13.

Volcanic ash is amorphorus, fine, dustlike particles thrown out of volcanoes. Ash falls on the surrounding land to form thick sediments for soil development (Fig. 10-14). Further from the volcano, ash provides a

Fig. 10-12 A loess deposit about 50 feet (15 meters) thick along the eastern side of the Mississippi River flood plain in western Tennessee.

Figure 10-13 Distribution of loess in the central United States.

Fig. 10-14 Diamond Head—the famous landmark near Honolulu—is an old volcanic cone composed mainly of ash. Note stratification and that the right side is highest because of prevailing wind from the left during the ash fall.

sprinkling of material that helps to rejuvenate soils.

Nature and Source of Organic Soil Parent Material

In locations where considerable quantities of plant material grow and decay is limited because of much water or low temperatures, a large accumulation of partially decayed vegetable matter gradually develops (Fig. 10-15). Such deposits are of wide occurrence and are not restricted to any given climatic zone. They are found in Europe, Asia, Africa, Canada, South America, the United States, and various other places, including the tropics. However, accumulations of this nature are more common in northern latitudes and occupy a larger percentage of the land surface in Norway, Sweden, Ireland, Scotland, northern Germany, Russia, and Holland than in countries lying farther south. In the tundra region organic deposits are of frequent and extensive occurrence.

Even when a rank growth of vegetation occurs each year and remains on the soil,

peat does not accumulate unless decay processes are very slow. The most common factor that limits decay is an excess of water. Accordingly, in moderate to warm climates, peat accumulates in shallow lakes and swamps. The topography, resulting from glaciation, in Minnesota, Wisconsin, and Michigan, and to a lesser extent in Maine, New York, and New Jersey, has given rise to innumerable small lakes and swamps. Thus conditions have been suitable for peat accumulation. Along the Atlantic and Gulf coasts many swamps have developed because of the slight elevation above sea level. Peat has accumulated in many of these areas.

Texture of Parent Material and Soil Properties

The textures of transported materials are related to their origin, and they may have great variability. Glacial and water-laid deposits range from sands to silty clays. Loessial deposits are high in silt, and many soils developed in loess have silt-loam A horizons. When the parent material is

Fig. 10-15 A lake that is rapidly being filled with vegetation and will soon become a peat swamp.

consolidated rock, the texture (size of mineral grains) of the rock becomes an important factor.

Granite and rhyolite are igneous rocks that have the same chemical composition. Rhyolite has the finer texture, or smaller mineral-grain size, because during formation it was subjected to more rapid cooling. This causes the rhyolite to weather more slowly and results in a finer-textured soil than that developed from granite. A similar comparison can be made between basalt and gabbro in young soils, but the textures of the old soils developed from these two materials are similar because all the minerals are weatherable.

Since the minerals in basalt weather more easily than those in granite, the finer-textured soil will develop from the basalt. Large areas of deep, clay-textured soils have developed from basalt in India and Australia. The complete weathering of minerals in basalt in humid tropic regions produces soils composed only of clay-sized mineral particles.

For some of the sedimentary rocks, a few generalizations can be drawn. Sandstones high in quartz weather to produce sandy soils. Soils developed from limestone and shale are usually fine-textured. Some cherty limestones, however, result in the formation of stony soils.

It is logical that the texture of the parent material will have a direct influence on the texture of the soil horizons in immature soils. The texture of resistant minerals will have a direct influence on the texture of even mature or old soils. Three additional ways in which the texture of the parent material influences soil development will be discussed in the following paragraphs. These are organic matter content, soil permeability (or the downward movement of water), and solum thickness.

Soils developed from fine-textured materials usually have a higher organic matter content than those formed from coarser-textured materials. The finer texture may enhance plant growth by providing a greater water and nutrient supply.

This results in a greater annual addition of organic matter to the soil. Fine-textured soils also tend to be less well aerated and have slightly lower average temperatures. This has the effect of retarding the rate of organic matter decomposition and thereby aiding its accumulation. In addition, certain organic compounds may combine with the clay to render soil organic matter resistant to decomposition, as discussed in Chapter 7.

The permeability will determine to a certain extent the quantity of precipitation that will run off and that which will infiltrate into the soil. In humid regions the development of acidity can readily occur in soils developing in calcareous materials, if they are permeable. The more water than moves through the soil, the more rapidly acidity develops, weathering proceeds, and colloidal materials are translocated. Certain soils of the coastal plains of the southeastern United States have developed in clay-textured marine sediments that are many thousands of years old. Where these parent materials are impermeable to water, some of the soils are still alkaline, even though the average annual precipitation exceeds 40 to 50 inches.

If the parent material is very coarse textured or gravelly, little surface is exposed to weathering and little water is retained for weathering and plant growth. In this case, very rapid permeability is associated with slow soil development.

It has been shown that fine-textured parent materials tend to retard leaching and the translocation of colloids. This contributes to the development of soils with thin solums. On sloping lands the fine-textured soils have greater runoff and, consequently, less water available for leaching. There is also more water active in erosion, which contributes to the development of a thin solum. Soils that develop in the coarser or permeable parent materials have the thicker solums. The relationships between texture and solum thickness are shown in Fig. 10-16.

Composition of Parent Material Is Important

The mineral composition of the parent material has much to do with the characteristics of the profile developed, at least until the soil becomes very old. If the material contains a large portion of aluminosilicate minerals, which decompose with relative ease, there will be much clay produced. Under suitable conditions, some of the clay will accumulate in the B horizon, thus making a finer-textured subsoil. On the other hand, if the parent material is composed almost entirely of minerals that weather slowly, there will be very little clay formation or clay accumulation in the B (illuviated) horizon. In nature we find all variations between these extremes.

It has been pointed out that soil acidity encourages mineral decomposition, translocation of colloids, and the overall development of the soil profile. Therefore, where parent materials are rich in lime, development of the soil is delayed, and it remains in the immature stage for a longer period of time.

Stratification of Parent Material Is Important

The discussion thus far has assumed that the parent material was uniform throughout its depth. Soils that develop in water-laid sediments, however, usually develop in stratified parent material. Stratification

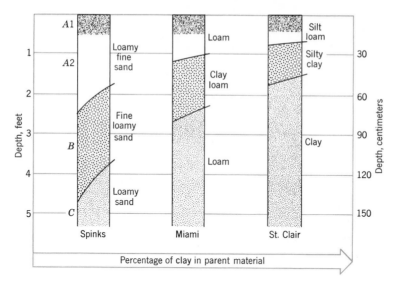

Fig. 10-16 Relationship between the texture of the parent material and the thickness and texture of the horizons of three forest soils of the northcentral United States (Alfisols).

could cause horizons of the same soil profile to develop in layers that have different textures and other characteristics. Where there is evidence of a lithological discontinuity, Roman numerals are used in horizon designation. The horizon sequence A1-A2-B1-IIB2-IIB3-IIIC would indicate that three different materials served as parent material for the horizons of the profile. Such a horizon sequence *could* indicate a situation where the upper three horizons developed from loess, the B2 and B3 developed from till, and the C horizon was disintegrated bedrock.

In the central United States there are large areas where a thin layer of loess was deposited over glacial till. Soils in such areas frequently have A and B horizons that have developed in loess. In cases where the loess is very thin, only the A horizon may have developed in loess, whereas the B and C horizons have developed from glacial till. The horizons formed in loess, compared to those developed in glacial till, frequently have more rapid permeability. This makes the thickness of the loess an important

factor in the design of terraces and the use of lister furrows for water-erosion control.

Where streams dissect an area underlain by strata of varying composition, various parent materials will be exposed along a traverse from the base to the top of a hill. This results in the development of a sequence of soils whose differences are due to differences in parent material. Such a sequence of soils is a *lithosequence* (Fig. 10-17).

Soil Development in Relation to Topography

Topography modifies soil-profile development in three ways: (1) by influencing the quantity of precipitation absorbed and retained in the soil, thus affecting moisture relations; (2) by influencing the rate of removal of the soil by erosion; and (3) by directing movement of materials in suspension or solution from one area to another. Since moisture is essential for the action of the chemical and biological processes of

Soil formed from outwash

Soil formed from sandstone

Outwash

Sandstone

Fig. 10-17 A roadcut showing outwash overlying sandstone. Strata of different materials can produce a lithosequence of soils on a hillside where the materials are exposed, as in the situation on the left side of the photograph.

weathering and effectively acts in conjunction with some of the physical forces, it is evident that a modification of moisture relationships within a soil will materially influence profile development. In a humid region it is noteworthy that, in similar parent material of intermediate or fine texture, the soil of steep ridges or hills differs from the soil on gentle slopes or on a level to undulating topography. In arid climates these soil differences associated with differences in slope are much less pronounced because of the absence of water tables near the surface in the more level areas.

Slope as a Factor in Soil Development

On steep slopes the continuous removal of surface soil by erosion keeps exposing the lower horizons and so modifies the profile. Consequently, the soils on steep slopes have thinner solums, less organic matter, and less conspicuous horizons than soils on level or undulating topography when the water table is well below the solum. These profile differences due to slope are least pronounced in soils developed in coarse-textured parent material in which internal drainage is very rapid.

Topography indirectly plays another part in profile development by influencing the supply of moisture available for plant growth. It also has a bearing on the agricultural value of the land because it is related not only to both external and internal drainage conditions but also to the ease of performing tillage operations.

Effect of Drainage Conditions on Soil Development

As pointed out before, drainage materially influences soil-forming processes. The accumulation of organic matter is usually facilitated because it is preserved by water (Fig. 10-18). Also, because of their low topographic position, poorly drained soils generally receive both organic and mineral matter from the adjacent slopes. In arid regions soluble salts also accumulate in areas that receive drainage water from sur-

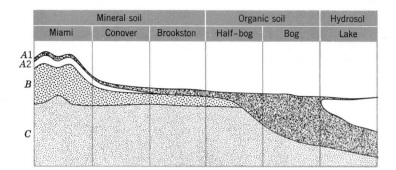

Mineral soil			Organic soil		Hydrosol
Miami	Conover	Brookston	Half-bog	Bog	Lake

Fig. 10-18 Topography, through its influence on drainage, is an effective factor in soil development in humid regions. The moderate humus accumulation, well-defined horizon of eluviation, and strongly developed B horizon of the Miami soils are superseded by an increase in humus, diminution in eluviation, and accumulation in the B horizon as poorer drainage becomes prominent. Finally, the bog and then the hydrosol or body of water itself are encountered. In the diagram the thickness of the A1 and A2 horizons is intentionally exaggerated.

rounding soils. Colors of the soil material at shallow depths are changed from yellows, reds, and browns, denoting good aeration and oxidizing conditions, to the drabs, grays, and mottled yellows that result from chemical reduction where drainage is poor. The horizon of eluviation is modified or may not be evident at all because of the slow or infrequent downward movement of water, and the B horizon is often replaced by a gray or bluish-gray horizon known as a Bg or gleyed layer. In this layer iron is reduced to the ferrous form in the presence of organic matter to produce gray colors.

As a result of differences in topography or drainage or both, the soil profiles developed in similar parent material of a like age and within a single zonal region vary appreciably. A group of soils developed under such conditions and showing such variations in profile charac-

teristics is designated as a *catena* or *toposequence of soils*. The Miami, Conover, and Brookston soils shown in Fig. 10-18 comprise a catena.

Locally, topography is perhaps the factor that most frequently causes soil differences. Parent material and to a lesser extent the vegetation also cause local soil differences (Fig. 10-19). The landscape in Fig. 10-19 is near the prairie-forest transition and trees occur mostly in the protected steep-sided valleys. Erosion has removed loess from the steeper slopes so the soils are developing from the underlying glacial till instead of from loess (Shelby soil). Parent material for the Adair and Clarinda has also been exposed by erosion. This material is a paleosol—an ancient soil that formed before the loess was deposited. The Edina soil is developing in a depression where runoff water from adjacent areas accumulates and is the most leached soil

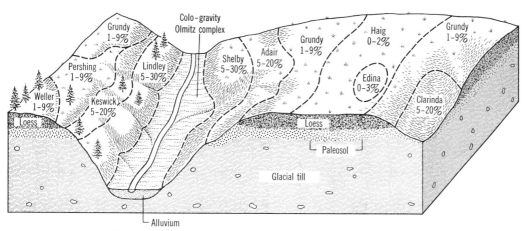

Fig. 10-19 Relationship of slope, vegetation, and parent material to soils of the Adair-Grundy-Haig soil association in southcentral Iowa. (From Oschwald, et al., 1965.)

because the water table is far below the surface.

Human Beings As a Factor of Soil Formation

People's use of the land for agriculture, forestry, grazing, and urbanization has produced extensive changes in soils. Many of these changes, including erosion, earth moving, drainage, salinity, depletion and addition of organic matter and nutrients, compaction, and flooding, have been discussed. Some activities can be judged harmful and others beneficial.

Urban soils are frequently highly modified by human beings, and homeowners sometimes face huge problems in growing lawn, shrubs, and gardens. The incorporation of lime containing mortar into soils from the laying of bricks and other masonary activities leaves many soils with a highly alkaline reaction that causes certain nutrient deficiencies. Sometimes the soil is little more than an accumulation of trash. Mapping the soils or making a soil survey of the area provides information useful in

planning sites for gardens, buildings, and the like (see Fig. 10-20). Soil scientists have mapped the soil in Washington, D.C. As a result of this soil survey, it was discovered

Fig. 10-20 A soil mapper identifies the soil by inspecting all the soil horizons to a depth of 5 feet (about 2 meters). Lines are then drawn around areas of similar soil (polypedons) on an aerial photograph to produce a soil map. (Photo USDA.)

that the cherry trees lining the Tidal Basin around the Jefferson Memorial are difficult to maintain because the water table is too high (poor aeration). Many of the trees have died and have been replaced. A soil survey before the trees were planted would have identified the problem and allowed necessary modification of soil for good cherry growth.

The Soil Survey

At the turn of the century there was an increasing awareness of the bonds between land and society. In an attempt to find the underlying causes of some agricultural problems and in an effort to build a solid foundation for future research, the United States Department of Agriculture in cooperation with the various state experiment stations began at that time a systematic investigation of our soil resources. This investigation assumed the form of a national inventory and survey.

At the present time soil surveying is the process of studying and mapping the earth's surface in terms of units called soil types. A soil survey report thus consists of two parts: (1) the soil map, which is accompanied by (2) a description of the area shown on the map.

Soil Mapping

The actual process of mapping or surveying consist of walking over the land at regular intervals and taking notes of soil differences and all related surface features, such as slope gradients, evidence of erosion, land use, vegetative cover, and cultural features. Boundaries are drawn directly on aerial photographs representing in most places changes from one soil type to another (see Figs. 10-20 and 10-21).

Types of Soil Surveys

Productive agricultural lands are inventoried in detail. A detailed map is one in which the scale is 2 inches or more to the mile. Four inches is the scale generally used at present, although occasionally an 8- or even a 12-inch scale is employed. In addition to soil-type boundaries, these maps show the location of gullies, railroads, houses, roads, streams, and other details

Table 10-2
Land Capability Classes

Land Suited to Cultivation and Other Uses
Class I.   Soils have few limitations that restrict their uses.
Class II.   Soils have some limitations that reduce the choice of plants or require moderate conservation practices.
Class III.   Soils have severe limitations that reduce the choice of plants, require special conservation practices, or both.
Class IV.   Soils have very severe limitations that restrict the choice of plants, require very careful management, or both.
Land Limited in Use—Generally Not Suited for Cultivation
Class V.   Soils have little or no erosion hazard but have other limitations impractical to remove that limit their use largely to pasture, range, woodland, or wildlife food and cover.
Class VI.   Soils have severe limitations that make them generally unsuited to cultivation and limit their use largely to pasture or range, woodland, or wildlife food and cover.
Class VII. Soils have very severe limitations that make them unsuited to cultivation and that restrict their use largely to grazing, woodland, or wildlife.
Class VIII. Soils and landforms have limitations that preclude their use for commercial plant production and restrict their use to recreation, wildlife, or water supply or to esthetic purposes.

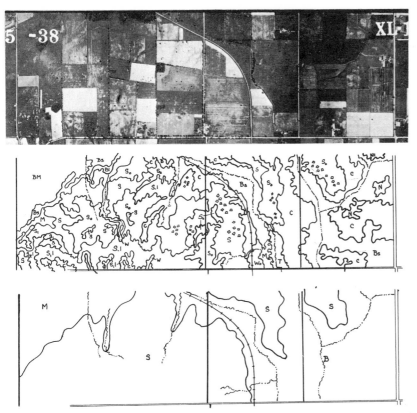

Fig. 10-21 *(Top)* Aerial photograph of good farming land in glaciated area. Such photographs are used extensively for a base map in making soil surveys. *(Center)* A detailed soil survey map of the same area. The symbols indicate soil types. *(Bottom)* A soil-association map of the area in which combinations of soil types occurring in close association are mapped as a unit.

needed in planning a soil-conserving practical farm- or ranch-management plan. This kind of work is expensive. Large expanses of land are too poor, however, to justify this cost. These areas are mapped on small scales, and the resulting study, called a reconnaissance survey, shows areas and regions that are dominated by soil associations in contrast to the detailed map where soil types are separated. Soil-association maps are useful in studying extensive land-use problems, and they possess the desirable characteristics of being made quickly at a comparatively low cost.

Purpose and Value of Soil Maps

Soil survey maps contain many types of information but perhaps that of greatest value is the soil type, slope, and degree of erosion that is recorded for each area delineated on the map. These maps form the basis to develop maps for a wide variety of uses. The areas can be grouped in land capability classes. Definitions of the land capability classes are given in Table 10-2. A land capability map can be constructed such as the one shown in Fig. 10-22. The Soil Conservation Service personnel use these maps to develop conservation plans

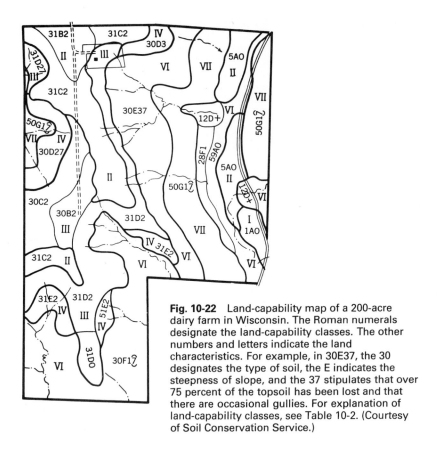

Fig. 10-22 Land-capability map of a 200-acre dairy farm in Wisconsin. The Roman numerals designate the land-capability classes. The other numbers and letters indicate the land characteristics. For example, in 30E37, the 30 designates the type of soil, the E indicates the steepness of slope, and the 37 stipulates that over 75 percent of the topsoil has been lost and that there are occasional gullies. For explanation of land-capability classes, see Table 10-2. (Courtesy of Soil Conservation Service.)

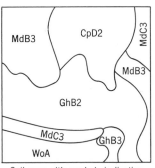

Soil map with symbols indicating soil type, slope, and erosion

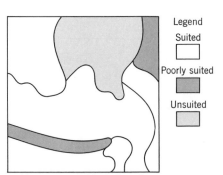

Map constructed from the soil showing suitability of land for grain and seed crops for wildlife management

Legend

Suited

Poorly suited

Unsuited

Fig. 10-23 Soil survey maps form the basis for the construction of many kinds of maps to serve a wide variety of needs: in this case for the wildlife management of a 40-acre tract. (Based on information from Allan et al., 1963.)

for farmers. The use of a soil survey map in wildlife management work is shown in Fig. 10-23.

Life insurance companies, banks, and other money-lending agencies use soil surveys in determining security for loans. Real estate companies and individuals interested in buying or selling land make extensive use of soil surveys. Taxes are based on soil types in some states. They are also used by highway and drainage engineers, by various kinds of manufacturers in selecting suitable locations for factories, and by merchandising and advertising companies in selecting areas for intensive campaigns. County agricultural agents and other extension workers find soil surveys helpful in their work. And, finally, the farmers themselves are making increasing use of soil survey maps and reports in planning their management programs and in interpreting modern agricultural research in terms of their own farm conditions.

References

Allan, P. F., L. E. Garland, and R. F. Dugan, "Rating Northeastern Soils for Their Suitability for Wildlife Habitat," *Trans. 28th Nor. Am. Wildlife and Nat. Res. Con.*, 1963.

Bidwell, O. W., and F. D. Hole, "Man as a Factor of Soil Formation," *Soil Science, 99*:65–72, 1965.

Boul, S. W., F. D. Hole, and R. J. McCracken, *Soil Genesis and Classification*, Iowa State Press, Ames, 1973.

Jenney, Hans, "A Study on the Influence of Climate Upon the Nitrogen and Organic Matter Content of the Soil," *Missouri Agr. Exp. Sta. Res. Bull., 152*, 1930.

Jenny, Hans, and C. D. Leonard, "Functional Relationships Between Soil Properties and Rainfall," *Soil Science, 38*:363–381, 1934.

Jenny, Hans, *Factors of Soil Formation*, McGraw-Hill, New York, 1941.

Muckenhirn, R. J., E. P. Whiteside, E. H. Templin, R. F. Chandler, and L. T. Alexander, "Soil Classification and the Genetic Factors of Soil Formation," *Soil Sci., 67*:93–105, 1949.

Oschwald, W. R., et al., "Principal Soils of Iowa," Ames, Iowa. Special Report No. 42, Department of Agronomy, Iowa State University, 1965.

Simonson, Roy W., "What Soils Are," in *Soil*, USDA Yearbook of Agr., Washington, D.C., 1957, pp. 17–31.

Simonson, Roy W., F. F. Riecken, and Guy D. Smith, *Understanding Iowa Soils*, Brown, Dubuque, Iowa, 1952.

Smith, G. D., W. H. Allaway, and F. F. Riecken, "Prairie Soils of the Upper Mississippi Valley," *Advances in Agronomy, 2*:157–205, 1950.

Smith, Horace, "Soil Survey of District of Columbia," *Soil Con. Service, USDA*, Washington, D.C., 1976.

Smith, R. M., George Samuels, and C. F. Cernuda, "Organic Matter and Nitrogen Build-ups in Some Puerto Rican Soil Profiles," *Soil Science, 72*:409–427, 1951.

Soil Survey Staff, *Soil Survey Manual*, USDA Handbook 18, Washington, D.C., 1951. (Also supplement May 1962.)

Soil Survey Staff, *Soil Taxonomy*, USDA Handbook 436, USDA, Washington, D.C., 1975.

Zack, Anne, "Soil Surveying the Nation's Capital," *Soil Conservation, 41*:12–15, 1975.

11

CLASSIFICATION AND GEOGRAPHY OF THE WORLD'S SOILS

Classification schemes of natural objects seek to organize knowledge so that the properties and relationships of the objects may be most easily remembered and understood for some specific purpose. The ultimate purpose of soil classification is maximum satisfaction of human wants that depend on use of the soil. This requires grouping soils with similar properties so that lands can be efficiently managed for crop production. Furthermore, soils that are suitable or unsuitable for pipelines, roads, recreation, forestry, agriculture, wildlife, building sites, and so forth can be identified. This chapter is designed as an introduction to the classification, nature, distribution, and use of the major soils in the world.

Soil Classification

A genetic classification was suggested about 1880 by the Russian scientist Dokuchaev, and it has been further developed by European and American workers. This system is based on the theory that each soil has a definite morphology (form and structure) that is related to a particular combination of soil-forming factors. This system reached its maximum development in 1949 and was in primary use (especially in the United States) until 1960. In 1960 the United States Department of Agriculture published *Soil Classification, A Comprehensive System.* This classification system places major emphasis on soil morphology and gives less emphasis to genesis or the soil forming factors as compared to previous systems. *Soil Classification, A Comprehensive System* has been improved since 1960 and was republished as *Soil Taxonomy* in 1975. Taxonomy is the classification of objects according to their natural relationships.

Diagnostic Horizons

The ABC nomenclature given in Chapter 10 is still used as the standard for describing and defining soil horizons. Soil Taxonomy has need for a more strict definition of soil horizons, and diagnostic horizons were developed to be used in defining most of the orders. Two kinds of diagnostic horizons, surface and subsurface, are recognized. The surface diagnostic horizons are called epipedons (Greek *epi*, over; and *pedon*, soil). A brief description of the diagnostic horizons is given in Table 11-1.

Soil Orders of Soil Taxonomy

The order is the highest category and there are 10 orders, each ending in sol (L. *solum* meaning soil). The orders, along with their meaning and approximate equivalents in great soil groups (1949), are given in Table 11-2. Entisols are very recent soils (see Table 11-2). Vertisols are soils high in clay that become "inverted" because of alternate swelling and shrinking. Inceptisols are young soils with just the beginning of genetic horizon development. Aridisols are soils

of arid regions. Mollisols are the grassland soils with thick, "soft," dark-colored surface horizons (mollic epipedons). Spodosols have spodic horizons (and are comparable to Podzols). Alfisol is derived from *pedalfer*—a word first used by Marbut to refer to humid region soils leached of lime and with a tendency for aluminum (Al) and iron (Fe) to accumulate in the subsoil. Ultisols are extremely leached soils, very low in bases. Oxisols are the red tropical soils rich in oxides of iron and aluminum and also 1 : 1 clays, that is, they have oxic horizons. The Histosols are bog soils composed mainly of plant tissue. The world distribution of orders is given in Fig. 11-1.

Suborders

The orders are divided into suborders primarily on the basis of chemical and physical properties that reflect either the presence or absence of waterlogging or genetic differences caused by climate and its partially associated variable, vegetation. The distribution of suborders in the United States is given in Fig. 11-2. For example, the Aqualfs are "wet" (aqu for aqua) Alfisols saturated with water sometime during the year. Borolls are Mollisols of the cool regions. The suborder names all have two syllables, with the last syllable indicating the order, such as *alf* for Alfisol and *oll* for Mollisol. Formative elements in names of suborders are given in Table 11-3.

Soil Orders—Properties, Distribution, and Use

Most of the people of the world make their living by tilling the soil. The soil directly influences their lives everyday in that it determines how they build their houses and

Table 11-1
Derivation and Major Features of Diagnostic Horizons

Horizon	Derivation	Major Features
Surface diagnostic horizons-epipedons		
Mollic	L. *mollis*, soft	Thick, dark colored, high base saturation, and strong structure so that the soil is not massive or hard when dry.
Umbric	L. *umbra*, shade	Same as mollic but highly H saturated and may be hard or massive when dry.
Ochric	Gr. *ochros*, pale	Thin, light-colored, and low in organic matter.
Histic	Gk. *histos*, tissue	Very high organic matter content and saturated with water at some time during the year unless artificially drained.
Anthropic	Gr. *anthropos*, man	Molliclike horizon that has a very high phosphate content resulting from long time cultivation and fertilization.
Plaggen	Ger. *plaggen*, sod	Very thick, over 20 in., produced by long continued manuring.
Subsurface diagnostic horizons		
Argillic	L. *argilla*, white clay	Illuvial horizon of silicate clay accumulation.
Natric	*Natrium*, sodium	Illuvial horizon of silicate clay accumulation, over 15% exchangeable sodium and columnar or prismatic structure.
Spodic	Gk. *spodos*, wood ash	Illuvial accumulation of free iron and aluminum oxides and organic matter.
Oxic	L. *oxide*, oxide	Altered subsurface horizon consisting of a mixture of hydrated oxides of iron, aluminum and 1:1 clays.
Cambic	L. *cambiare*, to change	An altered horizon due to movement of soil particles by frost, roots, and animals to such an extent to destroy original rock structure or aggregation into peds or both.
Agric	L. *ager*, field	An illuvial horizon of clay and organic matter accumulation just under the plow layer due to long-continued cultivation.

roads, and how they grow their crops. By affecting the amount and kinds of food they eat, soils affect their health. The order category is sufficiently general and yet sufficiently well defined to make a discussion of these and other aspects of the worlds' soils possible in one chapter. A broad schematic map of the soil orders of the world is presented in Fig. 11-1, and suborders for the United States appear in Fig. 11-2.

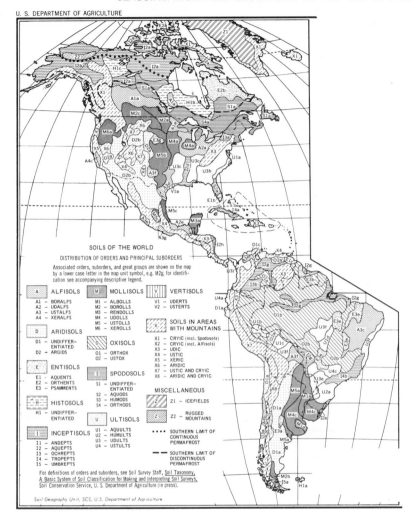

U. S. DEPARTMENT OF AGRICULTURE

SOILS OF THE WORLD

DISTRIBUTION OF ORDERS AND PRINCIPAL SUBORDERS

Associated orders, suborders, and great groups are shown on the map by a lower case letter in the map unit symbol, e.g. M2a; for identification see accompanying descriptive legend.

A ALFISOLS	**M** MOLLISOLS	**V** VERTISOLS
A1 – BORALFS	M1 – ALBOLLS	V1 – UDERTS
A2 – UDALFS	M2 – BOROLLS	V2 – USTERTS
A3 – USTALFS	M3 – RENDOLLS	
A4 – XERALFS	M4 – UDOLLS	
	M5 – USTOLLS	SOILS IN AREAS
D ARIDISOLS	M6 – XEROLLS	WITH MOUNTAINS
		X1 – CRYIC (incl. Spodosols)
D1 – UNDIFFER-	**O** OXISOLS	X2 – CRYIC (incl. Alfisols)
ENTIATED		X3 – UDIC
D2 – ARGIDS	O1 – ORTHOX	X4 – USTIC
	O2 – USTOX	X5 – XERIC
		X6 – ARIDIC
E ENTISOLS		X7 – USTIC AND CRYIC
	S SPODOSOLS	X8 – ARIDIC AND CRYIC
E1 – AQUENTS		
E2 – ORTHENTS	S1 – UNDIFFER-	MISCELLANEOUS
E3 – PSAMMENTS	ENTIATED	
	S2 – AQUODS	Z1 – ICEFIELDS
	S3 – HUMODS	
H HISTOSOLS	S4 – ORTHODS	Z2 – RUGGED
		MOUNTAINS
H1 – UNDIFFER-	**U** ULTISOLS	
ENTIATED		
	U1 – AQUULTS	•••• SOUTHERN LIMIT OF
I INCEPTISOLS	U2 – HUMULTS	CONTINUOUS
	U3 – UDULTS	PERMAFROST
I1 – ANDEPTS	U4 – USTULTS	
I2 – AQUEPTS		– – – SOUTHERN LIMIT OF
I3 – OCHREPTS		DISCONTINUOUS
I4 – TROPEPTS		PERMAFROST
I5 – UMBREPTS		

For definitions of orders and suborders, see Soil Survey Staff, Soil Taxonomy, A Basic System of Soil Classification for Making and Interpreting Soil Surveys, Soil Conservation Service, U. S. Department of Agriculture (in press).

Soil Geography Unit, SCS, U.S. Department of Agriculture

Fig. 11-1 Broad schematic map of the soil orders of the world.

SOILS OF THE WORLD

Distribution of Orders and Principal Suborders and Great Groups

1 : 50,000,000

A **ALFISOLS**—Soils with subsurface horizons of clay accumulation and medium to high base supply; either usually moist or moist for 90 consecutive days during a period when temperature is suitable for plant growth.

    A1 **Boralfs**—cold.
        A1a with Histosols, cryic temperature regimes common

        A1b with Spodosols, cryic temperature regimes

    A2 **Udalfs**—temperate to hot, usually moist.
        A2a with Aqualfs
        A2b with Aquolls
        A2c with Hapludults
        A2d with Ochrepts
        A2e with Troporthents
        A2f with Udorthents

    A3 **Ustalfs**—temperate to hot, dry more than 90 cumulative days during periods when temperature is suitable for plant growth.
        A3a with Tropepts
        A3b with Troporthents
        A3c with Tropustults

SOIL CONSERVATION SERVICE

Fig. 11-1—*cont.*

MAY 1972
USDA-SCS HYATTSVILLE MD 1972

A3d with Usterts
A3e with Ustochrepts
A3f with Ustolls
A3g with Ustorthents
A3h with Ustox
A3j Plinthustalfs with Ustorthents

A4 **Xeralfs**—temperate or warm, moist in
winter and dry more than 45 consecutive
days in summer.
A4a with Xerochrepts
A4b with Xerorthents
A4c with Xerults

D **ARIDISOLS**—Soils with pedogenic horizons,
usually dry in all horizons and never moist as
long as 90 consecutive days during a period
when temperature is suitable for plant growth.

D1 **Aridisols**—undifferentiated.
D1a with Orthents
D1b with Psamments
D1c with Ustalfs

D2 **Argids**—with horizons of clay
accumulation.
D2a with Fluvents
D2b with Torriorthents   *[continued overleaf*

E **ENTISOLS**—Soils without pedogenic horizons; either usually wet, usually moist, or usually dry.
- E1 **Aquents**—seasonally or perennially wet.
  - E1a Haplaquents with Udifluvents
  - E1b Psammaquents with Haplaquents
  - E1c Tropaquents with Hydraquents
- E2 **Orthents**—loamy or clayey textures, many shallow to rock.
  - E2a Cryorthents
  - E2b Cryorthents with Orthods
  - E2c Torriorthents with Aridisols
  - E2d Torriorthents with Ustalfs
  - E2e Xerorthents with Xeralfs
- E3 **Psamments**—sand or loamy sand textures.
  - E3a with Aridisols
  - E3b with Orthox
  - E3c with Torriorthents
  - E3d with Ustalfs
  - E3e with Ustox
  - E3f shifting sands
  - E3g Ustipsamments with Ustolls

H **HISTOSOLS**—Organic soils.
- H1 **Histosols**—undifferentiated.
  - H1a with Aquods
  - H1b with Boralfs
  - H1c with Cryaquepts

I **INCEPTISOLS**—Soils with pedogenic horizons of alteration or concentration but without accumulations of translocated materials other than carbonates or silica; usually moist or moist for 90 consecutive days during a period when temperature is suitable for plant growth.
- I1 **Andepts**—amorphous clay or vitric volcanic ash or pumice.
  - I1a Dystrandepts with Ochrepts
- I2 **Aquepts**—seasonally wet.
  - I2a Cryaquepts with Orthents
  - I2b Haplaquepts with Salorthids
  - I2c Haplaquepts with Humaquepts
  - I2d Haplaquepts with Ochraqualfs
  - I2e Humaquepts with Psamments
  - I2f Tropaquepts with Hydraquents
  - I2g Tropaquepts with Plinthaquults
  - I2h Tropaquepts with Tropaquents
  - I2j Tropaquepts with Tropudults
- I3 **Ochrepts**—thin, light-colored surface horizons and little organic matter.
  - I3a Dystrochrepts with Fragiochrepts
  - I3b Dystrochrepts with Orthox
  - I3c Xerochrepts with Xerolls

- I4 **Tropepts**—continuously warm or hot.
  - I4a with Ustalfs
  - I4b with Tropudults
  - I4c with Ustox
- I5 **Umbrepts**—dark-colored surface horizons with medium to low base supply.
  - I5a with Aqualfs

M **MOLLISOLS**—Soils with nearly black, organic-rich surface horizons and high base supply; either usually moist or usually dry.
- M1 **Albolls**—light gray subsurface horizon over slowly permeable horizon; seasonally wet.
  - M1a with Aquepts
- M2 **Borolls**—cold.
  - M2a with Aquolls
  - M2b with Orthids
  - M2c with Torriorthents
- M3 **Rendolls**—subsurface horizons have much calcium carbonate but no accumulation of clay.
  - M3a with Usterts
- M4 **Udolls**—temperate or warm, usually moist.
  - M4a with Aquolls
  - M4b with Eutrochrepts
  - M4c with Humaquepts
- M5 **Ustolls**—temperate to hot, dry more than 90 cumulative days in year.
  - M5a with Argialbolls
  - M5b with Ustalfs
  - M5c with Usterts
  - M5d with Ustochrepts
- M6 **Xerolls**—cool to warm, moist in winter and dry more then 45 consecutive days in summer.
  - M6a with Xerorthents

O **OXISOLS**—Soils with pedogenic horizons that are mixtures principally of kaolin, hydrated oxides, and quartz, and are low in weatherable minerals.
- O1 **Orthox**—hot, nearly always moist.
  - O1a with Plinthaquults
  - O1b with Tropudults
- O2 **Ustox**—warm or hot, dry for long periods but moist more than 90 consecutive days in the year.
  - O2a with Plinthaquults
  - O2b with Tropustults
  - O2c with Ustalfs

SOIL ORDERS—PROPERTIES, DISTRIBUTION, AND USE

S **SPODOSOLS**—Soils with accumulation of amorphous materials in subsurface horizons; usually moist or wet.
- S1 **Spodosols**—undifferentiated.
  - S1a cryic temperature regimes; with Boralfs
  - S1b cryic temperature regimes; with Histosols
- S2 **Aquods**—seasonally wet.
  - S2a Haplaquods with Quartzipsamments
- S3 **Humods**—with accumulations of organic matter in subsurface horizons.
  - S3a with Hapludalfs
- S4 **Orthods**—with accumulations of organic matter, iron, and aluminum in subsurface horizons.
  - S4a Haplorthods with Boralfs

U **ULTISOLS**—Soils with subsurface horizons of clay accumulation and low base supply; usually moist or moist for 90 consecutive days during a period when temperature is suitable for plant growth.
- U1 **Aquults**—seasonally wet.
  - U1a Ochraquults with Udults
  - U1b Plinthaquults with Orthox
  - U1c Plinthaquults with Plinthaquox
  - U1d Plinthaquults with Tropaquepts
- U2 **Humults**—temperate or warm and moist all of year; high content of organic matter.
  - U2a with Umbrepts
- U3 **Udults**—temperate to hot; never dry more than 90 cumulative days in the year.
  - U3a with Andepts
  - U3b with Dystrochrepts
  - U3c with Udalfs
  - U3d Hapludults with Dystrochrepts
  - U3e Rhodudults with Udalfs
  - U3f Tropudults with Aquults
  - U3g Tropudults with Hydraquents
  - U3h Tropudults with Orthox
  - U3j Tropudults with Tropepts
  - U3k Tropudults withTropudalfs
- U4 **Ustults**—warm or hot; dry more than 90 cumulative days in the year.
  - U4a with Ustochrepts
  - U4b Plinthustults with Ustorthents
  - U4c Rhodustults with Ustalfs
  - U4d Tropustults with Tropaquepts
  - U4e Tropustults with Ustalfs

V **VERTISOLS**—Soils with high content of swelling clays; deep, wide cracks develop during dry periods.
- V1 **Uderts**—usually moist in some part in most years; cracks open less then 90 cumulative days in the year.
  - V1a with Usterts
- V2 **Usterts**—cracks open more than 90 cumulative days in the year.
  - V2a with Tropaquepts
  - V2b with Tropofluvents
  - V2c with Ustalfs

X **Soils in areas with mountains**—Soils with various moisture and temperature regimes; many steep slopes; relief and total elevation vary greatly from place to place. Soils vary greatly within short distances and with changes in altitude; vertical zonation common.
- X1 Cryic great groups of Entisols, Inceptisols, and Spodosols.
- X2 Borolfs and cryic great groups of Entisols and Inceptisols.
- X3 Udic great groups of Alfisols, Entisols, and Ultisols; Inceptisols.
- X4 Ustic great groups of Alfisols, Inceptisols, Mollisols, and Ultisols.
- X5 Xeric great groups of Alfisols, Entisols, Inceptisols, Mollisols, and Ultisols.
- X6 Torric great groups of Entisols; Aridisols.
- X7 Ustic and cryic great groups of Alfisols, Entisols, Inceptisols, and Mollisols; ustic great groups of Ultisols; cryic great groups of Spodosols.
- X8 Aridisols, torric and cryic great groups of Entisols, and cryic great groups of Spodosols and Inceptisols.

Z **MISCELLANEOUS**
- Z1 Icefields.
- Z2 Rugged mountains—mostly devoid of soil (includes glaciers, permanent snow fields, and, in some places, areas of soil).

... Southern limit of continuous permafrost.

-- Southern limit of discontinuous permafrost.

Soil Geography Unit, SCS, U.S. Department of Agriculture   May 1972

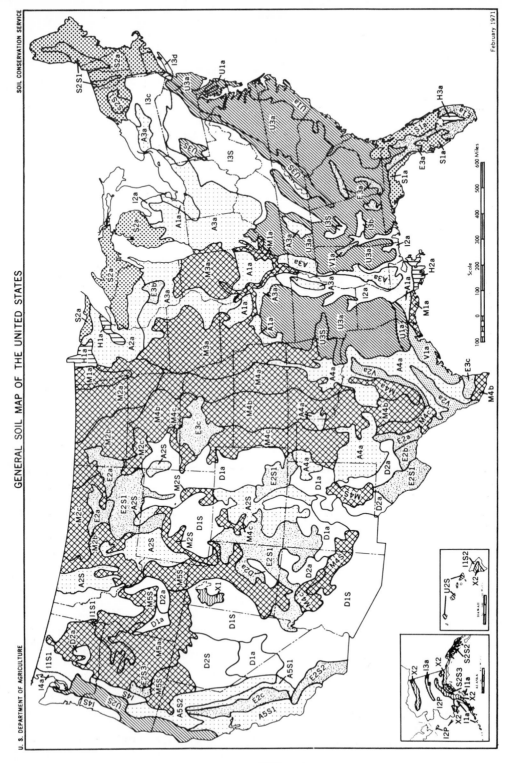

Fig. 11-2 General Soil Map of the United States

ALFISOLS

Aqualfs

A1a–Aqualfs with Udalfs, Haplaquepts, Udolls; gently sloping.

Boralfs

A2a–Boralfs with Udipsamments and Histosols; gently and moderately sloping.

A2S–Cryoboralfs with Borolls, Cryochrepts, Cryorthods, and rock outcrops; steep.

Udalfs

A3a–Udalfs with Aqualfs, Aquolls, Rendolls, Udolls, and Udults; gently or moderately sloping.

Ustalfs

A4a–Ustalfs with Ustochrepts, Ustolls, Usterts, Ustipsamments, and Ustorthents; gently or moderately sloping.

Xeralfs

A5S1–Xeralfs with Xerolls, Xerorthents, and Xererts; moderately sloping to steep.

A5S2–Ultic and lithic subgroups of Haploxeralfs with Andepts, Xerults, Xerolls, and Xerochrepts; steep.

ARIDISOLS

Argids

D1a–Argids with Orthids, Orthents, Psamments, and Ustolls; gently and moderately sloping.

D1S–Argids with Orthids, gently sloping; and Torriorthents, gently sloping to steep.

Orthids

D2a–Orthids with Argids, Orthents, and Xerolls; gently or moderately sloping.

D2S–Orthids, gently sloping to steep, with Argids, gently sloping; lithic subgroups of Torriorthents and Xerorthents, both steep.

ENTISOLS

Aquents

E1a–Aquents with Quartzipsamments, Aquepts, Aquolls, and Aquods; gently sloping.

Orthents

E2a–Torriorthents, steep, with borollic subgroups of Aridisols; Usterts and aridic and vertic subgroups of Borolls; gently or moderately sloping.

E2b–Torriorthents with Torrerts; gently or moderately sloping.

E2c–Xerorthents with Xeralfs, Orthids, and Argids; gently sloping.

E2S1–Torriorthents; steep, and Argids, Torrifluvents, Ustolls, and Borolls; gently sloping.

E2S2–Xerorthents with Xeralfs and Xerolls; steep.

E2S3–Cryorthents with Cryopsamments and Cryandepts; gently sloping to steep.

Psamments

E3a–Quartzipsamments with Aquults and Udults; gently or moderately sloping.

E3b–Udipsamments with Aquolls and Udalfs; gently or moderately sloping.

E3c–Ustipsamments with Ustalfs and Aquolls; gently or moderately sloping.

HISTOSOLS

Histosols

H1a–Hemists with Psammaquents and Udipsamments; gently sloping.

H2a–Hemists and Saprists with Fluvaquents and Haplaquepts; gently sloping.

H3a–Fibrists, Hemists, and Saprists with Psammaquents; gently sloping.

INCEPTISOLS

Andepts

I1a–Cryandepts with Cryaquepts, Histosols, and rock land; gently or moderately sloping.

I1S1–Cryandepts with Cryochrepts, Cryumbrepts, and Cryorthods; steep.

I1S2–Andepts with Tropepts, Ustolls, and Tropofolists; moderately sloping to steep.

Aquepts

I2a–Haplaquepts with Aqualfs, Aquolls, Udalfs, and Fluvaquents; gently sloping.

I2P–Cryaquepts with cryic great groups of Orthents, Histosols, and Ochrepts; gently sloping to steep.

Ochrepts

I3a–Cryochrepts with cryic great groups of Aquepts, Histosols, and Orthods; gently or moderately sloping.

I3b–Eutochrepts with Uderts; gently sloping.

I3c–Fragiochrepts with Fragiaquepts, gently or moderately sloping; and Dystrochrepts, steep.

I3d–Dystrochrepts with Udipsamments and Haplorthods; gently sloping.

I3S–Dystrochrepts, steep, with Udalfs and Udults; gently or moderately sloping.

[continued overleaf

Umbrepts

14a–Haplumbrepts with Aquepts and Orthods; gently or moderately sloping.

I4S–Haplumbrepts and Orthods; steep, with Xerolls and Andepts; gently sloping.

MOLLISOLS

Aquolls

M1a–Aquolls with Udalfs, Fluvents, Udipsamments, Ustipsamments, Aquepts, Eutrochrepts, and Borolls; gently sloping.

Borolls

M2a–Udic subgroups of Borolls with Aquolls and Ustorthents; gently sloping.

M2b–Typic subgroups of Borolls with Ustipsamments, Ustorthents, and Boralfs; gently sloping.

M2c–Aridic subgroups of Borolls with Borollic subgroups of Argids and Orthids, and Torriorthents; gently sloping.

M2S–Borolls with Boralfs, Argids, Torriorthents, and Ustolls; moderately sloping or steep.

Udolls

M3a–Udolls, with Aquolls, Udalfs, Aqualfs, Fluvents, Psamments, Ustorthents, Aquepts, and Albolls; gently or moderately sloping.

Ustolls

M4a–Udic subgroups of Ustolls with Orthents, Ustochrepts, Usterts, Aquents, Fluvents, and Udolls; gently or moderately sloping.

M4b–Typic subgroups of Ustolls with Ustalfs, Ustipsamments, Ustorthents, Ustochrepts, Aquolls, and Usterts; gently or moderately sloping.

M4c–Aridic subgroups of Ustolls with Ustalfs, Orthids, Ustipsamments, Ustorthents, Ustochrepts, Torriorthents, Borolls, Ustolls, and Usters, gently or moderately sloping.

M4S–Ustolls with Argids and Torriorthents; moderately sloping or steep.

Xerolls

M5a–Xerolls with Argids, Orthids, Fluvents, Cryoboralfs, Cryoborolls, and Xerorthents; gently or moderately sloping.

M5S–Xerolls with Cryoboralfs, Xeralfs, Xerorthents, and Xererts; moderately sloping or steep.

SPODOSOLS

Aquods

S1a–Aquods with Psammaquents, Aquolls, Humods, and Aquults; gently sloping.

Orthods

S2a–Orthods with Boralfs, Aquents, Orthents, Psamments, Histosols, Aquepts, Fragiochrepts, and Dystrochrepts; gently or moderately sloping.

S2S1–Orthods with Histosols, Aquents, and Aquepts; moderately sloping or steep.

S2S2–Cryorthods with Histosols; moderately sloping or steep.

S2S3–Cryorthods with Histosols, Andepts and Aquepts; gently sloping to steep.

ULTISOLS

Aquults

U1a–Aquults with Aquents, Histosols, Quartzipsamments, and Udults; gently sloping.

Humults

U2S–Humults with Andepts, Tropepts, Xerolls, Ustolls, Orthox, Torrox, and rock land; gently sloping to steep.

Udults

U3a–Udults with Udalfs, Fluvents, Aquents, Quartzipsamments, Aquepts, Dystrochrepts, and Aquults; gently or moderately sloping.

U3S–Udults with Dystrochrepts; moderately sloping or steep.

VERTISOLS

Uderts

V1a Uderts with Aqualfs, Eutrochrepts, Aquolls, and Ustolls; gently sloping.

Usterts

V2a–Usterts with Aqualfs, Orthids, Udifluvents, Aquolls, Ustolls, and Torrerts; gently sloping.

Areas With Little Soil

X1–Salt flats.

X2–Rock land (plus permanent snow fields and glaciers).

Slope Classes

Gently sloping–Slopes mainly less than 10 percent, including nearly level.

Moderately sloping–Slopes mainly between 10 and 25 percent.

Steep–Slopes mainly steeper than 25 percent.

Table 11-2

New Soil Orders and Approximate Equivalents in Great Soil Groups of 1949 System

Order	Formative Syllable	Derivation	Meaning	Approximate Equivalents
1. Entisol	ent	Coined syllable	Recent soil	Azonal soils and some Low Humic Gley soils
2. Vertisol	ert	L. *verto*, turn	Inverted soil	Grumusols
3. Inceptisol	ept	L. *inceptum*, beginning	Inception, or young soil	Ando, Sol Brun Acide, some Brown Forest, Low Humic Gley, and Humic Gley soils
4. Aridisol	id	L. *aridus*, dry	Arid soil	Desert, Reddish Desert, Sierozem, Solonchak, some Brown and Reddish Brown Soils, and associated Solonetz
5. Mollisol	oll	L. *mollis*, soft	Soft soil	Chestnut, Chernozem, Brunizem (Prairie), Rendzinas, some Brown, Brown Forest, and associated Solonetz and Humic Gley soils
6. Spodosol	od	Gk. *spodos*, wood ash	Ashy (Podzol) soil	Podzols, Brown Podzolic soils, and Ground-Water Podzols
7. Alfisol	alf	Coined syllable	Pedalfer (Al-Fe) soil	Gray-Brown Podzolic, Gray Wooded, Noncalcic Brown, Degraded Chernozem, and associated Planosols and Half-Bog soils
8. Ultisol	ult	L. *ultimus*, last	Ultimate (of leaching)	Red-Yellow Podzolic, Reddish-Brown Lateritic (of United States), and associated Planosols and Half-Bog soils
9. Oxisol	ox	F. *oxide*, oxide	Oxide soils	Laterite soils, Latosols
10. Histosol	ist	G. *histos*, tissue	Tissue (organic) soils	Bog soils

Entisols

Entisols are soils that tend to be of recent origin. They are characterized by youthfulness and are without natural genetic horizons or have only the beginnings of horizons. The central concept of Entisols is soils in deep regolith or earth with no horizons except perhaps a plow layer. Some Entisols, however, have plaggen, agric, or A2 (albic) horizons, and some have hard rock close to the surface.

Alluvial soils develop on alluvium of recent origin and have very weakly developed profiles. In many of them the color change from the A to C horizon is hard to see or is nonexistent. They are, in large part, soils in which most of the prop-

erties have been inherited. They are usually characterized by stratification. The texture is related to the rate at which the water deposited the alluvium. For this reason they tend to be coarse textured near the stream and finer textured near the outer edges of the flood plain. Mineralogically, they are related to the soils that served as a source for the alluvium.

Periodic flooding brings fresh minerals to the soils and they tend to remain fertile (see Fig. 11-3). The soil remains youthful because it is buried before maturity is reached. Alluvial soils played an important role in the development of early agriculture before the development of fertilizers and manuring systems. Soils of the Nile Valley

Table 11-3
Formative Elements in the Names of Suborders

Formative Elements	Derivation of Formative Element	Connotation of Formative Element
alb	L. *albus*, white	Presence of albic horizon (a bleached eluvial horizon)
and	Modified from Ando	Andolike
aqu	L. *aqua*, water	Characteristics associated with wetness
ar	L. *arare*, to plow	Mixed horizons
arg	Modified from argillic horizon; L. *argilla*, white clay	Presence of argillic horizon (a horizon with illuvial clay)
bor	Gk. *boreas*, northern	Cool
ferr	L. *ferrum*, iron	Presence of iron
fibr	L. *fibra*, fiber	Least decomposed stage
fluv	L. *fluvius*, river	Flood plains
hem	Gk. *hemi*, half	Intermediate stage of decomposition
hum	L. *humus*, earth	Presence of organic matter
lept	Gk. *leptos*, thin	Thin horizon
ochr	Gk. base of *ochros*, pale	Presence of ochric epipedon (a light surface)
orth	Gk. *orthos*, true	The common ones
plag	Modified from Ger. *plaggen*, sod	Presence of plaggen epipedon
psamm	Gk. *psammos*, sand	Sand textures
rend	Modified from Rendzina	Rendzinalike
sapr	Gk. *sapros*, rotten	Most decomposed stage
torr	L. *torridus*, hot and dry	Usually dry
trop	Modified from Gk. *tropikos*, of the solstice	Continually warm
ud	L. *udus*, humid	Of humid climates
umbr	L. *umbra*, shade	Presence of umbric epipedon (a dark surface)
ust	L. *ustus*, burnt	Of dry climates, usually hot in summer
xer	Gk. *xeros*, dry	Annual dry season

are a classic example. Most of the people in China live on floodplains. It has been estimated that one third of the world's population obtains its food from them.

Most soils in the world that developed from unconsolidated sediments were Entisols when they were young. Steep slopes where erosion occurs rapidly, an insufficient length of time, or movement of the material, as in the case of sand dunes, are the major causes for their existence. Unstabilized sand dunes develop into

Fig. 11-3 Lettuce being grown on highly productive Entisols (alluvial soils) with irrigation. Water from the river in the background (located where the line of trees are growing) is distributed over the field by gravity flow in the furrows between the rows. (Photo courtesy USDA.)

Entisols after vegetation has become established. Entisols are the dominant soils on stabilized sand dunes in the Nebraska Sandhills (Fig. 11-4).

Entisols developed from sand dunes have limited agricultural cropping value. The moderating influence of the lake on the climate has made it profitable to raise fruit on some of them in Michigan. Small areas of Entisols frequently exist on the steepest parts of cultivated fields and are effectively used with the surrounding zonal soils that comprise the major portions of the fields. They are low in organic matter content and are generally responsive to nitrogen fertilization. Many of them are neutral in reaction or calcareous at the surface.

Some Entisols have an A horizon resting directly on hard rock. An AR type of Entisol is shown in Fig. 1-5. Two very important factors that contribute to their development are the hardness of the rock and the steepness of the slope. The rate of rock disintegration does little more than keep pace with removal of material by erosion. Cracks in the underlying rock may

enable roots to penetrate much deeper than the A horizon. Where the A horizon is 1 to 2 feet thick, the land is profitably used for pasturing.

A/R Entisols are common in mountainous areas and give evidence to the fact that deep soils did not cover the land everywhere before agriculture began. Many deep, productive soils were once Entisols and, in a sense, these soils may be transitory in the development of well-differentiated profiles.

Fig. 11-4 Cattle grazing is the dominant land use on Entisols of the Nebraska sandhills. A grass cover is needed to prevent wind erosion.

Inceptisols

Inceptisol is derived from the Latin *inceptum*, meaning beginning. Development of genetic horizons is just beginning in Inceptisols, but they are still considered to be older than Entisols. Typically, Inceptisols have ochric epipedons and may have other diagnostic horizons, but show little evidence of eluviation or illuviation. Evidence of extreme weathering is lacking. They lack sufficient diagnostic features to be placed in any of the remaining eight soil orders.

Inceptisols occur in all climatic zones where there is some leaching in most years. On the soil-order map of Fig. 11-1, two large areas are shown that include the tundras of North America and Europe-Asia. Soils of the tundra are characterized by high organic matter content. The vegetation consists mostly of low-growing mosses, lichens, and sedges (see Fig. 11-5). These plants grow slowly, but the low soil temperature inhibits organic matter decomposition resulting in soils with a high content of organic matter. They usually have permafrost, are slightly to strongly acid, and have a surface microrelief caused by freezing and thawing. Most of the soils on the tundra show evidence of wetness or poor drainage.

Inceptisols of the tundra support a sparse population of nomadic hunters that live almost entirely on the products of the caribou. In recent years some of these Eskimos have been in the news for having high radioactivity. Testing of nuclear bombs in the Arctic produced radioactive fallout that was absorbed by the lichens. The caribou ate the lichens and the radioactivity was transferred to humans in the meat. Eskimos of the Brooks Range in Alaska were found to have about 100 times more radioactivity than persons in the "lower states."

Fig. 11-5 Dominant soils on the tundra above timberline in mountains are Inceptisols. Vegetation is sparse and the environment is very fragile.

Many Inceptisols are volcanic ash soils and represent a stage in the ultimate development of Ultisols and Oxisols in the humid tropics. They have amorphorus clays and are usually very acid. Many are intensively used for production of sugar cane, coffee, and other crops.

Inceptisols are found widely distributed in the world and some make good agricultural and grazing lands. The data in Table 11-4 show that Inceptisols occupy 15.8 percent of the land surface of the world and rank second in order of abundance. These soils, however, play a minor part in the production of the world's food.

Aridisols—The Dominant Soils of Desert Regions

Aridisols have an aridic soil moisture regime and are the dominant soils of desert regions. They are the most abundant soil order, making up nearly one fifth of the world's soils (Table 11-4). Desert shrubs

Table 11-4
Area of Soils of the World by Soil Order

Soil Order	Area in Thousands of Square Miles	Percentage of World Total	Rank
Alfisols	7,600	14.7	3
Aridisols	9,900	19.2	1
Entisols	6,500	12.5	4
Histosols	400	0.8	10
Inceptisols	8,100	15.8	2
Mollisols	4,600	9.0	6
Oxisols	4,800	9.2	5
Spodosols	2,800	5.4	8
Ultisols	4,400	8.5	7
Vertisols	1,100	2.1	9
Ice fields and rugged mountains	1,200	2.4	—
Islands, unclassified	200	0.4	—
Grand total	51,600	100.0	—

dominate the most arid regions, with shrubs giving way to desert (bunch) grasses with increasing moisture. Plants are widely spaced and use whatever soil moisture there is quite effectively. Perhaps one of the surprising things for persons who have lived all their lives in the humid region and then take a trip to the desert is the great diversity of plants and the considerable amount of vegetation (see Fig. 4-20). Many desert plants grow and function during the wetter seasons of the year and go dormant during the driest seasons. On the more humid or eastern edge of the arid region of the western United States, the desert bunch grasses give away to taller and more vigorous grasses and the Aridisols merge with the Mollisols.

Development and Properties of Aridisols. In arid regions the soil-forming processes are similar to those of humid regions, but the rate of soil formation is much slower in arid regions. The lesser amount of plant growth and the potential for organic matter decomposition produce soils with low organic matter contents. Winds play a major role in the development of Aridisols. Winds move dust about and occasional rains wash soluble nutrients from the dust on its transient journey across the desert. A more obvious role of the wind is the blowing away of fine soil particles, resulting in the formation of a concentration of gravel or the formation of desert pavement.

Water is less effective in leaching soluble salts and translocating colloidal material in arid regions because of the low amount of precipitation. Another factor is the torrential nature of much of the rainfall, which results in considerable runoff. A striking feature of most Aridisols is a zone at varying distances below the surface where calcium carbonate has been deposited by percolating water (calcic horizon). Many

Table 11-5

Some Properties of the Mohave Sandy Clay Loam—an Aridisol

Horizon	Depth, In.	Clay, %	Organic Matter, %	C/N	GEC. me/ 100 g	Exch. Na, %	pH	CaCO₃, %
A1	0–4	11	0.25	6	8	1.2	7.8	—
B1	4–10	14	0.19	6	15	2.0	7.4	—
B2t	10–27	25	0.24	7	22	2.5	8.5	—
B3ca	27–37	21	0.25	8	17	4.1	8.9	10
IICca	37–54	17	0.08	—	6	12.7	9.2	22

Adopted from profile 62 of "Soil Classification, a Comprehensive System," USDA, 1960.

Aridisols have well-developed argillic (Bt) horizons, which is evidence of considerable clay movement. The widespread occurrence of argillic horizons in many Aridisols suggests that many years ago a more humid climate existed than exists today. The Mohave is a common Aridisol from the southwestern United States and some data of a Mohave is given in Table 11-5 to illustrate features or properties commonly found in Aridisols. Note the presence of an argillic horizon (Bt), low content of organic matter and low carbon-nitrogen ratio of the organic matter, the presence of significant exchangeable sodium, high pH values, and accumulation of calcium carbonate (Ca) in the lower part of the profile (calcic horizon) (see color plate 1).

Aridisols are placed in suborders on the basis of the presence or absence of argillic horizons. The suborder Orthids includes Aridisols without argillic horizons and, by contrast, the suborder Argids includes Aridisols with argillic horizons. As we have already noted, the Mohave has an argillic horizon; therefore, the Mohave is an Argid. The general distribution of Orthids and Argids in the United States is shown in Fig. 11-2.

Relationship of Land Surfaces to Age of Aridisols. It is believed that Orthids are the "younger" Aridisols and that the Argids are the "older" Aridisols. There is evidence that Orthids in the United States have developed largely within the past 25,000 years in an arid climate. Orthids are mostly located where recent alluvium has been deposited. Argids are common on the older land surfaces in any landscape where there has been more time for the development of argillic horizons and greater likelihood that the soil has been influenced by a more humid climate more than 25,000 years ago. On the most recent sediments or land surfaces the Entisols are abundant. See Fig. 11-6 for a photograph of a desert landscape showing land surfaces of greatly different ages.

Land Use on Aridisols. The Aridisols of the western United States occur almost entirely within a region called the "western range and irrigated region." As the name implies, grazing of sheep and cattle and the production of crops by irrigation are the two major uses of land. The use of land for grazing is closely related to the precipitation, which largely determines the amount

Fig. 11-6 Grazing lands dominated by Aridisols. Soils in the area are closely related to age of land surface with Aridisols on the older surfaces and Entisols on the youngest surfaces. Note small irrigation reservoir and irrigated cropland near ranch headquarters. (USDA photo by Bluford W. Muir.)

of forage produced. Some areas are too dry for grazing, while other areas are more favorable and may also be able to take advantage of summer grazing on mountain meadows. As much as 75 acres or more in the drier areas is required per head of cattle, thus making large farms or ranches a necessity. Most ranchers supplement the range forage by producing some crops on a small acreage favorably located for irrigation (Fig. 11-6). The major hazard in the use of grazing lands is overgrazing, which results in the invasion of less desirable plant species and increased soil erosion.

Only about 1 or 2 percent of the land is irrigated because the production of crops by irrigation depends on a water supply. Most of the irrigated land is located on alluvial soils or Entisols along streams and rivers where the land is nearly level and the irrigation water can be distributed over the field by gravity (Fig. 11-7). In addition, the nearby rivers serve as a source of water from natural river flow and carry irrigation water released from water storage reservoirs. The alkaline nature of Aridisols may cause deficiencies of various micronutrients on certain crops. Major crops include alfalfa, cotton, citrus fruits, vegetables, and grain crops. In Arizona crop production only 2 percent of the land that is irrigated accounts for 60 percent of the total farm income. Grazing, by contrast, utilizes 80 percent of the land and accounts for only 40 percent of the total farm income.

Mollisols

Bordering many of the desert regions are areas of higher rainfall that support grasses that tend to cover the ground completely and produce an abundance of organic matter that decomposes within the soil. The rainfall, however, is sufficiently limited to prevent excessive leaching and base saturation remains high. Decomposition of abundant organic matter within the soil in the presence of calcium leads to the formation of mollic epipedons. The well-aggregated soil structure gives rise to the "softness" of the soil, which is neither massive nor very hard when dry. All Mollisols have mollic

Fig. 11-7 Aridisol landscapes on the Sonoran Desert of southern Arizona. *(Left)* Natural vegetation and the presence of desert pavement in the foreground. *(Right)* Similar land used for irrigated agriculture.

epipedons. Features of mollic horizons include (1) a thickness of 10 inches or greater, (2) dark color and at least 1 percent organic matter, and (3) over 50 percent base saturation. A soil with a mollic epipedon is shown in Fig. 11-8. In most Mollisols there has been sufficient clay migration to form a Bt or argillic horizon. As a group, Mollisols combine high soil fertility and fair-to-adequate rainfall so that they comprise perhaps the world's most productive agricultural soils.

Mollic epipedon

Fig. 11-8 A Mollisol with a mollic epipedon. The soil is the Aguilita clay loam from southwestern Puerto Rico and has developed from soft limestone.

Geographic Relationships of Mollisols.
Geographically, large areas of Mollisols and Aridisols share a common boundary. Illustrations shown in Fig. 11-1 include the great plains of North America, southern South America (Argentina), southern Soviet Union, and northeastern China. The drier Mollisols of the Great Plains near the Aridisol-Mollisol border have ustic moisture regimes and are called Ustolls. Ustolls typically have lime accumulation (Ca) layers. In eastern Nebraska the Ustolls border with Udolls–Mollisols with udic moisture regimes. Humid Mollisols or Udolls lack lime accumulation layers. In these areas increasing precipitation from Aridisols to the Mollisols results in gradual changes in soil properties, as illustrated in Fig. 11-9. With increasing precipitation there is a gradual increase in solum thickness, organic matter content, development of Bt horizon, and depth to the Ca layer. A soil profile near the Mollisol-Aridisol border in Colorado is shown in Fig. 11-10 and illustrates many of the dominant features of Mollisols (see also color plate 1).

Mollisols in the Palouse country of Washington, Idaho, and Oregon developed in loess (plus some volcanic ash) and are similar to Udolls in Iowa and Illinois. These western soils have a Mediterranean climate so that they are wet in the winter and dry in the summer. The soils are Xerolls. The cool Mollisols of the northern Great Plains are Borolls. Wet Mollisolls are Aquolls and are very important on the recently glaciated plains of the Midwest. See Fig. 11-2 for the distribution of the major areas of Mollisol suborders.

Land Use on Mollisols. The world's major grasslands lacked trees for lumber, readily available water supplies, and natural sites for protection against invaders. As a result, the areas were inhabited mainly by nomadic people until about 100 years ago. In fact, the virgin lands of the

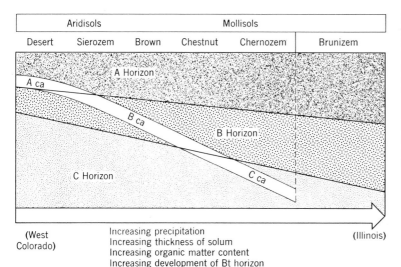

Fig. 11-9 Generalized relationships between Mollisols and Aridisols in many parts of the world showing a gradual change in soil properties with increasing precipitation. Specifically, the diagram illustrates soil relationships on the great plains of the United States.

Fig. 11-10 Ustoll near the Mollisol-Aridisol border in eastern Colorado. Obvious lime accumulation zone can be seen. (Caliche is a more or less cemented deposit of calcium carbonate.)

Soviet Union were opened to settled agriculture in 1954. Today the areas are characterized by having excellent soil for crops and low population density. Land use is importantly related to moisture supply and temperature.

Grazing is a major land use on Ustolls. The proportion of land planted to crops varies from virtually none in some areas to most in others. Dry-land farming is practiced extensively for wheat production (Fig. 11-11). Irrigation is utilized, particularly

SOIL PROFILES

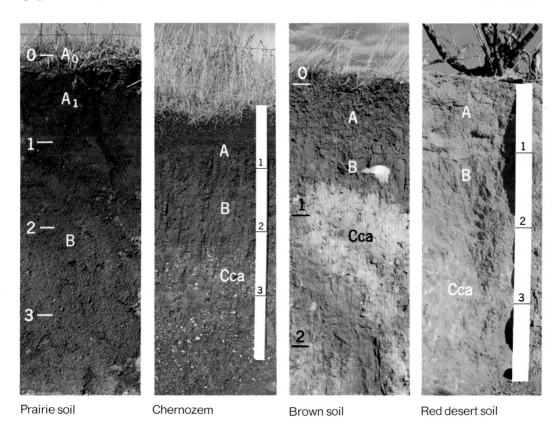

Prairie soil   Chernozem   Brown soil   Red desert soil

Soil Conservation Service,
U.S. Department of Agriculture

Podzol

Gray-brown
podzolic soil

Yellow podzolic soil

Latosol

Fig. 11-11 A common scene in the Ustoll region is wheat produced by fallowing for water conservation.

along the river valleys, and a wide variety of irrigated crops are grown. Fruits, vegetables, and sugar beets occupy smaller irrigated acreages.

As one travels east across the Great Plains, the rainfall increases and there is a decrease in grazing and an increase in wheat production. Near the Udoll border corn becomes a major crop. Udolls combine high natural soil fertility, adequate moisture, and warm summers that are ideal for corn and soybean production. The two most famous Udoll areas are the Corn Belt of the United States and the humid pampa of Argentina.

Most Borolls of the northern Great Plains and Canada have cryic soil temperature regimes. Winters are too cold for winter wheat (and summers are too cool for corn and soybeans). Winter wheat is planted in the fall and spring wheat is planted in the spring, giving winter wheat a longer growing season. Consequently, yields are greater in the winter wheat region. Winter wheat is also the major crop on the Xerolls of the Palouse hills (Fig. 11-12).

A very significant amount of the soil in the Midwest developed under the influence of poor drainage and have aquic moisture

regimes. The soils are Aquolls. Mollic horizons of these soils tend to be very thick, very dark-colored, and contain a high content of organic matter. Leaching has been minimal because of the presence of a high water table much of the time. They require drainage for crop production and, when properly drained, become some of the most productive soils for agriculture. In Illinois the most abundant single soil, the Drummer silty clay loam, is an Aquoll and much of the Corn Belt's reputation of corn production has resulted from the large acreages of Mollisols developed on flat lands under poor drainage (Fig. 11-13). Aquolls are also the dominant soils in the Red River Valley on the Minnesota-North Dakota border.

Wheat is the most important food crop in the world in terms of acreage and production. Corn is perhaps third in importance after rice. Compare the maps in Fig. 11-14, which show the production of wheat and corn in the world. The Soviet Union is the major wheat producer and the United States is the major corn producer. This correlates with the abundance of Udoll-type Mollisols in the United States and their general absence in the Soviet Union. In the United States corn production is centered in the Corn Belt on "moist" Udolls and the wheat production is concentrated on the "drier" Ustolls and cool Borolls. A similar comparison can be made for the pampa of Argentina. Over half of the world's production of corn and about 15 percent of the world's wheat is produced in the United States, which is a partial indication that a considerable amount of the Mollisols in the world are located in the United States. Mollisols occupy only 9 percent of the world's surface (Table 11-4), but produce a much larger percentage of the world's food.

Fig. 11-12 Landscape of Xerolls in the Palouse where winter wheat production is the dominant land use. (Photo courtesy Dr. H. W. Smith).

Spodosols

All Spodosols have a spodic horizon. Spodic horizons are illuvial subsurface horizons where amorphorus materials composed of organic matter, aluminum, and iron have accumulated and are about comparable to Bhir horizons. All Spodosols form in a

Fig. 11-13 Typical Aquoll-Udoll landscape in central Illinois with corn and soybeans as the major crops.

humid climate and mostly from sandy (siliceous) parent material. They are found from the tropics to the boreal regions, but the major areas in the world are just south of the tundras in North America and Europe-Asia. Here, glaciation left large areas of sandy parent materials and the climate is humid. No major area of Spodosols is shown in the southern hemisphere on the map in Fig. 11-1. Trees are the common vegetation (Fig. 11-15), although some of the most intensely developed Spodosols develop under heath vegetation. Ashy gray A2 horizons (albic) are a major feature of most Spodosols, but not a requirement (see Fig. 11-16).

Properties of Spodosols. Spodosols have solums that are very acid throughout, and they have low cation-exchange capacity (except where humus has accumulated) and low base saturation percentages. The base saturation of some horizons is frequently less than 10 percent. Spodosols have limited capacity to store water and are naturally infertile for most crops. Properties of a typical Spodosol are presented in Table 11-6 and color plate 2.

Land Use on Spodosols.
The unsuitability of Spodosols for agriculture in Colonial New England was described by C. L. W. Swanson, the former chief soil scientist of Connecticut.

The virgin soil under a long-established forest is not always good. When the settlers cleared New England forests 300 years ago, the topsoil they found was only 2 to 3 inches thick. Below this was sterile subsoil and when the plow mixed the two together, the blend was low in nearly everything a good soil should have. It was not the lavish virgin soil of popular fancy. Such soil could not support extractive agriculture which takes nutrients out of the soil and does not replace them. Many New England lands that were treated in this way soon went back to forest.

Similar conditions existed in other places, such as the Lake States. Large acreages of Spodosols were settled by farmers after logging removed the timber, but low soil fertility and droughtiness caused many farmers to abandon the land (Fig. 11-17). The first bulletin published by the Michigan Agricultural Experiment Station was devoted to solving the problems of farmers on the "sand plains" of north-central Michigan. Today the northern Spodosol regions are characterized by sparse farming generally, but in localized areas intensive production of fruits and vegetables occurs. The cool summers of the region attract many summer tourists, and many cities owe their existence to mining and lumbering.

Many Spodosols have developed from sandy parent material of marine origin along the southeastern coast of the United States and on the Florida peninsula. Use of modern technology has resulted in successful use of these soils for vegetable and cattle production.

Table 11-6
Some Properties of the Horizons of a Spodosol (Podzol)

Depth, in.	Horizon	Cation-Exchange Capacity, me/100 g	Percentage Base Saturation	pH	Percentage Organic Matter	Percentage Clay
+2–0	O2	—	—	3.6	45.6	—
0–4	A2	7.1	10	3.8	0.8	2
4–9	B2hir	14.3	4	4.4	7.2	4
9–15	B2ir	4.1	22	4.8	2.0	3
15–28	B3	6.5	9	5.2	0.8	2
28–50	C	4.6	4	5.0	0.1	7

Adapted from *Soil Survey Laboratory Memorandum* 1, 1952. Soil is profile No. 39, Worthington loam from Coos County, New Hampshire.

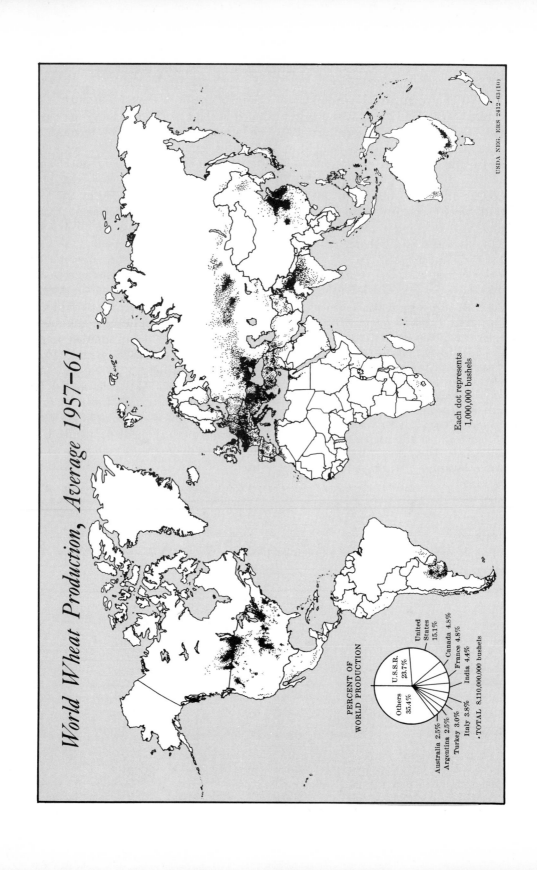

World Wheat Production, Average 1957–61

Each dot represents
1,000,000 bushels

PERCENT OF
WORLD PRODUCTION

Others 35.4%

U.S.S.R. 23.7%

United
States 15.1%

Canada 4.8%

France 4.8%

India 4.4%

Italy 3.8%

Turkey 3.0%

Argentina 2.5%

Australia 2.5%

TOTAL 8,110,000,000 bushels

USDA NEG. ERS 2412–63(10)

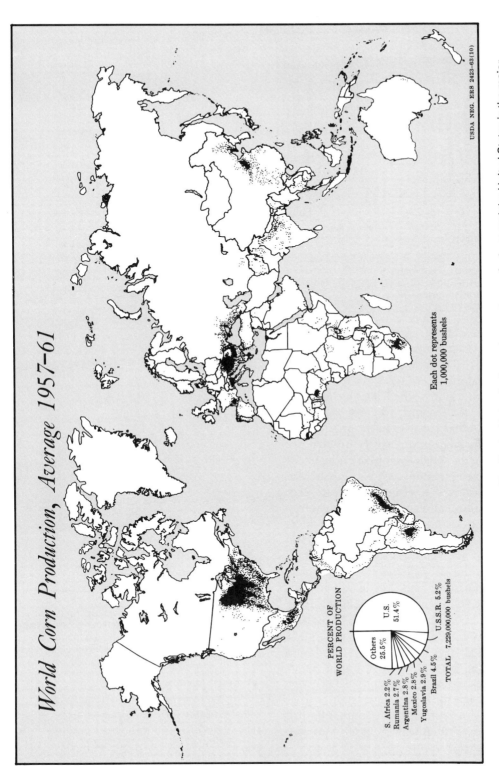

World Corn Production, Average 1957–61

PERCENT OF WORLD PRODUCTION

U.S. 51.4%

Others 25.5%

S. Africa 2.2%
Rumania 2.7%
Argentina 2.8%
Mexico 2.8%
Yugoslavia 2.9%
Brazil 4.5%

U.S.S.R. 5.2%

TOTAL 7,229,000,000 bushels

Each dot represents 1,000,000 bushels

USDA NEG. ERS 2423–63(10)

Fig. 11-14 World production of wheat and corn. The Soviet Union is the major producer of wheat and the United States is the major producer of corn. This reflects the large acreages of drier and cooler Mollisols in the Soviet Union and large acreages of more humid and warmer Mollisols in the United States.

Fig. 11-15 High rainfall as found in forest regions and quartz sand parent material are common where soils are Spodosols. Forestry and recreation are important land uses. (Photo courtesy Wayne Schmidt.)

Alfisols

The moister Mollisols (Udolls) occur in humid regions where trees are the natural vegetation. Many theories have been advanced to account for the extensive grasslands that exist in Iowa and Illinois. Along this wetter Mollisol boundary are large areas of soils developed under trees with ochric epipedons, argillic subsurface horizons (illuvial horizons of silicate clay accumulation), and similar or only slightly lower base saturation than nearby Mollisols. These soils are Alfisols.

Properties of Alfisols. Alfisols have argillic horizons and occur in regions where the soil is moist at least part of the year. The requirement for over 35 percent base saturation in the argillic horizon of Alfisols means that bases are being released in the soil by weathering about as fast as bases are being leached out. Thus, Alfisols rank only slightly lower than Mollisols for agriculture. Some physical properties of an Alfisol (Miami loam) are shown in Fig. 3-7 and color plate 2.

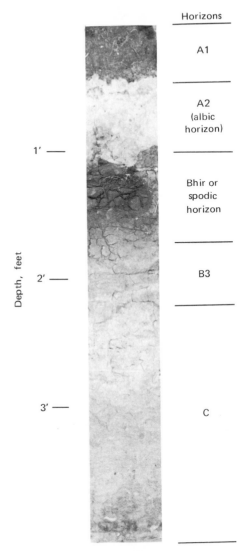

Fig. 11-16 A Spodosol profile showing a nearly "white" A2 horizon and an ortstein—a cemented Bhir or spodic horizon. All mineral horizons have a sandy texture, which is a common characteristic of Spodosols.

Land Use on Alfisols. Three large areas of Alfisols in the world are shown in Fig. 11-1. Most of these soils have udic

Fig. 11-17 These disintegrating buildings tell the story of forest removal, farming, and then abandonment of land in the Lake States with Spodosols of very low inherent fertility. It was not that the land was misused but that high production costs have made it noncompetitive for general farming.

moisture regimes and are Udalfs. These areas are in the northeastern United States, southeastern Canada, and northwestern Europe. Favorable climate and soils with fairly good fertility and physical properties make Alfisols one of the most productive of the soil orders for agriculture (Fig. 11-18). In all areas farm animals have played a very important role for power, food, and the production of manure that was carefully conserved and applied to the land to maintain soil fertility. Today, the availability of fertilizers has greatly reduced the farmers dependence on farm animals for manure. Wheat is a major crop on Alfisols in all three areas, but only in the United States is the climate warm enough for good corn pro-

duction (see Fig. 11-14). Potatoes are a major crop in Europe and rice and wheat are important crops in Asia. Cotton is grown on Udalfs developed from loess in western Tennessee and Louisiana.

When settlers moved across the United States in the eighteenth and nineteenth centuries, the farmers who settled on the Alfisols and Mollisols of the then "Northwest" found good soil and a surplus of agricultural products quickly followed. In 1827 Timothy Flint wrote:

Everyone who was willing to work had an abundance of the articles which the soil produced, far beyond the needs of the country, and it was a prevalent complaint in the Ohio Valley that this abundance greatly exceeded the chances for a profitable sale.

Fig. 11-18 A landscape in the Alfisol region of the eastern United States where much of the land is used for general farming and livestock production.

The farmers became agitators for the development of waterways that would permit shipment of excess food to markets. By the time the population of the United States reached 50 million (about 1880), enough food was being shipped to Europe to feed 25 million people.

A large area of Alfisols with ustic moisture regime exist in the Southwest, particularly in Texas and New Mexico. They occur in natural grasslands but do not have sufficient organic matter or dark color in epipedons to be Mollisols. Their use is similar to that of Ustolls in the same region. Alfisols with cryic temperature regime are Boralfs and are extensive east of the Mollisols on the northern Great Plains in the United States and Canada. Spring wheat is a major crop on Boralfs.

Xeralfs are Alfisols with xeric moisture regimes. They are important in the foothills and mountain ranges surrounding the central valley of California. Some like the San Joaquin on old terraces and alluvial fans, are highly developed (Fig. 3-6). Where slopes are favorable and irrigation water is available, a wide variety of crops are grown, including fruits, nuts, and cotton. A major citrus area of California is located on Xeralfs east of Fresno. A landscape with Xeralfs on the drier lower slopes and Xerults (Ultisols with xeric moisture regime) on the higher more humid slopes is shown in Fig. 11-19.

Fig. 11-19 Grazing is important on Xeralfs on the lower slopes of the Sierra Nevada Mountains in California. On the cooler and more humid upper slopes are Xerults (Ultisols with xeric moisture regime).

Ultisols

All of the soil orders discussed so far have *not* shown evidence of extreme weathering or age. Weathering and soil evolution is limited by precipitation, hardness of rock, high water table, or just the short period of time that has been available since the parent material was formed. Recent glaciation has played an important role in the existence of vast areas of slightly or only moderately weathered soils in the northern part of the northern hemisphere. As one approaches the humid tropics, ancient landscapes with very long periods of weathering coupled with abundant rainfall and high temperature have created two unique soil orders found in the humid tropics. These orders are Ultisol and Oxisol. The Oxisol order will be considered after Ultisol.

Properties of Ultisols. The word Ultisol comes from the Latin, *ultimus*, meaning last or, in the case of Ultisol, soils that are the most weathered and that show the ultimate effects of leaching. Ultisols have argillic horizons with low base saturation, being less than 35 percent. High amounts of exchangeable aluminum are usually present. They occur in the warmer parts of the world where the mean annual soil temperature is 47°F (8°C) or more and have a period each year when rainfall is considerably in excess of evapotranspiration. Few weatherable minerals usually exist in the soil to release bases and trees play a major role in transporting nutrients from the lower part of the soil to the upper part of the solum. Agriculture can be maintained only by shifting cultivation or by the use of fertilizers.

The gross morphology and horizon sequence is similar to the Gray-Brown Podzolics now classified as Alfisols (see Fig. 11-20). In fact, evidence points to the fact

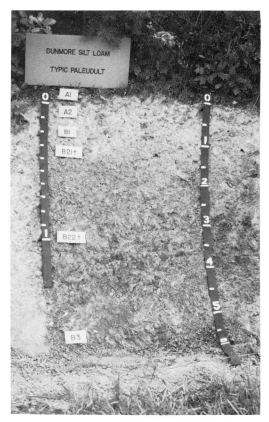

Fig. 11-20 Profile of an Ultisol (Udult with udic moisture regime). Scale in meters and feet.

that Ultisols can be Alfisols before they become sufficiently weathered to be Ultisols. Some properties of Ultisols located in the southeastern United States are presented in Table 11-7. Several things can be noted from the data: (1) the clay content shows the development of an argillic horizon; (2) the organic matter content of all horizons, except the very thin A1, is quite low; (3) cation-exchange capacity is relatively low, expressing the low organic matter content and presence of low cation-exchange capacity clays such as kaolinite; and (4) the amount of exchangeable bases and the percentage base saturation are very low, except for the very thin A1 horizon.

The addition and decomposition of residues from the vegetation play an important role in maintaining the higher base saturation in the upper part of the solum (A1 and A2 horizons). Generally, Ultisols have a very low fertility level for food crops, but respond well to fertilization because of their desirable physical properties (see color plate 2).

Land Use on Ultisols. Interestingly, the Ohio River is an approximate boundary between the Alfisolic and Mollisolic soils of the North and the Ultisolic soils of the South. An entirely different fortune befell the early settlers who moved south and east where Ultisols dominated the landscape. The plight of early Virginians and one farmer in particular, Edmund Ruffin, has been aptly described as follows.

Virginians by the thousands, seeing only a dismal future at home, emigrated to newer states of the South and Middle West. The rate of population growth dropped from 38 percent in 1820 to less than 14 percent in 1830, and then to a mere two percent in 1840. As Ruffin himself later described it, "All wished to sell, none to buy." So poor and exhausted were his lands and those of his neighbors that they averaged only ten bushels of corn per acre and even the better lands a mere six bushels of wheat.

Ruffin was determined to make good on the plantation he inherited from his father in Virginia. By chance he read Sir Humphrey Davy's book, *Elements of Agricultural Chemistry*, published in London in 1813. He was particularly intrigued by Davy's statement to the effect that a soil that contains salts of iron or any other acid matter could be ameliorated by the application of quick lime. Ruffin experimented with marl and found it beneficial. Not only were the Ultisols very low in fertility, the low pH

Table 11-7
Selected Characteristics of an Ultisol

Horizon	Depth, in.	Clay, %	Organic Matter, %	Cation-Exchange Capacity	Exchange-able Bases, me/ 100 g	Base Satura-tion, %
A1	0–1	9	5.2	14	4.1	29
A2	1–5	8	1.8	6	0.9	15
A3	5–7½	8	1.2	4	0.4	9
Blt	7½–12	17	0.4	5	0.4	8
B2, 3t	12–32	35	0.3	10	0.8	8
IIC	32–47	48	0.1	13	0.8	6

Data from profile 92 of "Soil Classification, a Comprehensive System," USDA, 1960. Soil from Jenkins County, Georgia that developed from coastal plain alluvium under mixed forest of pines, oak, sweetgum, and blackgum.

made manure ineffective through pH influence on nutrient availability and aluminum toxicity was a problem in the argillic horizon. Ruffin's discovery was a turning point in the use of very weathered Ultisols and led to the move away from soil exhaustion followed by abandonment to permanent agriculture. Today we can easily add fertilizers and alter soil pH. With the long year-round growing season and rainfall, this region that proved so troublesome at first is rapidly becoming one of the most productive agricultural and forestry regions in the United States. The Ultisols on the nearly level coastal plain have ideal topography for agricultural use. One big problem remaining to be solved is the development of low-cost methods of incorporating lime into subsoils to raise the pH and to reduce the Al toxicity.

One other major area of Ultisols is shown in Fig. 11-1 in southeastern Asia. Many smaller areas of Ultisols occur in regions where Oxisols are common. On many of these Ultisols shifting cultivation is still practiced.

Oxisols

All soils with oxic horizons belong to the Oxisol order. Oxic horizons are subsurface horizons consisting of a mixture of hydrated oxides of iron and/or aluminum and variable amounts of 1 : 1 lattice clays. Few other minerals exist in oxic horizons except some that are highly insoluble. More specifically, the oxic horizon has (1) a thickness of 12 inches or more, (2) a cation-exchange capacity of 16 milliequivalents or less for each 100 grams of clay, (3) none or only a trace of minerals that can weather to release bases, (4) little if any water-dispersible clay, and (5) diffuse boundaries with adjacent horizons. An Oxisol profile is shown in Fig. 11-21 that has very diffuse horizon boundaries (see also Fig. 1-8 and color plate 2).

Oxisols exist only on ancient land surfaces in the humid tropics and contain no reserve of bases beyond those on the exchange sites. Agriculture on Oxisol soils utilizes shifting cultivation similar to that on Ultisol soils. Ultisols and Oxisols commonly exist in the same landscapes and probably

Fig. 11-21 Oxisol profile and landscape south of Brasilia, Brazil. Note little change in soil properties with increasing soil depth—a common feature of Oxisols.

owe at least part of their differences to the tendency for Oxisols to develop from more basic rocks in which the minerals are more weatherable and with less tendency for silicate clays to form. As a result, Oxisols are richer in iron and have fewer weatherable minerals still remaining in the soil. The soil aggregates are very stable and the soils are very erosion resistant. Liming, fertilizers, irrigation, and other management practices have made some Oxisols some of the world's most productive soils, as illustrated in Fig. 11-22.

Fig. 11-22 Pineapple production on Oxisols in Hawaii.

Two very large areas of Oxisols exist: one in South America (including the Amazon Valley) and the other in Africa (including the Congo). Other large acreages exist in eastern India, Burma, and surrounding regions. The total acreage is comparable to that of Mollisols, but few people, by comparison, live on the Oxisols. The great Amazon Basin, the Sahara Desert, and the tundra region all have very low population densities of less than two persons per square mile.

Vertisols

Vertisols are mineral soils that (1) are over 20 inches (50 centimeters) thick, (2) have 30 percent or more clay in all horizons, and (3) have cracks at least 1-centimeter wide to a depth of 20 inches (unless irrigated) at some time in most years. Conditions that give rise to the development of Vertisols are parent materials high in, or that weather to form, large amounts of montmorillonitic (expanding) clay and a climate with a wet and dry season. The typical vegetation in natural areas is grass or herbaceous annuals, although some Vertisols support drought-tolerant woody plants.

Properties of Vertisols. The central concept of Vertisols is one of soils that crack widely in dry seasons (Fig. 11-23). After the cracks develop in the dry season, surface soil material sloughs off into cracks. The soil rewets in the wet season from water that quickly runs into the cracks and is held in the soil by impermeable underlying layers. Repeated drying or rewetting periods cause a "humping up" of areas between the cracks to produce a microrelief called *gilgai*. Repeated cycles of expansion and contraction cause a gradual inverting of the soil; thus, they are called Vertisols. Expansion and contraction in the subsoil with wetting and drying produces shiny ped

Fig. 11-23 Large cracks develop in Vertisols during the dry season. Photograph of Guanica clay in the Lajas Valley of south-western Puerto Rico.

surfaces, *slickensides* (see Fig. 11-24). The expansion and contraction causes a misalignment of fence and telephone posts. Pipelines may be broken and road and building foundations destroyed (see Fig. 11-25). G. W. Olson, soil scientist at Cornell University, found that the Mayans in Guatemala avoided Vertisol areas for the construction of temples.

Some properties of the Houston clay from the blackland prairie region of Texas are presented in Table 11-8 to illustrate some of the important properties of Vertisols relative to their nature and use. There is no evidence of clay migration and the content of clay is very high in all horizons. All horizons have a clay texture. The high content of expanding clay makes the soil very sticky when wet and very hard when dry. Permeability to water is very low when the soil is wet. The organic matter decreases gradually with increasing soil depth. Lime

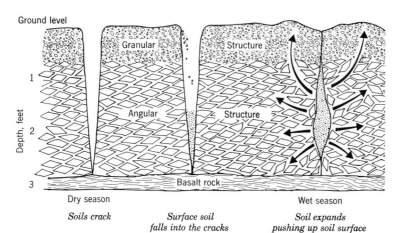

Fig. 11-24 Schematic drawing showing the formation of Vertisols. From left to right: (1) cracks develop in dry season; (2) loose material falls into cracks; and (3) wetting of the soil in the wet season causes expansion and movement of soil in the lower part of the soil to produce angular or wedge-shaped peds with shiny surfaces (slickensides) and a microrelief called gilgai. (Adapted from Boul, 1966.)

Table 11-8
Some Properties of the Houston Clay—a Vertisol

Horizon	Depth, in.	Clay, %	Organic Matter, %	CaCO₃, %	CEC, me/100 g
A11	0–18	58	4.1	17	64
A12	18–40	58	2.1	20	58
AC	40–60	58	1.0	26	53
C	60–78	59	0.4	32	47

Table is based on Kunze and Templin, 1956. Average of 5 profiles.

was present in the parent material, and the high amount of lime still remaining in the upper soil horizons is evidence of the "closed-system" nature of the soil and the limited opportunity for any soluble material to be leached out of the bottom of the profile. The high cation-exchange capacity reflects the high content of montmorillonitic clay.

Land Use on Vertisols. Vertisols are widely distributed in the world between 45° north and south latitude (see Fig. 11-1). The three largest areas of Vertisols in the world are in Australia (70 million acres), India (60 million acres), and the Sudan (40 million acres). Agriculturally the soils have great potential where power tools, fertilizers, and irrigation are available. The natural fertility level can be considered quite high, although the use of nitrogen and phosphorus is beneficial. Tillage of the soil is difficult with primitive tillage tools. The blacklands of Texas and Alabama are some of the best agricultural lands in the United States. Worldwide Vertisols are used mainly for cotton, wheat, corn, sorghum, rice, sugar cane, and pasture.

Fig. 11-25 Landslide on Vertisols produced by wet soil. Note that pipes are on top of ground to prevent breaking caused by expansion and contraction.

Histosols

Organic soils are classified as Histosols. Most Histosols are recognized by a histic epipedon that is over 12 inches thick, saturated with water at least 30 consecutive days a year, and contains at least 20 percent organic matter. Alaska and Minnesota are two states in the United States that have large areas of these soils. Perhaps the most famous area of Histosols in the United States is the Everglades in Florida. Histosols, however, are found throughout the world. Their total extent is less than 1 percent of the land surface of the world (Table 11-4).

Development of Histosols. Histosols develop where the soil is saturated from at least 1 month each year to continuous saturation. The characteristics of Histosols depend primarily on the nature of the vegetation that was deposited in the water and the degree of decomposition. In relatively deep water remains of algae and other aquatic plants give rise to highly colloidal material that shrinks greatly on drying. As the lake gradually fills, rushes, wild rice, water lilies, and similar plants flourish. The partially decayed remains of these plants are less slimy and colloidal. Gradually sedges, reeds, and eventually grasses are able to grow. Peat from such plants is much more fibrous than that produced from plants growing in deeper water. Shrubs and trees follow in time and produce a woody type of peat. Changes in water depth may cause a recurrence of deeper-water plants, and hence layers of more pulpy material may occur over fibrous peat and the like. The following plant succession in the filling of a Minnesota lake has been suggested by Soper.

1. Stonewort: waterweed stage.
2. Pondweed: water lily stage.
3. Rush: wild rice stage.
4. Bog: meadow stage.
5. Bog: heath stage.
6. Tamarack: spruce stage.
7. Pine association.

Land Use on Histosols. When organic soils are drained in such a way as to remove excess water rapidly yet maintain the water level at a relatively shallow depth, they may be used for very intensive types of crop production. In the northern states these soils are used for the production of onions, celery, mint, potatoes, cabbage, cranberries, carrots, and other root crops. Corn is produced to some extent, and considerable areas are used as pasture. Late spring or summer and early fall frosts are the greatest hazard to crop growth in the temperate region. Other hazards include fires and wind erosion (Fig. 11-26). A great variety of special crops is grown on the organic soils of the South and East. Special methods of tillage, coupled with careful application of

Fig. 11-26 Histosol or organic soil landscape. Crop on the left is grass sod to be used for landscaping. The lightness of the soil makes it ideal for sod production, but this also contributes to its susceptibility to wind erosion. Note windbreak of trees at far end of field.

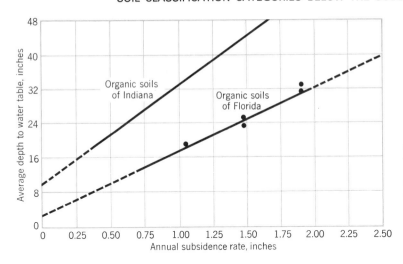

Fig. 11-27 This shows that the subsidence rate for organic soils depends on the depth to the water table. The lower the water table, the greater the soil loss. Subsidence is greatest in Florida where temperature is highest.

fertilizer, are required to bring these soils to their highest state of productivity. Drainage is required, and this results in subsidence because of oxidation and dehydration of organic matter. Comparative rates of subsidence in Florida and Indiana are shown in Fig. 11-27. Subsidence not only results in an eventual loss of the Histosol for crop production, but also creates engineering problems (see Fig. 11-28).

Soil Classification Categories Below the Suborders

The discussion of the soil orders permitted an introduction to soil geography on a worldwide basis. Within each order there is much diversity and a need for much more precise definition and classification of use to permit wise utilization. These additional categories will be considered next.

Fig. 11-28 Organic soil being removed and replaced with sand to create a stable roadbed.

Great Groups and Subgroups

Suborders are divided into great groups. Great group names are coined by prefixing one or more additional formative elements to the appropriate suborder name. The prefixes are used to indicate the presence or absence of certain diagnostic horizons. The formative elements, with their meanings and connotations, are shown in Table 11-9. As an example, a Fragiaqualf is a "wet" Alfisol with a fragipan. Subgroup names indicate to what extent the central concept of the great group is expressed. A typic Fragiaqualf is a soil that is typical for the Fragiaqualf great group.

Family and Series

The families indicate features that are important to plant growth such as texture, mineralogical composition, or temperature. The series gets down to the individual soil and the name is that of a natural feature or place near where the soil was first recognized. Familiar series names include Amarillo, Carlsbad, and Fresno and obviously refer to soils located in Texas, New Mexico, and California, respectively.

Table 11-9
Formative Elements for Names of Great Groups

Formative Element	Connotation	Formative Element	Connotation
acr	Extreme weathering	moll	Mollic epidedon
agr	Agric horizon	nadur	See *Natr* and *Dur*
alb	Albic horizon	natr	Natric horizon
and	Ando-like	ochr	Ochric epipedon
anthr	Anthropic epipedon	pale	Old development
aqu	Wetness	pell	Low chroma
arg	Argillic horizon	plac	Thin pan
calc	Calcic horizon	plag	Plaggen horizon
camb	Cambic horizon	plinth	Plinthite
chrom	High chroma	quartz	High quartz
cry	Cold	rend	Rendzina-like
dur	Duripan	rhod	Dark-red colors
dystr, dys	Low base saturation	sal	Salic horizon
eutr, eu	High base saturation	sider	Free iron oxides
ferr	Iron	sphangno	Sphagnum-moss
frag	Fragipan	torr	Usually dry
fragloss	See *frag* and *gloss*	trop	Continually warm
gibbs	Gibbsite	ud	Humid climates
gloss	Tongued	umbe	Umbric epipedon
hal	Salty	ust	Dry climate, usually hot in summer
hapl	Minimum horizon		
hum	Humus	verm	Wormy, or mixed by animals
hydr	Water	vitr	Glass
hyp	Hypnum moss	xer	Annual dry season
luo, lu	illuvial	sombr	A dark horizon

References

Anonymous, "Fallout in the Food Chain," *Time*, September 13, 1963, p. 63.

Boul. S. W., "Soils of Arizona," *Ariz. Agr. Exp. Sta. Tech. Bull.*, *117*, 1966.

Dudal, R., and D. L. Bramao, "Dark Clay Soils of Tropical and Subtropical Regions," *FAO Agr. Dev. Paper*, *83*, 1965.

Kellogg, Charles E., "Soils," *Sci. Am.*, *183*:30–39, July 1950.

Kunze, G. W., and E. H. Templin, "Houston Black Clay, the Type Grumusol: II. Mineralogical and Chemical Characterization," *Soil Sci. Soc. Am. Proc.*, *20*:91–96, 1956.

Lotspeich, Frederick B., and Henry W. Smith, "Soils of the Palouse Loess: I. The Palouse Catena," *Soil Science*, *76*:467–480, 1953.

Oakes, Harvery, and James Throp, "Dark-Clay Soils of Warm Regions Variously Called Rendzina, Black Cotton Soils, Regur and Tirs," *Soil Sci. Soc. Am. Proc.*, *15*:347–354, 1951.

Rourke, John D., "Soils of the World-Probable Occurrence of Orders and Suborders," USDA, Washington, D.C., May 1968.

Ruffin, Edmund, *An Essay on Calcareous Manures*, Belknap Press, Cambridge, Mass., 1961. Original book published in 1832.

Smith, H. N., *Virgin Land*, Vintage, New York, 1957.

Soil Survey Staff, *"Soil Survey Laboratory Memorandum 1,"* USDA, Washington, D.C., 1952.

Soil Survey Staff, *Soil Classification, A Comprehensive System*, USDA, Washington, D.C., 1960.

Soil Survey Staff, *Soil Taxonomy*, USDA Handbook 436, Washington, D.C., 1975.

Soper, E. K., "The Peat Deposits of Minnesota," *Minn. Geol. Sur. Bull.*, *16*, 1919.

Stephens, John C., "Drainage of Peat and Muck Lands," in *Water*, USDA Yearbook, Washington, D.C., 1955, pp. 539–557.

Swanson, C. L. W., "The Road to Fertility," *Time*, January 18, 1954.

Tuan, Yi-Fu, *China*, Aldine, Chicago, 1969.

12
SOILS AND MINERAL NUTRITION OF PLANTS

The growth and development of plants are determined by numerous factors of soil and climate and by factors inherent in the plants themselves. Some of these factors are under the control of human beings, but many are not. People have little control over air, light, and temperature, for example, but can influence the supply of plant nutrients in the soil. They may increase the supply of available nutrients by modifying soil conditions or by making additions in the form of fertilizers. Anyone dealing directly with the growth of plants is particularly concerned with their nutrient requirements. The emphasis in this chapter will be the amounts and forms of nutrients in soils and their uptake by plants.

The Essential Elements

If a soil is to produce crops successfully, it must have, among other things, an adequate supply of all the necessary nutrients that plants take from the soil. Not only must required nutrient elements be present in forms that plants can use, but there should also be a rough balance between them in accordance with the amounts needed by plants. If any of these elements is lacking or if it is present in improper proportions, normal plant growth will not occur. Elements required by plants are *essential elements*.

Characteristics of an Essential Element

For many centuries people knew that substances such as manure, ashes, and blood had a stimulating effect on plant growth. The effect was found to result basically from the essential elements contained in the materials. As recently as 1800, however, it was not known which elements removed from the soil were indispensable. Discovery of the chemical elements and techniques for their determination were prerequisites for determining which nutrients were essential for plant growth. Two of the criteria commonly used in establishing the essentiality of a plant nutrient are (1) its necessity for the plant to complete its life cycle, and (2) its direct involvement in the nutrition of the plant apart from possible effects in correcting some unfavorable condition in the soil or culture medium.

Concentrations of Essential Elements in Plants and Functions

The elements generally required by plants are divided into two groups based on the *amount* that plants require. *Macronutrients* are *necessary* in relatively large amounts, usually over 500 parts per million in the plant. *Micronutrients* are necessary only in extremely small amounts, usually less than 50 parts per million in the plant. A list of the macro- and micronutrients and their major roles in plant growth are given in Table 12-1.

Quantities of Nutrients in Crops

The nutrient content of several crops is given in Table 12-2. In considering these quantities of nutrients, it is important to keep in mind that they are taken from the more readily available supply in the soil. Furthermore, the quantities removed by a single crop may seem rather small in some instances; but, when the quantities contained in all the crops of a rotation are totaled or when the amounts removed by crops for several years are considered, the necessity of supplying plant nutrients in the form of fertilizer and manure to maintain soil fertility is apparent.

The quantities of elements given in Table 12-2 do not represent the total quantities that crops require during growth but, instead, the quantities contained in the harvested material. Roots and other portions of the plant that may not be harvested require considerable quantities of nutrients. Fertilizers are used to make up the difference between the amount of nutrients in crops and the amount of nutrients supplied by the soil.

Nutrient Deficiency Symptoms

When plants are starving for any particular nutrient, characteristic symptoms usually appear on these plants. If crops are not vigorous and healthy, it is important to know and understand the cause. If the unhealthy appearance is due to disease, it may be possible to save the crop by spraying or, if it is a nutrient deficiency, fertilizers may be applied as a top-dressing on the soil or on the foliage as a spray in time to save the crop. These deficiency symptoms appear only when the supply of a particular element is so low that the plant can no longer function normally, and then it will

NUTRIENT DEFICIENCY SYMPTOMS

Table 12-1
Essential Mineral Elements and Role in Plants

Element	Role in Plants
Macronutrients	
Nitrogen (N)	Constituent of all proteins, chlorophyll, and in coenzymes and nucleic acids.
Phosphorus (P)	Important in energy transfer as part of adenosine triphosphate. Constituent of many proteins, coenzymes, nucleic acids, and metabolic substrates.
Potassium (K)	Little if any role as constituent of plant compounds. Functions in regulatory mechanisms as photosynthesis, carbohydrate translocation, protein synthesis, etc.
Calcium (Ca)	Cell wall component and plays role in the structure and permeability of membranes.
Magnesium (Mg)	Constituent of chlorophyll and enzyme activator.
Sulfur (S)	Important constituent of plant proteins.
Micronutrients	
Boron (B)	Somewhat uncertain, but believed important in sugar translocation and carbohydrate metabolism.
Iron (Fe)	Chlorophyll synthesis and in enzymes for electron transfer.
Manganese (Mn)	Controls several oxidation-reduction systems, formation of O_2 in photosynthesis.
Copper (Cu)	Catalyst for respiration, enzyme constituent.
Zinc (Zn)	In enzyme systems that regulate various metabolic activities.
Molydbenum (Mo)	In nitrogenase needed for nitrogen fixation.
Cobalt (Co)	Essential for symbiotic nitrogen fixation.
Chlorine (Cl)	Activates system for production of O_2 in photosynthesis.

Compiled from many sources.

usually be profitable to apply fertilizer long before the symptoms indicating acute starvation actually appear.

Nutrient Mobility in Plants and Deficiency Symptoms

Translocation of nutrients within the plant is an ever continuing process. In this regard there is considerable difference in the mobility of the various nutrients. When a shortage of a mobile nutrient occurs, it is removed from the older, first-formed tissues and translocated to the growing points. This causes the symptoms to appear on the lower leaves. Nitrogen is very mobile in plants, and deficient plants have yellow-colored lower leaves and green upper leaves. Other nutrients that are mobile in the plant include phosphorus, potassium,

Table 12-2
Approximate Pounds per Acre of Nutrients Contained in Crops

Crop	Acre Yield	Nitrogen	Phosphorus as P	Potassium as K	Calcium	Magnesium	Sulfur	Copper	Manganese	Zinc
Grains										
Barley (grain)	40 bu.	35	7	8	1	2	3	0.03	0.03	0.06
Barley (straw)	1 ton	15	3	25	8	2	4	0.01	0.32	0.05
Corn (grain)	150 bu.	135	23	33	16	20	14	0.06	0.09	0.15
Corn (stover)	4.5 tons	100	16	120	28	17	10	0.05	1.50	0.30
Oats (grain)	80 bu.	50	9	13	2	3	5	0.03	0.12	0.05
Oats (straw)	2 tons	25	7	66	8	8	9	0.03	—	0.29
Rice (rough)	80 bu.	50	9	8	3	4	3	0.01	0.08	0.07
Rice (straw)	2.5 tons	30	5	58	9	5	—	—	1.58	—
Rye (grain)	30 bu.	35	5	8	2	3	7	0.02	0.22	0.03
Rye (straw)	1.5 tons	15	4	21	8	2	3	0.01	0.14	0.07
Sorghum (grain)	60 bu.	50	11	13	4	5	5	0.01	0.04	0.04
Sorghum (stover)	3 tons	65	9	79	29	18	—	—	—	—
Wheat (grain)	40 bu.	50	11	13	1	6	3	0.03	0.09	0.14
Wheat (straw)	1.5 tons	20	3	29	6	3	5	0.01	0.16	0.05
Hay										
Alfalfa	4 tons	180	18	150	112	21	19	0.06	0.44	0.42
Bluegrass	2 tons	60	9	50	16	7	5	0.02	0.30	0.08
Coastal Bermuda	8 tons	185	31	224	59	24	—	0.21	—	—
Cowpea	2 tons	120	11	66	55	15	13	—	0.65	—
Peanut	2.25 tons	105	11	79	45	17	16	—	0.23	—
Red clover	2.5 tons	100	11	83	69	17	7	0.04	0.54	0.36
Soybean	2 tons	90	9	42	40	18	10	0.04	0.46	0.15
Timothy	2.5 tons	60	11	79	18	6	5	0.03	0.31	0.20

Crop	Yield									
Fruits and vegetables										
Apples	500 bu	30	5	37	8	5	10	0.03	0.03	0.03
Beans, dry	30 bu.	75	11	21	2	2	5	0.02	0.03	0.06
Cabbage	20 tons	130	16	108	20	8	44	0.04	0.10	0.08
Onions	7.5 tons	45	9	33	11	2	18	0.03	0.08	0.31
Oranges (70 pound boxes)	800 boxes	85	13	116	33	12	9	0.20	0.06	0.24
Peaches	600 bu.	35	9	54	4	8	2	—	—	0.01
Potatoes (tubers)	400 bu.	80	13	125	3	6	6	0.04	0.09	0.05
Spinach	5 tons	50	7	25	12	5	4	0.02	0.10	0.10
Sweet potatoes (roots)	300 bu.	45	7	62	4	9	6	0.03	0.06	0.03
Tomatoes (fruit)	20 tons	120	18	133	7	11	14	0.07	0.13	0.16
Turnips (roots)	10 tons	45	9	75	12	6	—	—	—	—
Other crops										
Cotton (seed and lint)	1500 lbs.	40	9	13	2	4	2	0.06	0.11	0.32
Cotton (stalks, leaves, and burs)	2000 lbs.	35	5	29	28	8	—	—	—	—
Peanuts (nuts)	1.25 tons	90	5	13	1	3	6	0.02	0.01	—
Soybeans (grain)	40 bu.	150	16	46	7	7	4	0.04	0.05	0.04
Sugar beets (roots)	15 tons	60	9	42	33	24	10	0.03	0.75	—
Sugarcane	30 tons	96	24	224	28	24	24	—	—	—
Tobacco (leaves)	2000 lbs.	75	7	100	75	18	14	0.03	0.55	0.07
Tobacco (stalks)	—	35	7	42	—	—	—	—	—	—

Reprinted from *Plant Food Review*, 1962, pp. 22–25, publication of The Fertilizer Institute.

Fig. 12-1 Manganese deficient bean plants. Note the healthy lower leaves and the light-colored intervein areas of the upper leaves. A progression of more severe symptoms occurs from the bottom to the top of the plant that is related to the immobility of manganese in plants.

and magnesium. Those with limited mobility that produce symptoms on the new leaves or growing points include calcium, boron, iron, copper, and manganese (see Fig. 12-1).

Deficiency Symptoms

It should be pointed out that nutrient deficiency symptoms are not always easily diagnosed. Some of them might be mistaken for discoloration or abnormal characteristics produced by diseases, or they may be due to a deficiency of some other element or factor of plant growth. But information concerning these symptoms has been accumulating rapidly, and they have become a valuable aid in determining the need for certain nutrients, especially when used in conjunction with tissue and soil tests.

Nutrient Uptake from Soils

During seed germination nutrients are supplied from the store of nutrients con-

tained in the seed. The nutrients in the seed are eventually depleted and the plant becomes dependent on the soil for nutrients. As roots elongate through soil, an increasing amount of nutrients becomes *positionally available* to the plant. As the root system enlarges and ramifies a greater soil volume, there is an increase in the ability of the plant to absorb nutrients from the soil. The extent and distribution of roots in soils are discussed in Chapter 2. This discussion will emphasize the processes important in the movement of nutrients to root surfaces, the nutrient absorption process, and the factors that affect nutrient absorption from soils.

Role of Mass Flow and Diffusion in Nutrient Uptake

Two important points must be kept in mind to understand how plants utilize nutrients in soils so effectively. As we have noted, root extension through the soil continuously exposes roots to "new" supplies of nutrients (and water). Second, after roots have invaded a region of the soil, *mass flow* and *diffusion* play an important role in moving nutrients over short distances to root surfaces. As water is absorbed, a water-tension gradient in the soil is established, and water slowly moves to root surfaces. Nutrients dissolved in the water are carried along in the water by mass flow. The amount of nutrients moved to roots by mass flow depends on the amount of water moved to the root and the concentration of nutrients in the water.

The range of concentration for some nutrients in soil water is given in Table 12-3. The concentration of calcium ranges from 8 to 450 parts per million. For a concentration of calcium of 8 parts per million in soil water and 2200 parts per million of calcium in the plant, it would

Table 12–3
Relation Between Concentration of Ions in the Expressed Soil Solution and Concentration Within The Corn Plant

| | Concentration, ppm | | | Ratio of Corn Plant Content to Lowest and Highest Soil Solution Contents | |
| | Soil Solution | | Corn-Plant[a] | | |
	Low	High	Ave.	Low	High
Calcium	8	450	2200	275	4.9
Potassium	3	156	20,000	6666	128.0
Magnesium	3	204	1800	600	8.8
Nitrogen	6	1700	15,000	2500	8.8
Phosphorus	0.03	7.2	2000	66,666	278.0
Sulfur	118	655	1700	155	2.6

Adapted from S. A. Barber, "A Diffusion and Mass Flow Concept of Soil Nutrient Availability," *Soil Sci., 93*:39–49, 1962.
Used by permission of the author and The Williams and Wilkins Co.
[a] Dry-weight basis.

require that the plant absorb 275 times more water than the plant's weight to move the amount of calcium needed to the roots by mass flow. Stated in another way, if the transpiration ratio is 275 and the concentration of calcium in soil water is 8 parts per million, enough calcium will be moved to the surface of roots to supply plant needs. Transpiration ratios are more commonly about 500, resulting in the expectation that more calcium is moved to root surfaces by mass flow than plants need.

The situation for phosphorus is very different. The phosphorus concentration in the soil solution is usually low, and a transpiration ratio of over 60,000 is sometimes needed to move enough phosphorus to the roots, according to the data in Table 12-3. From this illustration and others that could be drawn from the data, it is evident that some situations exist where a mechanism other than mass flow is needed to account for the movement of nutrients to the roots. This mechanism or process is *diffusion.*

Diffusion includes the movement of nutrient ions through the soil water and from exchange site to exchange site along the surfaces of soil particles. When there is insufficient movement of nutrients to the root surface by mass flow, diffusion plays an important role. In these cases the uptake of the ion reduces the concentration of that ion at the root surfaces, establishing a diffusion gradient outward from the root surface and causing the diffusion of ions toward the root. From the data in Table 12-3 it can be concluded that mass flow usually plays the dominant role in movement of calcium and sulfur to root surfaces and that diffusion plays the dominant role in movement of phosphorus to root surfaces. For potassium, magnesium, and nitrogen it appears that both mass flow and diffusion are important depending, of course, on the particular concentration in the soil solution and the transpiration ratio. At certain points along root surfaces there is an intimate contact with soil surfaces so that ions are exchanged directly from the soil particle to the root surface by a process called *contact exchange*.

The Process of Nutrient Uptake

Two well-established phenomena are known about nutrient absorption by plants. First, metabolic energy is required. If root respiration is curtailed, the net uptake of nutrients is minor, even from a concentration solution. Second, the process is selective. Plants have a capacity to selectively absorb certain ions over a wide range of conditions while effectively excluding others. Any theory of ion absorption must take into account these two phenomena.

The roots of plants are more or less surrounded by the soil solution and are in intimate contact with soil particles at many sites. Root cells have an "outer" space into which ions from the soil solution and from the exchange sites of the soil can diffuse. Diffusion of ions into these spaces is reversible and occurs without regard to the plants metabolic activities. It is a passive activity so far as the plant is concerned. This is a prelude to the irreversible transport of the ions across a seemingly impermeable membrane that requires an expenditure of energy.

The interior surface of the "outer" space has binding sites where carrier molecules are believed to be located. The carriers combine with the ions and together they migrate across a membrane that is impermeable to the ion alone. Once across the membrane, the carrier molecule and ion separate as the ion is deposited in the "inner" space of the cell, commonly called the vacuole. The carrier transport requires energy that is derived from respiration and the process enables the cell to achieve an ionic concentration in the cell that may be many times that of the external soil solution (see Fig. 12-2).

This theory explains the selective absorption of ions, since the carriers are specific for a given ion or group of ions. Nitrate and phosphate are both anions but require separate carriers. In fact, different carriers are required for $H_2PO_4^-$ and HPO_4^{2-}. Calcium and magnesium are transported by different carriers. The carrier for potassium can also transport cesium, while still another can transport sodium and lithium. Anions and cations are absorbed by the same mechanism but using different carriers.

The surfaces of leaves and stems, as well as the roots, can absorb nutrients. Any exposed plant surface appears to be able to function in this regard. Carbon dioxide absorption by the leaves is the major avenue for obtaining carbon. The processes involved in the transport of a nutrient ion from the soil environment into the root and its translocation and distribution within the plant are complex and interrelated.

Factors Affecting Nutrient Uptake

The factors that affect metabolism, and thereby the availability of respiratory energy, will directly affect nutrient uptake. These include the supply of respiratory substrate, temperature, and oxygen supply. The oxygen supply can be significantly altered by management practices. Soil compaction can reduce nutrient uptake

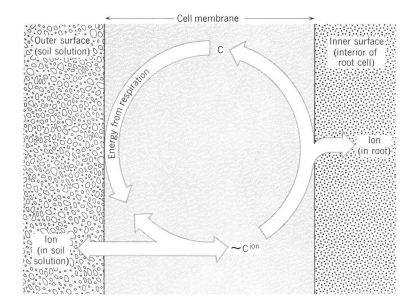

Fig. 12-2 A diagrammatic representation of the carrier theory of nutrient uptake. An organic carrier, C, links up with an ion from the soil solution at the exterior surface of the root cell. The ion is carried across the cell's membrane and deposited in the interior of the cell in the roots. (From Hanson, 1967. Reprinted from *Plant Food Review*, September 1967, p. 8, publication of Fertilizer Institute.)

through its affect on the availability of oxygen for root respiration. It is interesting that not all nutrients are reduced to the same degree. Lawton established that soil compaction reduced the uptake of potassium more than it did phosphorus or nitrogen, and that calcium was least affected in the case of corn. Plants grown in water culture usually obtain maximum growth when air is bubbled through the water.

An increase in the concentration of the nutrient in the environment external to the root will favor an increase in nutrient uptake when the initial concentration is low.

The moisture content of the soil is important because it influences the rate of movement and diffusion of ions into the "outer" spaces of the root cells. For example, it has been observed that drying the soil reduces phosphorus uptake. This is expected, because phosphorus is so slightly soluble and has such low mobility in the soil.

The density and distribution of roots in the soil is also important. Roots that enter a zone with a high nitrogen and phosphorus level tend to proliferate. It seems that when phosphate fertilizer is placed with the nitrogen in the soil, greater uptake of phosphorus occurs. The density of roots and extent of root surfaces become more important as the mobility of the nutrient in the soil decreases. Thus, root proliferation is expected to be less important for nitrate uptake than for phosphate uptake. Crops with deeply penetrating root systems generally require less fertilization than those with shallow root systems.

Nitrogen

We live in an ocean of nitrogen, but the supply of food for human beings and other animals is more limited by nitrogen than any other nutrient. The atmosphere is 79 percent nitrogen (by volume) as inert N_2

gas that resists reacting with other elements to create a form of nitrogen that most plants can use. Increasing the soil nitrogen supply for plants consists essentially of increasing the amount of biological fixation or adding fertilizer nitrogen. It is paradoxical that the nutrient absorbed from the soil in greatest quantity by plants is the nutrient most limiting in supply.

The Soil Nitrogen Cycle

During the first billion years of the earth's history, a large amount of reduced nitrogen was released into the atmosphere from the earth's interior. Green plants evolved that produced oxygen and microorganisms oxidized nitrogen to N_2 gas, and an atmosphere was formed similar to that existing today. The atmosphere now contains over 99 percent of the nitrogen currently in the earth's nitrogen cycle. Ever

since the modern atmosphere was formed, there has tended to be a nitrogen cycle consisting of an equilibrium between the nitrogen in the atmosphere and the nitrogen in the soils and oceans.

Atmospheric N_2 is characterized by strong attraction between the nitrogen atoms and great resistance to react with other elements. Higher plants are incapable of using N_2. The process of converting N_2 into usable forms that vascular plants can use is *nitrogen fixation*. Natural fixation is due to microorganisms (mainly bacteria in soils and algae in water) and certain atmospheric pheonomena, including lighting. Denitrifing bacteria in soils convert available soil nitrogen back into N_2 in a process called *denitrification*. These two processes, fixation and denitrification, are shown as processes 1 and 5 in Fig. 12-3. Fixation and denitrification are approximately equal and responsible for a quasi-

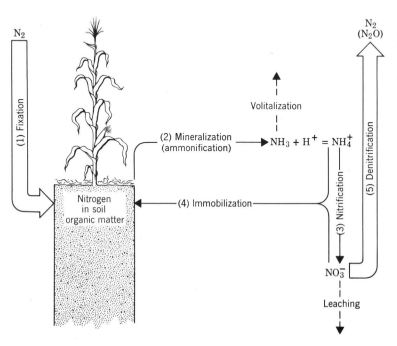

Fig. 12-3 The nitrogen cycle is composed of two subcycles. One subcycle consists of the addition of nitrogen to soils by fixation (1) and loss by denitrification (5). The other subcycle consists of cycling of nitrogen within the soil involving mineralization (2), nitrification (3) and immobilization (4). Nitrogen can be lost from the soil by leaching and volatilization.

equilibrium between the nitrogen in the atmosphere and the nitrogen in the lands and oceans.

A subcycle exists in the soil involving nitrogen in soil organic matter and the soil organisms consisting of processes 2, 3, and 4, as shown in Fig. 12-3. *Mineralization* of organic nitrogen results in available nitrogen as ammonium (NH_4^+). Nitrification produces available nitrogen as nitrate (NO_3^-). *Immobilization*, uptake of nitrogen by roots and microorganisms, incorporates the nitrogen back into organic form. The discussion that follows will consider the major processes of the nitrogen cycle in the number sequence presented in Fig. 12-3.

Nitrogen Fixation

There is virtually an inexhaustible supply of nitrogen in the atmosphere since, at sea level, there are about 34,500 tons of nitrogen in the air over an acre. It takes about 1 million years for the nitrogen in the atmosphere to move through one cycle.

Some nitrogen is fixed by electrical discharge (lightning) and other ionizing phenomena of the upper atmosphere, and the nitrogen is added to the soil as a component of precipitation. Most of the nitrogen naturally added to soils is added through biological fixation-symbiotic and nonsymbiotic. Biological nitrogen fixation is a reduction reaction requiring energy that is supplied by adenosine triphosphate (ATP). Nitrogen-fixing microorganisms contain the enzyme nitrogenase that combines with a nitrogen molecule, N_2. Pyruvic acid is the hydrogen donor, and fixation occurs in a series of steps that reduces the N_2 to NH_3, as shown in Fig. 12-4. Molybdenum is a part of nitrogenase and essential for biological nitrogen fixation. The organisms that fix nitrogen also require cobalt, which is the only known need of cobalt by plants.

Fig. 12-4 Simplified series of steps in nitrogen fixation. (Adapted from Delwiche, in *The Science Teacher*, Vol. 36, No. 3, p. 19, March 1969.)

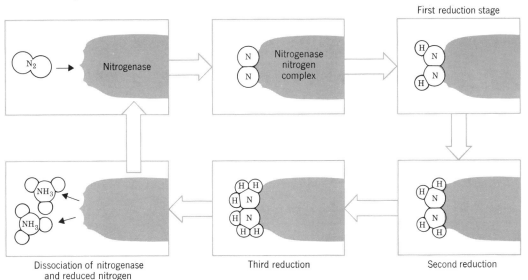

Symbiotic Nitrogen Fixation. The most important bacteria, from the agricultural point of view, capable of utilizing the free nitrogen of the air are those that cause the formation of nodules on the roots of legumes. These organisms, when growing in the nodules of legume plants, derive their food and minerals from the legume and, in turn, they supply the legume with some of its nitrogen. This growing together for a mutual benefit is called *symbiosis*, and hence the organisms are designated symbiotic nitrogen-fixing bacteria. It has been estimated that nearly 2 million tons of nitrogen are fixed annually by legume bacteria in the United States.

Rhizobium and Nitrogen Fixation. Legume plants form a symbiotic relationship with heterotrophic bacteria of the genus *Rhizobium*. The root of the host plant appears to secrete a substance that activates *Rhizobium* bacteria. When the bacteria make contact with a root hair, the root hair curls. An infection thread is formed in the root through which the bacteria migrate to the center of the root (Fig. 12-5). Once inside the root, the bacteria rapidly multiply and are transformed into swollen, irrgular-shaped bodies called *bacteroids*. An enlargement of the root occurs and, eventually, a gall or nodule is formed. The bacteroids receive food, nutrients, and probably certain growth compounds from the host plant. The legume host plant is benefited by the N_2 fixed in the nodule. Some of the fixed nitrogen is transported from the nodules to various parts of the host plant.

Quantity of Nitrogen Fixed Symbiotically. The quantity of nitrogen added to the soil through growth of legumes varies greatly according to conditions, such as the kind of legume, the nature of the soil, the effectiveness of the bacteria present, and seasonal conditions. It appears that the intimate relations existing between nodule bacteria and their host plants are determined mainly by the carbohydrate supply in the host plants. Any environmental condition affecting the pro-

Figure 12-5 Early stages in the formation of a nodule. *(a)* Response of the bacteria to a product of the host plant; organism moves toward root hair. *(b)* Curling of the root hair. *(c)* Early penetration of the infection thread. (From P. W. Wilson, "The Biochemistry of Symbiotic Nitrogen Fixation," p. 74, The University of Wisconsin Press, Copyright © 1940. Used by permission of P. W. Wilson and The University of Wisconsin Press.)

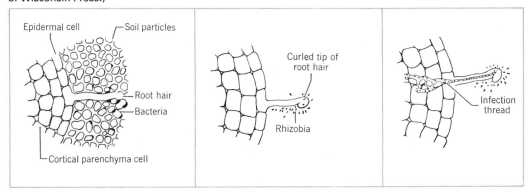

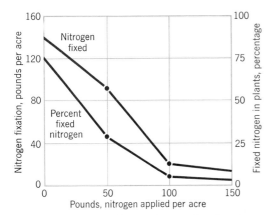

Fig. 12-6 Amount of total nitrogen fixed and percent of nitrogen in soybean plants from symbiotic fixation in relation to the application of nitrogen fertilizer. (Data from Weber, 1966.)

duction of carbohydrates in the plant would automatically affect the quantity of nitrogen fixed by good strains of legume bacteria. It has also been found that symbiotic nitrogen fixation is inhibited by an abundance of available soil nitrogen.

One method used to study the quantity of nitrogen fixed is to compare the amount of nitrogen in nodulated and nonnodulated plants. Weber used this method and found that as much as 142 pounds of nitrogen was fixed when soybeans were grown. This is shown in Fig. 12-6, which also shows that symbiotically fixed nitrogen represented 75 percent of the total nitrogen in the tops of the soybean plants. Furthermore, as the amount of available nitrogen in the soil was increased by the addition of fertilizer, the total amount and percentage of nitrogen fixed by bacteria was markedly decreased. From these and other data it is reasonable to conclude that 50 to 200 pounds of nitrogen are commonly fixed per acre per year when properly inoculated legume crops are grown. Inoculated alfalfa and clover are capable of fixing so much nitrogen that

nitrogen fertilizers do not increase plant growth but only reduce the amount of nitrogen fixed. Peas and navy beans fix less nitrogen, and it is a standard practice to use nitrogen fertilizer to increase yields. Many trees are legumes and fix nitrogen. Black locust is an example of a nitrogen-fixing legume tree.

In actual farm practice the amount of nitrogen added to a soil by legume bacteria is determined by the methods of disposing of the legume crop. If the crop is turned under as a green manure, the total quantity of nitrogen taken from the air is added. If the crop is cut for hay and sold off the farm, little or no gain is realized; with some legumes there may even be a net loss of nitrogen. And, if the crop is cut for hay and fed on the farm, about one half the nitrogen that was taken from the air by the legume bacteria can be returned to the soil if special care is exercised in handling the manure to prevent loss. It is generally assumed (although not necessarily true for all legumes) that the amount of nitrogen in the roots and stubble equals the amount of nitrogen taken from the soil; this would mean that the quantity of nitrogen removed in the harvested crop is equal to the nitrogen obtained from the air.

Nonlegume Symbiotic Nitrogen Fixation. In many wild land ecosystems there are few or no legumes to fix nitrogen. Now it is known that many nonlegume species have root nodules and fix nitrogen symbiotically. This means that symbiotic nitrogen fixation is important in natural range and forest soils as well as in agroecosystems. Red alder is an example of a nonlegume capable of symbiotic nitrogen fixation. This feature makes red alder a good pioneer specie for invasion of freshly exposed parent materials and burned-over lands where soils have little nitrogen-supplying capacity because of low organic matter content. Nonlegumes known to fix nitrogen symbiotically are listed in Table 12-4. The organisms causing formation of nodules and fixation of nitrogen are believed to be actinomycetes. The contribution of nitrogen to the earth's terrestrial ecosystems by symbiotic nonlegume fixation is several times that of herbaceous legumes.

Very recent research in Brazil suggests that some tropical grasses are involved in symbiotic nitrogen fixation. This has raised hopes that the major food crops of the world, like wheat and corn, may eventually be inoculated and made to fix nitrogen. A large international research effort has been launched to study the problem. If this research leads to the effective inoculation of cereal crops, it will be hailed as one of the greatest scientific advances of recent time. The cost of producing the world's food would be reduced, or the same resources would be able to produce more food.

Nonsymbiotic Nitrogen Fixation. There are certain groups of bacteria living in the soil independently of higher plants that have the ability to use atmospheric nitrogen in the synthesis of their body tissues. Since these bacteria do not grow in association (mutual relationship) with higher plants, they are termed nonsymbiotic. A dozen or more different bacteria have been found that fix N_2 nonsymbiotically. However, the two organisms that have been studied the most belong to the genus

Table 12-4

Distribution of Nodulated Nonlegumes

Family	Genus	Species Nodulated	Geographical Distribution
Betulaceae	Alnus	15	Cool regions of the northern hemisphere
Elgaeagnaceae	Elasagnus	9	Asia, Europe, North America
	Hippophae	1	Asia and Europe, from the Himalayas to the Arctic Circle
	Shepherdia	2	Confined to North America
Myricaceae	Myrica	7	Temperate regions of both hemispheres
Coriariaceae	Coriaria	3	Widely separated regions, chiefly Japan, New Zealand, Central and South America, and the Mediterranean region
Rhamnaceae	Ceanothus	7	Confined to North America
Casuarinaceae	Casuarina	12	Tropics and subtropics, extending from East Africa to the Indian Archipelago, Pacific Islands, and Australia

Reproduced from Stevenson, F. J., "Origin and Distribution of Nitrogen in Soil," *Soil Nitrogen,* Agronomy Monograph #10, 1965, by permission of The American Society of Agronomy.

Azotobacter and the genus *Clostridium.*

Azotobacter are widely distributed in nature. They have been found in soils (of pH 6.0 or above) in practically every locality where examinations have been made. The greatest limiting factor affecting their distribution in soils appears to be the soil reaction. These organisms may exist in soils below pH 6.0 but, as a rule, they are not active, as far as nitrogen fixation is concerned, under such conditions. *Azotobacter* are favored by good aeration, abundant organic matter (particularly of a carbonaceous nature), the presence of ample available calcium and sufficient quantities of available nutrient elements, especially phosphorus, and proper moisture and temperature relations.

The anaerobic bacteria, *Clostridia,* are much more acid tolerant than most members of the aerobic group and perhaps, for that reason, are more widespread. It is believed that these organisms can be found in every soil and that under suitable conditions they fix some nitrogen. It is not necessary that soils be waterlogged in order for anaerobic bacteria to function. A soil in good tilth may contain considerable areas within the granules favorable for the activities of these anaerobic nitrogen-fixing bacteria.

A question that naturally comes to mind is "How much nitrogen is fixed per acre per year by these nonsymbiotic microorganisms under favorable field conditions?" This question cannot be answered definitely because of the many difficulties encountered in making such a measurement under field conditions. From laboratory studies it is known that the nitrogen fixers utilize nitrate and ammonium nitrogen, which are normally present in the soil. To the extent that soil nitrogen is used, symbiotic fixation of atmospheric nitrogen

is inhibited. Furthermore, larger amounts of carbohydrates are used in relation to the amount of nitrogen fixed. It is estimated that only 5 to 20 pounds of nitrogen is fixed per 1000 pounds of organic matter decomposed. Considering the amount of decomposing organic matter available in soils, the large number of competing organisms, and the inhibitory effect of the available soil nitrogen, it appears that nonsymbiotic nitrogen fixation is a minor or unimportant factor in crop production. For the natural ecosystem, the small quantity fixed each year over thousands of years is undoubtedly important.

Mineralization

Symbiotically fixed nitrogen is utilized within plants and eventually appears as nitrogen in dead tissue of plants and animals or animal feces and incorporates into soil organic matter and humus. At any one time over 99 percent of the soil nitrogen is in organic matter. About 2 to 3 percent of the total organic nitrogen is mineralized in a single year, resulting in one complete turnover of soil nitrogen every 30 to 50 years. Many different kinds of heterotrophic organisms are engaged in organic matter decomposition with the subsequent minerlization of nitrogen to ammonia (NH_3) (see number 3 of Fig. 12-3). Mineralization is also called *ammonification* because the end product is ammonia. Some NH_3 produced at the very surface of the soil escapes by volatilization, especially when soil pH is 8 or more.

Most of the ammonia produced in the soil quickly forms ammonium (NH_4^+). There is a strong tendency for ammonium to form because of the presence of hydrogen ions in the soil and the strong bond formed between the ammonia and

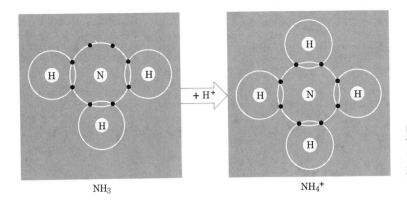

Fig. 12-7 Diagram of formation of ammonium (NH_4^+) showing arrangement of atoms. Valence of nitrogen is -3.

hydrogen from electron sharing (Fig. 12-7). The ammonium ion has a charge of $+1$ and is available to plants. There is evidence that ammonium is the major form of nitrogen used by plants in forests and in rangelands. Ammonium is adsorbed on the cation-exchange complex and is held against leaching. Some ammonium is fixed in illitic clay minerals where it occupies the space in the crystal lattice normally occupied by potassium.

An acre furrow slice of soil has about 1000 pounds of nitrogen for each percent of organic matter. This is based on the fact that an acre furrow slice weighs 2 million pounds and organic matter is 5 percent nitrogen. If the mineralization rate is 2 to 3 percent per year, there will be approximately 20 to 30 pounds of nitrogen mineralized each year for each percent of organic matter in the surface soil. Organic matter also exists below the surface plow layer, and this organic matter is slowly mineralized as well. It is important to realize that the nitrogen-supplying power of a soil is intimately related to the *organic matter content* and *mineralization rate*. Sandy soils low in organic matter are poor in ability to supply available nitrogen. Desert soils, because of low organic matter content, are

likely to supply only 5 percent of the nitrogen needs of a crop produced with irrigation. Organic soils that have been drained, on the other hand, have maximum potential for supplying nitrogen to plants, since the soils are mostly organic matter. Little nitrogen is mineralized in water-saturated organic soils, however, because of oxygen deficiency for aerobic heterotrophic decomposers. In swamps some plants snare insects to obtain nitrogen.

Nitrification

If ammonium is not absorbed by roots or microorganisms or fixed in clay, ammonium is commonly oxidized to nitrate. This process, nitrification, is a two-step process with nitrite as the intermediate product. Specific autotrophic bacteria are involved. The reactions and organisms are as follows.

$$(1) \quad NH_4^+ + 1\tfrac{1}{2}O_2 \xrightarrow{\text{Nitrosomonas}}$$

$$NO_2^- + 2H^+ + H_2O + \text{energy}$$

$$(2) \quad NO_2^- + \tfrac{1}{2}O_2 \xrightarrow{\text{Nitrobacter}}$$

$$NO_3^- + \text{energy}$$

Nitrification results in the presence of available nitrogen in soils as an anion. This is of no particular concern, since most plants utilize nitrate about as readily as ammonium. Paddy rice prefers ammonium and rice plants in rice paddies feed mainly on ammonium because anaerobic conditions inhibit the nitrification of the ammonium.

Nitrate is stable in well-aerated soils and readily moves with soil water (mass flow) to root surfaces. Nitrate is also readily leached from soils, and this has important implications for economy of use and nitrate pollution of groundwater. A product called N-Serve has been developed to inhibit nitrification and keep nitrogen in the ammonium form in soil to protect it from leaching and denitrification. In well-aerated soils the ammonium in fertilizers is rapidly converted to nitrate, making the typical plant response to ammonium nitrate fertilizer the same as that of a nitrate fertilizer. Nitrification results in the production of hydrogen ions and is a potential for increasing soil acidity.

Immobilization

Both ammonium and nitrate are available forms of nitrogen for roots and microorganisms, and their use results in the conversion of mineral forms of nitrogen into organic form. The process is immobilization as shown by number 4 in Fig. 12-3. Immobilized nitrogen is "safe" in the soil and subject to repeated cycling through the nitrogen subcycle in the soil, which involves mineralization, nitrification, and immobilization. Below the root zone or zone of biological activity nitrate will not be immobilized or denitrified. It is natural for some nitrate to be leached to the water table in well-drained soils of humid regions and become a component of the groundwater.

Denitrification

Just as it is natural for nitrogen to be added to soils by fixation, it is natural for nitrogen to be lost from soils by denitrification. Denitrification is the reduction of nitrate to gaseous nitrogen and its escape from the soil. In fact, in a climax forest or grassland where the organic matter content of the soil remains constant from year to year and the amount of nitrogen cycling in the soil remains about constant from year to year, the addition of nitrogen by fixation approaches the losses of nitrogen from the soil by denitrification. Thus, denitrification is one of the most significant processes in the nitrogen cycle and accounts for significant losses of nitrogen from soils.

Denitrification is carried out by facultative anaerobic organisms that use nitrate in place of oxygen in respiration, as follows.

$$C_6H_{12}O_6 + 4NO_3^- \rightarrow$$

$$6CO_2 + 6H_2O + 2N_2 \text{ (plus NO, N}_2\text{O, and NO}_2\text{)}$$

Denitrification occurs under anaerobic conditions in water-saturated soils and may occur in the interior of moist aggregates in soils considered to be well-drained.

Normally, denitrification is detrimental to agriculture, since nitrogen is lost, as when nitrate fertilizer is applied to poorly aerated soils. Denitrification, however, is an important process than can be used to help prevent excess nitrates from building up in the groundwater of irrigated valleys where large amounts of nitrogen fertilizer are used. In a large, irrigated valley, excess nitrate is commonly leached to a shallow water table and gradually flows to the lower end of the valley in the groundwater. This takes a considerable period of time. In route, microorganisms gradually denitrify the nitrate, which escapes as nitrogen gas.

This enables the drainage water from the irrigated valley to have a low value of nitrates, which is not detrimental to other water users downstream. Denitrification has also been used as a means to remove nitrate from manure wastes and sewage effluent.

Human Intrusion in the Nitrogen Cycle

Mineralization of soil organic matter is the major source of available nitrogen for plants. Mineralization of 50 pounds of nitrogen per acre per year is realistic for many soils. By contrast a 150-bushel quantity of corn (minus roots) contains 235 pounds of nitrogen (Table 12-2). It is obvious that the natural sources of nitrogen in the soil are small compared to the needs of 150-bushel corn crop. The United States produces over 50 percent of the corn produced in the world; this is accomplished by using a large amount of nitrogen fertilizer. The balance sheet for the earth's nitrogen cycle in Fig. 12-8 shows that the amount of industrially fixed nitrogen was equivalent to the terrestrial fixation and about two times that fixed by legume crops. The biosphere is now receiving annually about 9 million metric tons more nitrogen per year than is being lost. It is obvious that humans have become important intruders in the earth's nitrogen cycle because of industrial nitrogen fixation. The long-time effects of a buildup of nitrogen in the biosphere are unknown, but the buildup represents potential danger for nitrate pollution of groundwater and eutrophication of lakes. It is important to realize that adding more nitrogen to the soil as fertilizer does not necessarily result in more leaching of nitrate to the water table. This results from the fact that greatly increased plant growth demands more nitrogen uptake. Nitrate losses are increased,

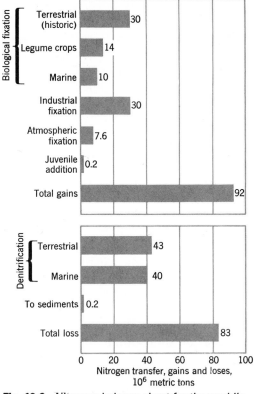

Fig. 12-8 Nitrogen balance sheet for the world's biosphere. Nine million metric tons (92-83) more nitrogen is being added than is being removed by denitrification (and loss to sediments). (From "The Nitrogen Cycle, by C. C. Delwicke, copyright © 1970 by *Scientific American*, Inc. All rights reserved.)

however, when the immobilization capacity of the soil is exceeded.

Effect of Nitrogen on Plant Growth

An abundance of nitrogen promotes rapid growth with a greater development of dark green leaves and stems. Although one of the most striking functions of nitrogen is the encouragement of above-ground vegetative growth, this growth cannot take place except in the presence of adequate quantities of available phosphorus, pot-

Nitrogen deficiency on corn. Lower leaves turn yellow along midrib, starting at the leaf tip.

Left is nitrogen-deficient section of corn leaf and cut open section of stem that remains uncolored when treated with diaphenylamine, indicating low nitrate level in plant sap. On right the leaf is dark green and diaphenylamine produces a dark blue color indicative of high nitrate level in plant sap.

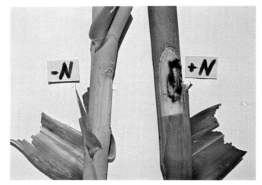

Potassium deficiency on alfalfa (and clovers) shows as a series of white dots near leaf margins; in advanced stages entire leaf margin turns white. (Courtesy American Potash Institute.)

Potassium deficiency on corn. Lower leaves have yellow margins. (Courtesy American Potash Institute.)

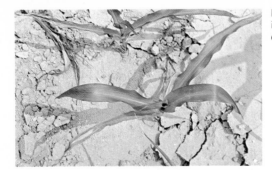

Phosphorus deficiency on corn is indicated by purplish discoloration.

FE DEFICIENCY
PIN OAK

Iron deficiency on pin oak. The leaf veins remain green as the intervein areas lose their green color.

Iron deficiency on roses. Green veins with yellow intervein areas, showing most on newest leaves.

Manganese deficiency on kidney beans. Leaf veins remain green as the intervein areas loose their green color and turn yellow.

Zinc deficiency of navy beans grown on calcareous soil. The small unfertilized zinc deficient plants stand in marked contrast to the taller plants that were fertilized with zinc. A case where zinc fertilization is necessary to produce a crop.

Boron deficiency on sugar beets causes heart rot. Most advanced symptom is on left. (Courtesy American Potash Institute.)

Magensium deficiency on coffee. Veins remain green and intervein areas turn yellow.

Manganese-treated plants in rear showing no deficiency. Plants in foreground are unfertilized and difference in degree of manganese deficiency symptoms is indicative of varietal response to limited soil manganese.

Excess soluble salt symptoms on geranium. The leaf margins turn yellow and become necrotic.

asium, and other essential elements.

An ample supply of available nitrogen during the early life of the plant may stimulate growth and result in earlier maturity. However, the presence of an excess of nitrogen throughout the growing season of the plant frequently prolongs the growth period. This effect is especially significant for certain crops in regions having a short growing season or in areas where an early fall freeze may do great damage to fruit trees whose season's growth period has been prolonged.

A large supply of available nitrogen encourages the production of soft, succulent tissue that is susceptible to mechanical injury and the attack of disease. Either effect may decrease the quality of the crop. However, the development of softness in the tissues may be desirable or undesirable, depending on the kind of crop. For vegetables used for their leaves, pronounced succulence, tenderness, and crispness are desired. Other vegetables and some fruits may have their keeping and shipping qualities impaired when they are grown with an excess of available nitrogen. An excess of nitrogen may encourage lodging in grains, which frequently decreases the quality, but a normal amount of nitrogen usually increases plumpness in grains.

Nitrogen Deficiency Symptoms

The need for more nitrogen is indicated by a light green to yellow appearance of the leaves. As a rule, the older bottom leaves start to turn light green, then turn yellow at the tip. The entire leaf may turn yellow, even though the tissues are alive and turgid. In the corn plant the yellowing extends up the midrib of the leaf, with the outer edges remaining green the longest (see Fig. 12-9). A nitrogen-starved cucumber may

Fig. 12-9 Nitrogen deficiency symptoms of corn. Lower leaves turn yellow at tip and then along midrib. The lowest or oldest leaves loose nitrogen first and then die. A 15-bushel-yield reduction has been observed for each nitrogen-deficient leaf at silking.

have a small or pointed blossom end; a deficiency of nitrogen may cause the kernels of cereals to become shriveled and light in weight. In fruit trees the early shedding of leaves, death of lateral buds, poor set of fruit, and development of unusually colored fruit are indications of a lack of nitrogen. Colored plates showing nitrogen deficiency are found in plate 1.

Phosphorus

Phosphorus plays an indispensable role as a universal fuel for all biochemical work in living cells. High energy adenosine triphosphate (ATP) bonds release energy for work when converted to adenosine diphosphate (ADP). Phosphorus is also an important element in bones and teeth. The relation of phosphorus in soils and plants to animal health and the extensive occurrence of phosphorus deficiency in grazing animals are well known. Here the emphasis will be on development of an understanding of the nature of phosphorus in soils and the conditions that control the uptake of phosphorus from soils by plants. Such knowledge helps answer questions like "Is phosphorus likely to be deficient in my garden? If so, what can be done about it?" or "What are the problems of increasing or improving the world food supply that relate to soil phosphorus?"

Phosphorus Cycle in Soils

The earth's crust contains about $\frac{1}{10}$ percent phosphorus and a bushel of grain contains about $\frac{1}{10}$ pound of phosphorus. On this basis the phosphorus in an acre furrow slice of an average soil could produce 20,000 bushels of grain. This does not include the phosphorus that could be absorbed by roots at depths below the plow layer. Phos-

phorus, however, commonly limits plant growth. The major problem in phosphorus uptake from soils by plants is the low solubility of most phosphorus compounds and the resulting very low concentration of available phosphorus in the soil solution at any one time.

Most of the phosphorus in igneous rocks and soil parent materials occurs as apatite (Fig. 12-10). Fluorapatite ($Ca_{10}(PO_4)_6 F_2$) is the most common apatite mineral. Fluorapatite contains fluorine (F), which contributes to a very stable crystalline structure that resists weathering. The structure is similar to teeth dentine, and fluoridation of water is designed to incorporate fluorine into teeth to increase resistance to decay. Apatite weathers slowly and the available phosphorus occurs mostly as $H_2PO_4^-$ in the soil solution. The $H_2PO_4^-$ is immobilized by plants and microorganisms, and a significant amount of the phosphorus in soils is converted into the organic form during soil formation. As with nitrogen, the organic phosphorus is mineralized (process 3) to

Fig. 12-10 Major processes in the soil phosphorus cycle. The availability of phosphorus to plants is determined by the amount of phosphorus in the soil solution.

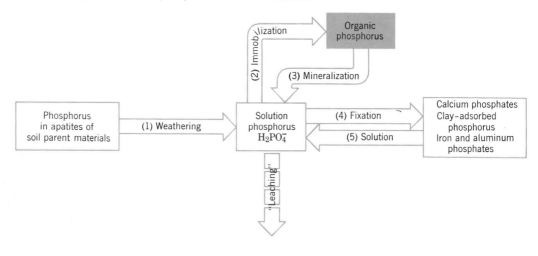

complete a subcycle of the overall phosphorus cycle.

A major difference between the nitrogen and phosphorus cycles in the soil is that the available forms of nitrogen (ammonium and nitrate) are relatively stable ions that remain available for plant use. The $H_2PO_4^-$, by contrast, reacts quickly with other ions in the soil solution to become much less soluble or unavailable to plants. Reactions with calcium, iron, and aluminum are the most common. Phosphate is also strongly adsorbed on clay surfaces. An equilibrium is established between the concentration of $H_2PO_4^-$ in the soil solution and the fixed mineral forms, as represented by reactions 4 and 5 in Fig. 12-10. The concentration of phosphorus in the soil solution is primarily a function of the solubility of the fixed forms of phosphorus. The result is a *very low concentration of phosphorus in the soil solution at any one time because of the low solubility of the fixed forms.* In general, there is decreasing solubility or availability in the order calcium phosphates, clay-absorbed phosphate, and iron and aluminum phosphates.

Changes in Fixed Phosphorus Over Time

The forms in which the phosphorus occurs in soils change over time. Some of the mineral phosphorus is converted to organic phosphorus, as we have already noted. There is a shift over time of the mineral forms to compounds of less solubility. Even the first precipitated calcium phosphates may slowly change from relatively soluble tricalcium phosphate ($Ca_3(PO_4)_2$) to apatite forms. The freshly precipitated or raw oxides of iron and aluminum phosphate slowly crystallize into strengite (FePO·2 H_2O) and variscite (AlPO$_4$·2H$_2$O), which expose less surface area and dissolve more

slowly. In highly weathered tropical soils some of the phosphorus may be encased or coated with iron and aluminum oxides and protected from solution. This encased or "occuled" phosphorus is the least soluble of the fixed forms.

Youthful soils where most of the fixed phosphorus is in calcium phosphate have higher available phosphorus than strongly weathered tropical soils (Oxisols). The change in the kinds of fixed phosphorus in soils as a function of soil age is shown in Fig. 12-11. Strongly weathered tropical soils maintain a very low equilibrium concentration of phosphorus in the soil solution because phosphorus is mainly in minerals of very slow solubility—not because the soils have a low total phosphorus content.

Fig. 12-11 Percentage distribution of four forms of phosphorus as related to weathering or soil age. (Adapted from S. C. Chang and M. L. Jackson, *Journal of Soil Science* (1958). By permission of the Oxford University Press.)

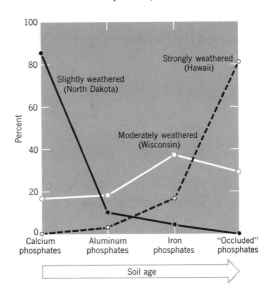

Plant Uptake of Phosphorus From Soils

The dominant form of phosphate available to plants is $H_2PO_4^-$. The presence of water is important for phosphorus absorption in soils. Plants absorb about 500 pounds of water per pound of growth. The phosphorus in 500 pounds of soil water, however, is very inadequate to meet the plant needs if water and phosphate are absorbed in the ratio that they exist in soils. For analysis consider:

transpiration ratio $= 500$

Phosphorus in plant tissue $= 0.3$ percent

Phosphorus content of soil solution

$= 0.03$ parts per million

It would require 100,000 pounds of water to contain the phosphorus needed for 1 pound of plant tissue, calculated as follows.

$$\frac{0.03 \text{ pounds phosphorus}}{1 \text{ million pounds water}}$$

$$= \frac{0.003 \text{ pounds phosphorus}}{X \text{ pounds water}}$$

$X = 100{,}000$ pounds of water

The 100,000 pounds of water is 200 times greater than the transpiration ratio. Thus, during the time that plants absorb water, they must in effect deplete the soil solution of its phosphorus and have the phosphorus replaced through 200 cycles. This example shows the great dependence of plants on diffusion of phosphorus on the surfaces of soil particles into and through water films to plant roots to supply the phosphorus needs of plants. The greater the concentration of phosphorus in the soil water, the easier it is for plants to satisfy the phosphorus requirements and, in effect, the greater is phosphorus availability in the soil.

Effect of pH on Phosphorus Availability

The ions in the soil solution are a function of pH. As the pH goes below 5.5, soluble iron and aluminum increase considerably. This causes fixation of phosphorus as iron and aluminum phosphates. Best availability of phosphorus is in the range of 6 to 7. Calcium phosphates begin to precipitate at about pH 6.0. Above pH 7.0, the tendency for apatite formation again reduces phosphorus solubility or availability.

Increases in pH above 7 create sufficient OH^- to react with $H_2PO_4^-$ to form $HPO_4^=$ and water to cause the latter form of phosphorus to become the most abundant. Since $HPO_4^=$ is less readily taken up by plants than $H_2PO_4^-$, one can conclude that part of the reduced availability of phosphorus in alkaline soils is due to the presence of hydroxyl ions and formation of $HPO_4^=$.

Phosphorus Movement and Loss From Soils

The very low concentration of phosphorus in the soil solution at any one time means that leaching removes little phosphorus from soils. The tendency for phosphate ions in the soil solution to become fixed makes it difficult for plants to satisfy their phosphorus needs but, on the other hand, it decreases leaching of phosphorus from soils, which may produce eutrophication in lakes. Removal of phosphorus from soils into local waters tends to be associated most with removal of adsorbed phosphorus on soil particles by erosion than by the leaching of phosphorus in solution to the underground water table.

Phosphorus fertilizers are mainly soluble calcium and ammonium phosphates. They dissolve into the soil solution and are subject to fixation. For this reason the phosphorus from fertilizers may move about 1 inch from place of application or, for practical purposes, the fertilizer phosphorus does not move after placement in the soil. An obvious consequence is that surface-applied phosphorus is much less effective than phosphorus applied within the soil where roots are more abundant and more water is available for its solution. Another important consequence of phosphorus fixation is that only about 10 to 20 percent of the phosphorus applied to soils is used by the next crop or used within the next year. This has caused several times more phosphorus to be added to the soils of the United States than is removed in crops.

Anthropic Horizons—Indicators of Human Activity

Anthropic horizons are epipedons that have greatly increased phosphorus content because of human activity. Anthropic horizons are thick and dark colored like mollic horizons but, in addition, have over 250 part per million of phosphorus as P_2O_5 soluble in dilute acid. Some European farmers have produced anthropic horizons because of the long-continued use of large applications of organic matter plus materials containing phosphorus. Some of the phosphorus came from bones that were collected on battlefields. No such horizons have been produced in the United States from farming, only increased phosphorus content.

Human occupancy at a campsite or village results in the disposal of refuse containing phosphorus. This is particularly true where the refuse contains many bones. The phosphorus in bones (and teeth) is a calcium phosphate similar to apatite; it accumulates as such in soils or some other insoluble fixed form. Accumulations of phosphorus in soils as a result of human activity makes a determination of the phosphorus content of soil an important tool for discovering the locations of ancient campsites, villages, and roads (see Fig. 12-12).

Fig. 12-12 Students removing artifacts from Koster Indian site near mouth of Illinois River. During the past 8500 years there were 12 periods of occupancy and formation of anthropic horizons. The anthropic horizons are separated by layers of sterile soil representing periods of abandonment and deposition of eroded soil from the slopes above the site.

Effect of Phosphorus on Plant Growth

The effects of too little or too much phosphorus on plant growth are less striking than those of nitrogen or potassium. It appears to hasten maturity more than most nutrients, as an excess stimulates early maturation. Phosphorus deficiency is characterized by stunted plants having about equally affected root and top growth. Many soils produce forage deficient in phosphorus in terms of the nutritional requirement of animals, and fertilization with sufficient phosphorus to increase the phosphorus content of the forage improves forage quality in these cases.

Phosphorus Deficiency Symptoms

If phosphorus is deficient, cell division in plants is retarded and growth is stunted. A dark green color associated with a purplish coloration in the seedling stage of growth is a symptom of phosphorus deficiency (see color plate 2). Later, plants become yellow. Occasionally a pale or yellowish green color develops when the lack of phosphorus inhibits the utilization of nitrogen by the plant. Bronze or purple leaves sometimes

are observed at the top of new shoots of phosphorus-starved apple trees. In the absence of sufficient phosphorus, general maturity of the crop and seed formation are usually delayed. With corn, poor pollination frequently is associated with phosphorus starvation. Perhaps the most characteristic symptom of phosphorus deficiency, among plants in general, is the stunted growth.

Potassium

Many soils have an abundance of available potassium, and plants do not respond to potassium fertilizers even though plants generally use more potassium from soils than any nutrient except nitrogen. This is in stark contrast to what we have just noted regarding the general need of nitrogen and phosphorus fertilizers in agroecosystems. Basically, the potassium in soils is in minerals that weather and release potassium ions. The ions are adsorbed on the cation exchange and are readily available for plant uptake. The available potassium accumulates in soils with ustic or drier moisture regimes in the absence of leaching. Such

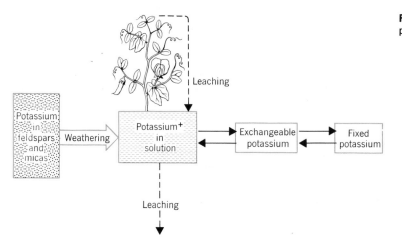

Fig. 12-13 The potassium cycle in soils.

soils are generally neutral or alkaline, do not require lime, and do not need potassium fertilizers even for high crop yields. Leaching in humid regions removes the available potassium and creates a need for potassium fertilizer when moderate or high crop yields are desired. Organic soils are notoriously deficient in potassium because they contain few minerals that contain potassium. Our discussion of potassium will emphasize the nature of the potassium in soils and factors that affect soils ability to meet the potassium needs of plants.

The Potassium Cycle in Soils

The earth's crust has an average potassium content of 2.6 percent. Parent materials and youthful soils could easily contain 40,000 to 50,000 pounds of potassium per acre furrow slice. The potassium content of the soil at depths below the plow layer could be similar. About 95 to 99 percent of this potassium is in the lattice of the following minerals.

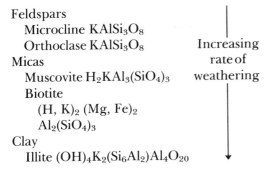

Feldspars
    Microcline $KAlSi_3O_8$
    Orthoclase $KAlSi_3O_8$
Micas
    Muscovite $H_2KAl_3(SiO_4)_3$
    Biotite
      $(H, K)_2 (Mg, Fe)_2$
      $Al_2(SiO_4)_3$
Clay
    Illite $(OH)_4K_2(Si_6Al_2)Al_4O_{20}$

Increasing rate of weathering

The micas weather faster and release their potassium more readily than the feldspars. These minerals exist primarily in the sand and silt fractions.

During weathering, the potassium ion, K^+, is released into the soil solution (Fig. 12-13). Plants absorb potassium as K^+ mainly from the soil solution and a little by contact exchange from the cation-exchange surfaces. A few pounds of K^+ exists in the soil solution and up to a few hundred pounds of K^+ per acre furrow slice exist on the cation-exchange sites in most mineral soils.

An equilibrium exists between the solution potassium and the exchangeable potassium, as shown in Fig. 12-13. Consider that weathering is occurring in a soil when plants are dormant. The concentration of potassium in the soil solution increases, which forces more potassium onto exchange positions by mass action. During this time, the release of potassium exceeds plant uptake and the exchangeable or available potassium increases. During periods of rapid growth, plants may remove potassium from soil faster than it is released by weathering, and the equilibrium is shifted to the left. As plants absorb the potassium from the soil solution, it dissociates from cation-exchange sites in an effort to maintain the equilibrium. This sequence of events is typical of the annual changes that occur from winter to summer in available potassium in soils.

An equilibrium also exists between the exchangeable and fixed potassium. Fixation occurs by migration of K^+ into vacant positions of the mineral lattice from which a K^+ had been removed by weathering. Weathering begins at the edges of mineral particles and progresses inward (Fig. 12-14). Along the edges the potassium is weathered out, leaving vacant spaces in the lattice, while the interior of the particle is still fresh and unweathered. Loss of potassium along the edges removes the potassium bridges that hold adjacent layers of the crystals together, and layers separate or expand along the edges. Potassium fixation is the reverse of weathering out of the potassium from the lattice (in Fig. 12-15).

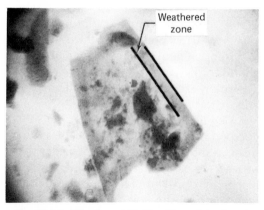

Fig. 12-14 Weathered biotite silt particle showing loss of potassium from the darkened area around the edges. The loss of potassium results in expansion of the 2:1 (mica) layers. The weathered zone is vermiculite. (Photo courtesy Dr. M. M. Mortland.)

Fixation and release is a reversible process dependent on the concentration of K^+ on the exchange sites which, in turn, is dependent on the concentration of K^+ in the soil solution. Complete loss of potassium from between the layers causes a complete separation of the mineral layers and loss of potassium-fixing capacity. Potassium fixation in soils conserves potassium that might otherwise be lost by leaching when potassium release by weathering exceeds plant uptake. Fixation also allows some potassium from fertilization to be temporarily stored in a safe, unavailable position until plant uptake has reduced the amount of potassium on the exchange complex. Occasionally a soil has such high potassium-fixing capacity that most of the potassium from fertilizer use goes into satisfying the fixing capacity instead of increasing the uptake of potassium and increasing yield. Ammonium ions are similar in size to K^+ and are fixed in the same lattice spaces as potassium.

The discussion thus far has stressed the association between potassium and the

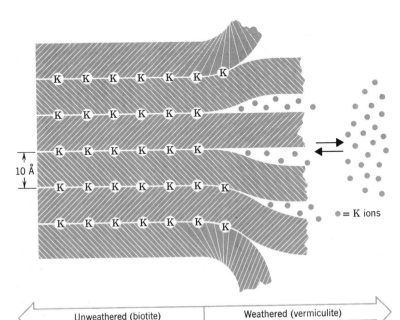

Fig. 12-15 Diagram showing the separation and curling of mica layers resulting from the loss of K^+ during weathering of biotite. Fixation of K^+ results in an uncurling and collapse of the layers or a reversal of weathering. Ions in the soil solution diffuse into and out of the expansion space between the layers.

= K ions

10 Å

Unweathered (biotite)   Weathered (vermiculite)

mineral components of the soil. Apparently most of the potassium in plants does not form an integral part of plant and microbial tissue in the way that nitrogen is incorporated into protiens. In fact, a lot of potassium is leached from the leaves of plants during rains. Consequently, the organic matter is not a significant source of potassium for plants. Potassium is immobilized and mineralized, but it does not accumulate in the soil organic fraction. Soils with the lowest available potassium are acid organic soils. Soils with the highest available potassium tend to be fine-textured soils that are neutral or alkaline.

Forms of Soil Potassium versus Uptake and Plant Growth

Both the solution and exchangeable potassium are considered plant available. Removing the exchangeable potassium removes the available potassium (solution potassium is also removed) and plants must depend on release of fixed potassium or potassium weathered from minerals. Removing the exchangeable potassium from a silt loam soil in Wisconsin reduced the yields of corn and oats to 62 percent of the normal or untreated soil (Table 12-5). The removal of exchangeable and fixed potassium lowered yields to about 20 percent of normal. The addition of 800 pounds of potassium per acre in fertilizer to soil from which the exchangeable and fixed potassium had been removed resulted in corn and oats yields greater than that of the normal soil. The experimental results confirm the validity of the distinctions made in regard to the various forms of potassium in soils and their availability to plants.

The plant uptake data in Table 12-5 show decreases that parallel growth for removal of exchangeable potassium and removal of exchangeable potassium plus fixed potassium. Uptake as a result of the application of 800 pounds of potassium per acre to soil from which the exchangeable and fixed potassium had previously been removed, however, resulted in *luxury con-*

Table 12-5
Effect of Different Forms of Potassium on Plant Growth and Uptake of Potassium

| | Percentage of Untreated Soil | | | |
| | Yield | | Potassium Uptake | |
Soil Treatment	Corn	Oats	Corn	Oats
Normal or untreated soil	100	100	100	100
Exchangeable potassium removed	62	62	41	28
Exchangeable + fixed potassium removed	20	18	23	18
Exchangeable and fixed potassium removed + 800 pounds of potassium as fertilizer per acre	149	116	959	452

Adapted from Attoe and Truog, 1945.

sumption of potassium. This is evidence that the fertilizer caused an overabundance of solution and exchangeable potassium and resulted in excessive potassium uptake before most of the fertilizer potassium was dissipated by plant uptake, fixation, or leaching. Luxury consumption of potassium limits the amount of potassium fertilizer that can be applied at one time for efficient use of the fertilizer and maintain a desirable magnesium/potassium ratio in grass forages.

Role of Diffusion in Plant Uptake From Soils

As with phosphorus, there is less potassium in the soil water than needed for plant growth if water and potassium are taken up by plants in the same ratio that they normally exist in soils. Diffusion of potassium from cation-exchange sites through water films to root surfaces is very important in uptake of potassium from soils by plants. About 90 percent of the uptake was found to be due to diffusion, and the remainder was caused by mass flow and root interception in an experiment with soybeans. Root interception can be expected to account for little of the uptake, since only about 1 percent of the surface soil volume is made up of roots, and uptake from root interception is closely related to root volume.

Potassium Movement and Loss From Soils

Potassium is intermediate between nitrogen and phosphorus in regard to mobility in soils. Some potassium is leached from soils in humid regions, but the losses do not appear to have any environmental consequences. Many soils have argillic horizons with considerable capacity to hold potassium in the exchange and fixed positions. Some of the potassium leached from the surface soil is held in the B horizon and returned to the surface by plant roots. Long-time losses of potassium by leaching result in gradual decreases in the potassium content of soils and development of soils with a limited supply of available potassium for crops in humid regions. By the time soils become Ultisols and Oxisols, the original potassium minerals have been essentially weathered, and deep rooting of trees is important to bring up potassium from less weathered soil, where the supply of potassium is greater.

Effects of Potassium on Plant Growth

Potassium has a counterbalancing effect on the results of a nitrogen excess. It enhances the synthesis and translocation of carbohydrates, thereby encouraging cell wall thickness and stalk strength. A deficiency is sometimes expressed by stalk breakage or lodging. It also increases the sugar content of sugar beets and sugar canes. Highest dry matter yields of these two crops can be obtained with very high rates of nitrogen fertilization, but the greatest production of sugar results from moderate nitrogen applications and a sufficient level of available potassium. Root crops like potatoes also have a high potassium requirement. Less succulent foliage is promoted by good supplies of potassium and reduces disease. There is some evidence that indicates that alfalfa is less susceptible to frost injury when it is well fertilized with potassium.

Potassium Deficiency Symptoms

A deficiency of potassium usually shows up as a "leaf scorch" in most plants. Corn indicates a need for potassium by a yellowing of the tips and margins of the lower leaves (color plate 1). This coloration does not move up the midrib, as with a nitrogen deficiency, but gradually spreads upward and inward from the leaf tip and edges. This leaf scorch is frequently mistaken for "burning" or "firing" and is ascribed to a deficiency of moisture during dry weather. When insufficiently supplied with potassium, alfalfa frequently develops a series of white spots near the margin of the older leaves (color plate 1). Sometimes this spotting effect is accompanied by a yellowing of the leaf edges; at times the leaf margins turn yellow without the formation of white spots. The edges of the leaves finally dry up and curl under. Potato plants indicate a potassium deficiency by a marginal scorch of the lower leaves; frequently the areas between the veins of potato leaves bulge out, giving a wrinkled appearance. A cucumber starved for potassium grows with a small stem end. Symptoms for soybeans are yellowing of leaf margins.

Calcium and Magnesium

There are many similarities between the behavior of calcium, magnesium, and potassium in soils. They are all available as exchangeable cations, and the amount available is importantly related to mineral weathering and degree of leaching. All three are absorbed as cations, calcium and magnesium as Ca^{+2} and Mg^{+2}. Some important calcium minerals include calcium feldspars, apatite, calcite, dolomite, gypsum, and amphibole. Important magnesium minerals include biotite, dolomite, augite, serpentine, hornblende, and olivine.

The cations set free in weathering are adsorbed on the cation-exchange sites. An equilibrium is established between the exchangeable and solution forms. Diffusion to roots surfaces is the most important process in uptake from soils. In contrast to potassium, there is no significant fixation into unavailable forms.

The amount available is related to cation-exchange capacity and percentage saturation. Calcium deficiency for plants has occasionally been observed on very acid soils with low calcium saturation. Compared to calcium, magnesium is less strongly adsorbed to cation-exchange sites, much less exchangeable magnesium exists in soils, and magnesium deficiencies have been observed much more frequently. Acid sandy soils with less than 75 pounds of exchangeable magnesium per acre furrow slice are likely to be magnesium deficient for corn in Michigan. Some soils formed from parent materials rich in serpentine have abundant magnesium for plant growth, very low calcium saturation, and deficiencies of other nutrients, which produces stunted growth; these soils are referred to as "serpentine barrens."

A deficiency of calcium is characterized by malformation and disintegration of the terminal portion of the plant. Calcium is not readily removed from the older tissues to be used for new growth when a deficiency occurs. The deficiency symptoms have been established for many plants, by the use of greenhouse methods, but they are seldom seen in the field.

Magnesium is a constituent of chlorophyll. As with several other nutrient elements, a deficiency of magnesium results in a characteristic discoloration of the leaves. Sometimes a premature defoliation of the plant results from magnesium deficiency. The chlorosis of tobacco, known as "sand drown," is due to a magnesium deficiency. Cotton plants suffering from a lack of this element produce purplish red

Fig. 12-16 Magnesium deficiency symptoms on corn. The veins are green and the intervein areas are yellow.

leaves with green veins. Leaves of sorghum and corn become striped; the veins remain green, but the areas between the veins become purple in sorghum and yellow in corn (Fig. 12-16). The lower leaves of the plant are affected first. In legumes the deficiency is shown by chlorotic leaves. For coffee, see color plate 4.

Sulfur

Sulfur (S) exists in soil minerals and is immobilized into important plant compounds and eventually accumulates in soil organic matter. Sulfur, similar to phosphorus, is made available in soils by both mineral weathering and mineralization. Plants obtain most of their sulfur from soils as sulfate, (SO_4^{-2}), but some is absorbed through leaves as SO_2. Sulfates are reduced in water-logged soils to hydrogen sulfide (H_2S gas) and elemental sulfur. Many reactions of sulfur are similar to those of nitrogen in that both elements undergo changes in oxidation and reduction, are leached from soils as anions, and significant amounts are added to the soil in precipitation.

When coal and oil are burned, sulfur is released into the atmosphere as SO_2. There are many areas where enough sulfur is washed out of the atmosphere by rain to satisfy plant needs. Many studies have shown a relationship between addition of sulfur to soils in precipitation and closeness to industrial centers. In fact, so much sulfur is added to some soils that all vegetation nearby has been killed.

Areas in the United States where plant response to sulfur has been obtained are shown in Fig. 12-17. Plants with the greatest sulfur need include cabbage, turnips, cauliflower, onions, radishes, and asparagus. Intermediate sulfur users are legumes, cotton, tobacco, and alfalfa. The grasses have the lowest sulfur requirements. Sulfur is mobile in plants and deficiency symptoms are similar to nitrogen. Plants are stunted and light green to yellow in color.

Micronutrients

Micronutrients function largely in plant enzyme systems and are required in very small amounts. The factors that determine the amounts available to plants are importantly related to soil conditions and plant species. Iron, for example, is one of the

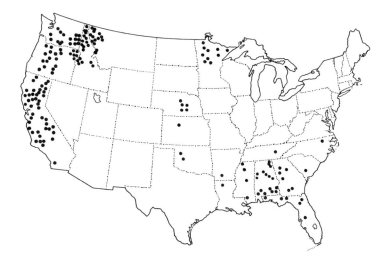

Fig. 12-17 Locations where crops have responded to sulfur fertilizers. (From Jordan and Reisenauer, 1957.)

most abundant elements in soils. Some Oxisols are high in iron content and yet need iron fertilization for profitable production of pineapple. Four of the seven micronutrients are utilized as cations and three as anions. The cations will be considered first.

Iron and Manganese

Iron (Fe) and manganese (Mn) are weathered from minerals and appear as divalent cations in the soil solution; as such, they are available to plants. Reaction with hydroxyl and biological oxidation convert iron and manganese to insoluble forms that are unavailable. Generally, in acid soils, sufficient Fe^{+2} and Mn^{+2} exist in the soil solution to meet the needs of plants. In some very acid soils there are toxic concentrations of iron and manganese. Deficiencies are common in alkaline soils for many plants because of the insolubility of the hydroxide and oxide forms of iron and manganese. Plants particularly susceptible to iron deficiency include roses, pin oak (Fig. 9-8), azalea, rhododendron, and many fruits

and ornamentals. Cereals and grasses, including sugar cane, tend to have a manganese deficiency in alkaline soils.

Deficiency symptoms for iron are striking and are commonly seen on plants growing on calcareous or alkaline soils. Iron-deficient plants have a light-yellow leaf color, which is more evident on the younger leaves. The intervein areas are most affected, and the veins retain a darker color. *Iron-chlorosis* is the name given to this condition, as shown in color plate 2.

In the absence of sufficient manganese, tomato, bean, oat, tobacco, and various other plants are dwarfed. Associated with this dwarfing is a chlorosis of the upper leaves of the plant. The veins, however, remain green (Fig. 12-1, and color plate 3). The "gray speck" of oats has been attributed to a shortage of this element in some soils.

Copper and Zinc

Copper (Cu) and zinc (Zn) are released in weathering as Cu^{+2} and Zn^{+2}, absorbed by plants, and adsorbed on cation-exchange sites. Copper complexes with organic matter, and there is evidence that the complexing can reduce the availability of copper to plants in soils with high organic matter content. Copper availability also decreases with increase in pH. Freshly developed organic soils and leached sandy soils are most likely to be copper deficient for some plants.

Copper is immobile in plants and deficiency symptoms are highly variable from plant to plant. Gum pockets under the bark and twig dieback occur on fruit trees. Older leaf tips become necrotic in small grains.

Zinc deficiency was first discovered on plants growing on organic soils in Florida. Deficiencies are common on calcareous soils where high pH reduces zinc availability and on acid sandy soils where zinc has been leached from the soil. Organic soils are also low in zinc reserves and plants are commonly deficient. In some cases zinc deficiency is caused by high phosphorus fertilization.

Widespread zinc deficiencies occur in the United States on corn, sorghum, citrus and deciduous fruit, beans, vegetable crops, and ornamentals. Pecan rosette, the yellows of walnut trees, the mottle leaf of citrus, the little leaf of the stone fruits and grapes, white bud of corn, and the bronzing of the leaves of tung trees are all ascribed to zinc deficiencies. In tobacco plants a zinc deficiency is characterized by a spotting of the lower leaves; in extreme cases, almost total collapse of the leaf tissue may occur. Zinc deficiency frequently delays the maturity of white beans when grown on alkaline soil in Michigan.

Boron

Most of the boron (B) in soils is in the mineral tourmaline and is released in weathering as the borate ion, BO_3^{-3}. The borate ion (or H_3BO_3) is absorbed by plants, and boron accumulates in soil organic matter. Mineral and organic forms of boron are both important in supplying boron to plants. Dry weather that limits decomposition of organic matter in the surface soil has caused boron deficiency in alfalfa. Fixation at high pH and leaching of boron from acid soils results in maximum availability near pH 7.

Many physiological diseases of plants, such as the internal cork of apples, yellows of alfalfa, top rot of tobacco, cracked stem of celery (color plate 3), and heart rot and internal black spot of beets are associated with a deficiency of boron. In sugar beets, boron deficiency appears as a stunting and curling or twisting of the petioles, associated with a crinkling of the heart leaves. They have unusually dark green and thicker leaves that wilt more rapidly under drought conditions. The older leaves frequently become chlorotic, and rotting of the beets, starting in the crown, may occur. Girdle or canker of table beets occurs as a cracking of the outer skin of the beets near the soil surface, followed by a breakdown of the root tissue.

Chlorine as a Unique Fertilizer Element

The chlorine requirement of crops is very small, even though chlorine may be one of the most abundant anions in plants. It is unique in that there is little probability that it will ever be needed as a fertilizer because of its addition to the atmosphere from ocean spray and, consequently, widespread addition of it to soils in precipitation. The presence of chlorine in most of the potassium fertilizer used today is another important manner in which chlorine is added to soils.

Molybdenum

Weathering of minerals releases molybdenum (Mo), and it is probably absorbed by plants as the molybdate ion, MoO_4^{-2}. Molybdenum accumulates in soil organic matter and is adsorbed as an anion by the clay fraction. In soils with low pH, molybdenum is fixed by reaction with iron. Liming acid soils has commonly removed plant molybdenum deficiencies. Some calcareous soils in the western United States developed from high molybdenum parent materials, and high levels of available molybdenum result in toxic concentrations in forage for grazing animals.

This element is needed for nitrogen fixation in legumes; when it is deficient, legumes show symptoms of nitrogen deficiency. When molybdenum causes other metabolic disturbances in the plant, other symptoms are seen. A deficiency of molybdenum in cauliflower causes a cupping of the leaves. It is caused by the reduced rate of expansion near the leaf margin compared to that in the center of the leaf. The leaves also tend to be long and slender, giving rise to the symptom called "whiptail." Interveinal chlorosis, stunting of the plant, and general paleness are also exhibited, depending on the kind of plant.

References

Allison, Franklin E., "Nitrogen and Soil Fertility," in *Soil*, USDA Yearbook, Washington, D.C., 1957, pp. 85–94.

Attoe, O. J., and E. Truog, "Exchangeable and Acid-Soluble Potassium as Regards Availability and Reciprocal Relationships," *Soil Sci. Soc. Am. Proc.*, 10:81–86, 1945.

Barber, S. A., "A Diffusion and Mass Flow Concept of Soil Nutrient Availability," *Soil Sci.*, 93:39–49, 1962.

Black, C. A., *Soil-Plant Relations*, Wiley, New York, 1968.

Cook, R. L., and C. E. Millar, "Plant Nutrient Deficiencies," *Mich. Agr. Exp. Sta. Spec. Bull.*, 353, 1953.

Delwiche, C. C., "The Nitrogen Cycle," *Sci. Am.*, September 1970, pp. 137–146.

Hanson, J. B., "Roots, Selectors of Plants Nutrients," *Plant Food Rev.*, Spring 1967.

Hodges, T. K., "Ion Absorption by Plant Roots," *Advances in Agronomy*, 25:163–207, 1973.

Jordan, Howard V., and H. M. Reisenauer, "Sulfur and Soil Fertility," in *Soil*, USDA Yearbook, Washington, D.C., 1957, pp. 107–111.

Lawton, Kirk, "The Influence of Soil Aeration on the Growth and Absorption of Nutrients by Corn Plants," *Soil Sci. Soc. Am. Proc.*, 10:263–268, 1945.

Lucas, R. E., and Bernard D. Knezek, "Climatic and Soil Conditions Promoting Micronutrient Deficiencies in Plants," in *Micronutrients in Agr.*, Soil Sci. Soc. Am., Madison, Wis., 1972, pp. 265–288.

Marx, Jean L., "Nitrogen Fixation in Maize," *Science*, 189:368, 1975.

Mortland, M. M., "Kinetics of Potassium Release from Biotite," *Soil Sci. Soc. Am. Proc.*, 22:503–508, 1958.

Oliver, S., and S. A. Barber, "An Evaluation of the Mechanisms Governing the Supply of Ca, Mg, K, and Na to Soybean Roots," *Soil Sci. Soc. Am. Proc.*, 30:82–86, 1966.

Olsen, Sterling R., and Maurice Fried, "Soil Phosphorus and Fertility," in *Soil*, USDA Yearbook, Washington, D.C., 1957, pp. 94–100.

Ray, Peter Martin, *The Living Plant*, Holt, Rinehart and Winston, New York, 1972.

Richards, B. N., *Introduction to the Soil Ecosystem*, Longman, New York, 1974.

Smith, Rex L., et al., "Nitrogen Fixation in Grasses Inoculated with *Spirillum* lipoferum," *Science*, 193:1003–1005, 1976.

Stevenson, F. J., "Origin and Distribution of Nitrogen in Soil," in *Soil Nitrogen*, Agronomy 10, Am. Soc. Agron., Madison, Wis., 1965, pp. 1–42.

Struever, Stuart, "The Koster Expedition and the New Archeology," *The Science Teacher*, *42*:26–32, 1975.

Subbaroa, Y. V., and Roscoe Ellis, Jr., "Reaction Products of Polyphosphates and Orthophosphates with Soils and Influence on Uptake of Phosphorus by Plants," *Soil Sci. Soc. Am. Proc.*, *39*:1085–1088, 1975.

Viets, Frank G., "The Plant's Need for and Use of Nitrogen," in *Soil Nitrogen*, 10, Am. Soc. Agron., Madison, Wis., 1965, pp. 503–549.

Viets, F. G., C. E. Nelson, and C. L. Crawford, "The Relationship Among Corn Yields, Leaf Composition and Fertilizer Applied," *Soil Sci. Soc. Am. Proc.*, *18*:297–301, 1954.

Walker, Richard B., "Factors Affecting Plant Growth on Serpentine Soils," *Ecology*, *35*:259–266, 1954.

Weber, C. R., "Nodulating and Nonnodulating Isolines," *Agron. Jour.*, *58*:43–46, 1966.

Youngberg, C. T., and A. G. Wollum II, "Nonleguminous Symbiotic Nitrogen Fixation," in *Tree Growth and Forest Soils*, Proc. of 3rd North Am. Forest Soils Conf., Corvallis, Ore., 1968.

13

COMPOSITION, MANUFACTURE, AND USE OF FERTILIZERS

The history of humans, in one sense, has been a record of efforts to increase the food supply by increasing the available nutrient supply. For thousands of years people have used lime, marl, ashes, bones, manures, mud, and legumes to add nutrients to soils. In Chapter 7 the use of trees for nutrient accumulation under shifting cultivation was discussed. Rapid progress in the development of chemical fertilizers occurred after the discovery of the major essential plant nutrients a little over a century ago. Now, it has been estimated that at least a fourth of our total food supply can be attributed to the use of chemical fertilizer. The present importance of fertilizers has been well summarized by Lamer who wrote,

Under present economic and political conditions, in all countries of the world, fertilizers are one of the most important strategic weapons of modern agriculture. Agriculture history has passed through various stages in its development; at present, it is in the fertilizer epoch.

This chapter presents information for gaining an understanding of the nature, manufacture, and efficient use of fertilizers—one of the most important means of increasing the food supply in a world where the number of underfed people increases daily.

Fertilizer Materials

Fertilizers, in a broad sense, include all materials that are added to soils to supply certain elements essential to the growth of plants. However, the term *fertilizer* usually refers to manufactured fertilizers. Fertilizers do not contain plant nutrients in elemental form as nitrogen, phosphorus, or potassium, but the nutrients exist in compounds that provide the ionic forms of nutrients that plants can absorb. The major kinds of nitrogen fertilizers will be discussed first.

Nitrogen Fertilizer Materials

The atmosphere is about 79 percent nitrogen and, as noted in Chapter 12, there are about 34,500 tons of nitrogen over every acre. Yet it is ironic that our food supply is more limited by a lack of nitrogen than any other plant nutrient. The major source of essentially all industrial nitrogen (including fertilizer nitrogen) results from the fixation of atmospheric nitrogen, according to the following generalized reaction.

$$N_2 + 3H_2 \xrightarrow[\text{pressure and catalysts}]{\text{proper temperature}}$$

$$2NH_3 \text{ (anhydrous ammonia)} \qquad (1)$$

The source of hydrogen is usually natural gas (CH_4). A process flow chart of ammonia synthesis is presented in Fig. 13-1.

Ammonia (called anhydrous ammonia in the fertilizer trade) is the most concentrated nitrogen fertilizer, being 82 percent nitrogen. At normal temperature and pressure anhydrous ammonia is a gas and is stored and transported as a liquid under pressure. After direct application of NH_3 to the soil, the NH_3 absorbs a hydrogen ion and is converted to NH_4^+, a stable ammonium ion.

About 98 percent of the nitrogen fertilizer produced in the world is ammonia or one of its derivatives. The manufacture of ammonia and five of its derivatives is shown diagrammatically in Fig. 13-2, and a brief summary of some properties of nitrogen fertilizers is given in Table 13-1.

Frequent attempts have been made to determine the relative efficiency of nitrogen fertilizers by applying equal quantities of nitrogen per acre, in the various materials, for a given crop. Since so many factors, such as temperature and moisture conditions, soil reaction, leaching, kind of crop, and time and method of application, affect the action of any nitrogen fertilizer, relative fertilizing values so obtained may be misleading.

Nitrogen, in the nitrate form, is readily soluble in water and is quickly used by most crops. Nitrate is easily leached from the soil by rains because of its high solubility and because nitrate is not held in the soil to any appreciable extent. Although the ammonium form of nitrogen is soluble in water, it is not leached out as readily as nitrate because large quantities of ammonium can be adsorbed and held by the cation-exchange complex. Plants can use ammoniacal nitrogen, but most of it is converted to nitrates as a result of nitrification before plants take it up. Consequently, nitrate and ammonium forms of nitrogen are of about equal value for most crops and most situations.

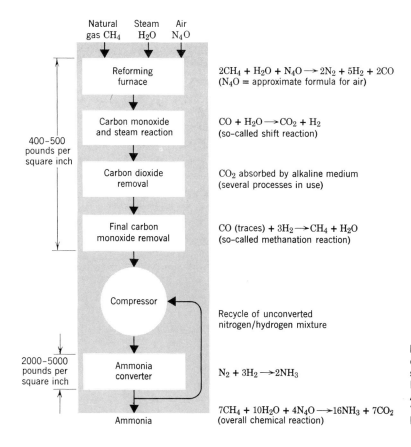

$$2CH_4 + H_2O + N_4O \longrightarrow 2N_2 + 5H_2 + 2CO$$
(N_4O = approximate formula for air)

$$CO + H_2O \longrightarrow CO_2 + H_2$$
(so-called shift reaction)

CO_2 absorbed by alkaline medium
(several processes in use)

$$CO \text{ (traces)} + 3H_2 \longrightarrow CH_4 + H_2O$$
(so-called methanation reaction)

Recycle of unconverted
nitrogen/hydrogen mixture

$$N_2 + 3H_2 \longrightarrow 2NH_3$$

$$7CH_4 + 10H_2O + 4N_4O \longrightarrow 16NH_3 + 7CO_2$$
(overall chemical reaction)

Fig. 13-1 Process flow chart of ammonia synthesis. (From President's Science Advisory Committee, "The World Food Problem," May 1967.)

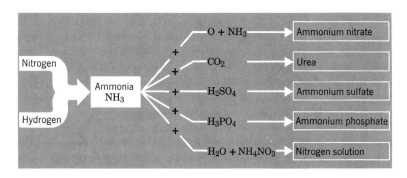

Fig. 13-2 Diagrammatic representation of the manufacture of ammonia and some of its derivatives.

Table 13-1
The Principal Fertilizers Supplying Nitrogen

Nitrogen Carrier	Nitrogen, %	Remarks
Anhydrous ammonia	82	Used directly, or for ammoniation, nitrogen solutions, etc.
Ammonium nitrate	33	Conditioned to resist absorption of water
Ammonia liquor (aqua ammonia)	24	Formed when ammonia is absorbed in water
Nitrogen solutions	Variable	Many kinds formed from solutions of aqua ammonium, ammonium nitrate, urea, etc.
Urea	45	$CO(NH_2)_2$, is hydrolyzed to ammonium in soils
Urea-formaldehyde	35–40	Contains insoluble, slowly available nitrogen
Ammonium sulfate	20	Also produced as a by-product of the coking of coal.
Sodium nitrate	16	Also a naturally occurring salt in Chile
Sulfur-coated urea	39	Slow release, also contains 10 percent sulfur

Urea hydrolyzes rapidly in warm, moist soils to form ammonium carbonate. The ammonium may be used directly by plants or may be converted to nitrate and then used as nitrate. Urea-formaldehyde is one of the more recently developed nitrogen fertilizers and is not water soluble. The nitrogen in urea-formaldehyde is released in available form slowly to provide a continual supply of nitrogen through the growing season. Its high cost greatly limits its use.

Sulfur-coated urea shows good potential for the slow release of nitrogen for crops such as sugar cane and pineapples, which take up to 2 years to mature. Granular urea is sprayed with liquid sulfur and coated with a wax microbiocide. The slow release of nitrogen reduces leaching losses and results in less frequent applications and more efficient use.

Phosphorus Fertilizer Materials

Historically, bones served as a source of phosphorus, yet as recently as 1840 the value of bones as a fertilizer was found to result largely from their phosphorus content. At about this same time, Liebig suggested that bones be treated with sulfuric acid to increase the solubility or availability of the phosphorus. This marked the beginning of the modern fertilizer industry because it led, in 1842, to the patenting of a process for the manufacture of superphosphate by the treatment of mineral rock phosphate with sulfuric acid.

Huge deposits of rock phosphate ore are located in Florida east of Tampa, where 73 percent of the phosphate ore in the United States is mined. The deposits were formed 10 to 15 million years ago and are buried under about 20 to 40 feet of sand (Fig. 13-3). The rock phosphate is in a matrix of sand and clay. Mining consists of removal of the overburden and use of an hydraulic gun to break up the ore to form a slurry (see Fig. 13-4). The slurry is pumped to a recovery plant where the rock phosphate pebbles are separated from the sand and clay by a washing and screening process. The rock phosphate pebbles are ground to form a powder called rock phosphate that is sometimes used directly as a fertilizer. Rock phosphate is very insoluble, so the phosphorus has low availability for plant growth.

The only important source of mineral phosphate used to manufacture fertilizers is rock phosphate. The production of ordinary superphosphate by the acidulation of rock phosphate with sulfuric acid is shown in Equation 2.

$$3[Ca_3(PO_4)_2] \cdot CaF_2 + 7H_2SO_4 \rightarrow$$

rock phosphate   sulfuric acid   (2)

$$3Ca(H_2PO_4)_2 + 7CaSO_4 + 2HF$$

monocalcium phosphate   gypsum   hydrogen fluoride

The ordinary superphosphate produced by reaction 2 consists of about half

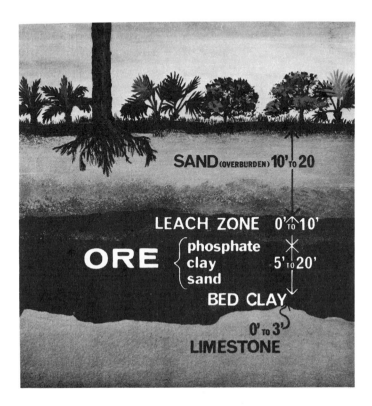

Fig. 13-3 Location and nature of Florida rock phosphate deposits. (By permission of the Florida Phosphate Council, Inc.)

Fig. 13-4 In mining rock phosphate the overburden is removed and the ore matrix is dumped into a "sump" where a hydraulic gun forms a slurry that is pumped to the recovery plant. (Photo courtesy International Minerals and Chemical Corporation.)

monocalcium phosphate and about half gypsum. The hydrogen fluoride can be recovered and, in some cases, is used to fluorinate water. Ordinary superphosphate has a phosphorus content of about 9 percent phosphorus or 20 percent P_2O_5.

Under proper conditions the reaction of rock phosphate with sulfuric acid will yield phosphoric acid. By treating rock phosphate with phosphoric acid, a more concentrated superphosphate can be produced as follows.

$$\text{Rock phosphate} + 14H_3PO_4 \rightarrow$$

$$10Ca(H_2PO_4)_2 + 2HF \qquad (3)$$

The same phosphorus compound is produced with sulfuric acid and phosphoric acid, but without the production of any gypsum when phosphoric acid is used. This more concentrated superphosphate contains about 20 percent phosphorus or the equivalent of 45 percent P_2O_5. Concentrated superphosphate is commonly called triple superphosphate. Both types of superphosphates are of about equal quality as fertilizers when the same amount of phosphorus is applied.

Ammonium phosphates are produced by neutralizing phosphoric acid with ammonia. The two popular kinds produced are monoammonium phosphate and diammonium phosphate. Ammonium phosphates are good sources of both phosphorus and nitrogen, the phosphorus being water soluble. Some ammonium phosphate is produced as a by-product of the coking industry by using the ammonia produced in the coking of coal to neutralize phosphoric acid.

Basic slag, sometimes referred to as Thomas phosphate, is produced as a by-product of the iron industry. Phosphorus is contained in certain iron ores, and steel made from them is brittle if most of the phosphorus is not removed. The slag is produced by oxidizing the phosphorus in molten iron by a blast of air blown through it. The molten iron is contained in a converter lined with lime, and the oxidized phosphorus combines with this lime. The slag so produced rises to the surface and is drawn off, cooled, and ground so finely that most of it will pass a 100-mesh screen. The phosphorus in slag is soluble in citric acid and is considered available to crops. A summary of these and other phosphate materials appears in Table 13-2.

Potassium Fertilizer Materials

Settlement of the eastern seaboard of the United States was stimulated in the early 1600s by a need in England for several important products, including woodashes or potash. To obtain potash, the ashes of trees were leached with water to remove the potassium compounds (K_2CO_3), dried and then calcined to produce potassium oxide. Later, huge mineral deposits of potassium salts were discovered in Germany, and Germany supplied the world market until after World War I. During the 1920s the Carlsbad, New Mexico potash mines were opened and, more recently, in 1959 the

Table 13-2

The Principal Phosphatic Materials

Material	Percentage Available		Remarks
	P_2O_5	P	
Rock phosphate	25–35[a]	11.0–15.4[a]	Effectiveness depends on degree of fineness, soil conditions, and crop grown
Superphosphate (ordinary)	20	8.7	Made by treating ground phosphate rock with H_2SO_4
Triple super-phosphate	46	20	Made by treating ground phosphate rock with liquid H_3PO_4
Monoammonium phosphate	48	21	Made by neutralizing H_3PO_4 with ammonia
Diammonium phosphate	53	23	Made by neutralizing H_3PO_4 with ammonia
Basic slag	5–20	2.2–8.8	By-product obtained in the manufacture of steel
Bone meals	17–30[a]	7.5–13.2[a]	Includes raw as well as steamed bone meals
Colloidal phosphate	18–23[a]	7.9–10.1[a]	A finely divided, relatively low-grade rock phosphate or phosphatic clay

[a] Total phosphoric acid instead of amount that is available.

Fig. 13-5 A continuous mining machine at work more than half a mile underground at Esterhazy, Sask., Canada. (Courtesy of American Potash Institute and International Minerals and Chemical Company.)

mining of potash was begun in Saskatchewan, Canada (see Fig. 13-5).

Common minerals in the potash deposits include sylvite (KCl), sylvinite (mixture of KCl and NaCl), kainite ($MgSO_4KCl \cdot 3H_2O$), and langbeinite ($K_2SO_4 \cdot 2MgSO_4$). Processing of the ore consists of separating KCl from the other products in the ore. The flotation method is commonly used to separate KCl from the ore mixture. The ore is ground, suspended in water, and treated with a flotation agent that adheres to the KCl crystals. As air is passed through the suspension, the KCl crystals float to the top and are skimmed off (Fig. 13-6). After further purification, the nearly pure KCl is dried and screened for particle size. The fertilizer material is called muriate of potash (KCl) and is about 60 percent K_2O.

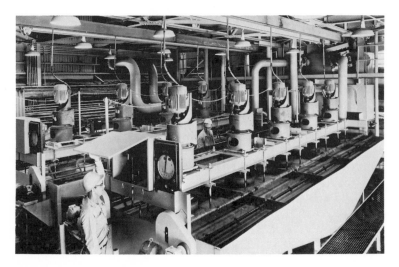

Fig. 13-6 A flotation plant for the recovery of potassium minerals. (Courtesy of American Potash Institute and the International Minerals and Chemical Company.)

Although the great bulk of the potassium comes from mines and about 95 percent of the potassium fertilizer is KCl, several other sources and materials are worthy of mention. Some potassium is obtained from brine lakes at Searles Lake in California and Salduro Marsh in Utah. If needed, the sea represents an "endless" supply of potassium, since each cubic mile of sea water contains about 1.6 million tons of potassium. The second most widely used potassium fertilizer, although used to only a minor extent, is K_2SO_4, which is about 40 percent potassium or 48 percent K_2O. A very small amount of KNO_3 is also produced.

All the potash fertilizer salts are soluble in water and are considered readily available. In general, it can be said that there is very little difference in their effects on crop production except in rather special cases. Discrimination is sometimes made against carriers having a high percentage of chlorine for special crops like potatoes and tobacco. The sulfate (or carbonate if available) is usually preferred for tobacco, especially where large amounts are to be added because a crop of superior burning quality is produced. The muriate is just as efficient as the sulfate for potato production, according to most of the experimental evidence.

Materials Containing Micronutrients

Although the bulk of fertilizer consists of carriers containing nitrogen, phosphorus, and/or potassium, micronutrients are sometimes added to fertilizers to supply one or more of the micronutrients. Some common micronutrient carriers are listed in Table 13-3.

Mixed Fertilizers

Soils vary greatly in their ability to supply crops with available nutrients, and the mineral requirements of different crops are also quite variable. In order to supply nutrient deficiencies in soils and to meet the various requirements of different crops, fertilizers containing two or more essential elements are prepared in many different grades. They are known as *mixed fertilizers* and are made by mixing two or more of the separate fertilizer carriers.

Preparation of Mixed Fertilizers

The preparation of mixed fertilizers can be a relatively simple operation, especially if the mixture is to be of a low grade (i.e., a fertilizer containing a comparatively low percentage of nutrients). It consists essentially in mixing suitable materials in the correct proportion to give the desired grade or analysis. There are many bulk

Table 13-3
Some Common Micronutrient Carriers

Carrier	Nutrient Composition
Borax ($Na_2B_4O_7 \cdot 10H_2O$)	11% B
Copper sulfate ($CuSO_4 \cdot 5H_2O$)	25% Cu
Ferrous-sulfate ($FeSO_4 \cdot 7H_2O$)	20% Fe
Manganese oxide (MnO)	48% Mn
Manganese sulfate (variable hydration)	23–25% Mn
Zinc sulfate ($ZnSO_4 \cdot 7H_2O$)	35% Zn

mixing plants in the United States that follow this simple procedure. On the other hand, many of the larger fertilizer factories follow much more involved processes that require careful chemical and temperature control, particularly in the preparation of some of the carriers of materials used in making the final mix.

After the acidulation of rock phosphate in the manufacture of superphosphate, the product is allowed to stand in a large pile for a considerable time to "cure" before being ground. If these phosphates are not properly "cured," the chemical reactions involved are not completed and the fertilizer made from them will harden in the bags. Mixtures of superphosphate and potash carriers are often made and stored for a considerable time before being used in the final fertilizer mixture.

Fertilizers in storage tend to assume an unfavorable mechanical condition, chiefly the result of "setting up," which is essentially a cementation, as in plaster of paris. It may also be due to the surface tension effects of moisture, which forms films around the particles of fertilizer.

The drilling qualities of certain mixed fertilizers made from fertilizer salts that take up moisture readily can be appreciably increased by adding an organic filler such as muck or diatomaceous earth. Fertilizers are commonly formulated today so that the only material in them classified as filler is what is needed to insure a good physical condition.

The Fertilizer Grade or Guaranteed Analysis

The fertilizer grade or guaranteed analysis expresses the nutrient content. In all states the grade is expressed in the order $N-P_2O_5-K_2O$. The total nitrogen is expressed as elemental nitrogen (N); the phosphorus is expressed as available P_2O_5; and the potassium is given as water-soluble K_2O.

Considerable interest has developed in recent years in changing the basis for expressing the fertilizer grade. In 1955 the Soil Science Society of America passed a resolution favoring a change from the $N-P_2O_5-K_2O$ basis for expression to the $N-P-K$ or elemental basis. They outlined the following advantages for such a change. First, greater uniformity would exist in the expression of the nutrient content of fertilizers, of soils, and of plants. Plant composition in present-day literature is usually expressed on the elemental basis. The same is true in the field of animal nutrition and biochemistry, where feeding standards are usually expressed in terms of elemental amounts of phosphorus and calcium. Second, it would result in greater simplicity. The elemental basis requires fewer symbols for expression. It is a common practice to refer to $N-P-K$ fertilizer but to refer to the calculations of the amounts of plant nutrients on the $N-P_2O_5-K_2O$ basis. This leads to confusion and creates an obstacle to learning in education programs for farmers. Third, accuracy of expression of the true ratio of the major nutrient elements in a fertilizer requires that the elemental basis be used. Reference to a 10—20—10 fertilizer tends to lead one to believe that the ratio of $N-P-K$ is 1—2—1. The ratio of $N-P_2O_5-K_2O$ is 1—2—1, but the actual $N-P-K$ ratio is 1.0—0.88—0.83. A scale for converting P_2O_5 to P and K_2O to K is given in Fig. 13-7.

Calculation of Fertilizer Formulas

In calculating formulas for mixtures it is necessary to decide first what percentages of nitrogen, available phosphoric acid, and water-soluble potash are desired in the fer-

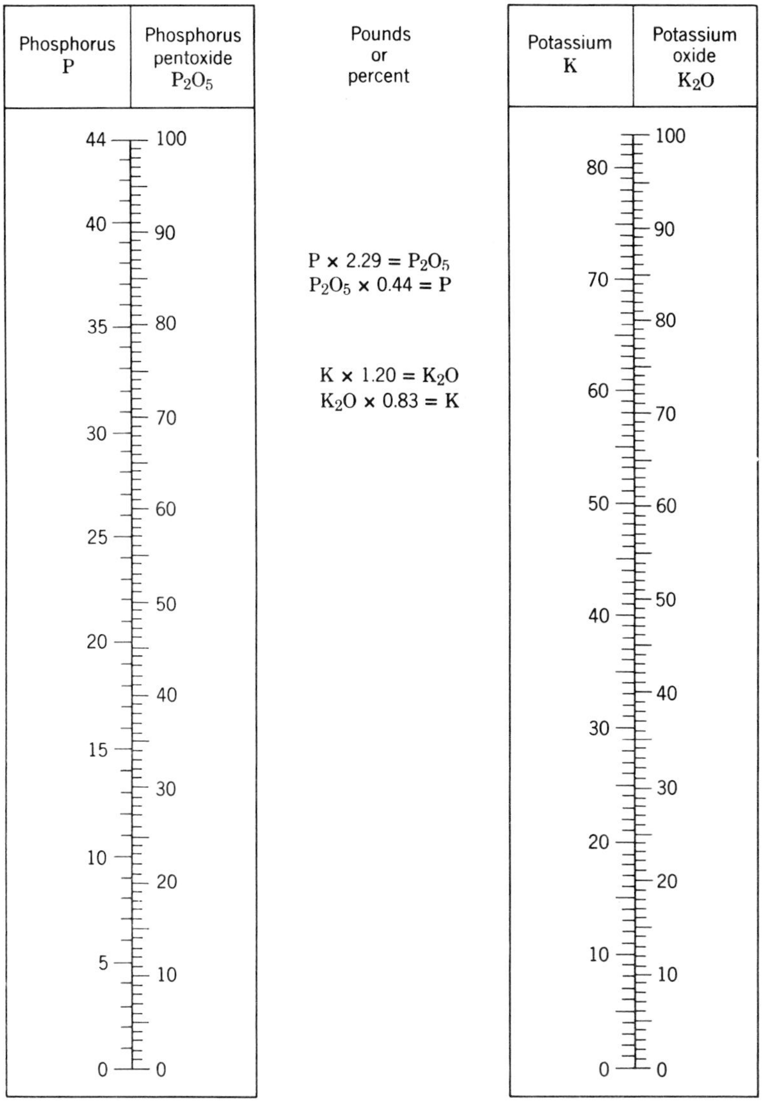

Phosphorus P	Phosphorus pentoxide P_2O_5	Pounds or percent	Potassium K	Potassium oxide K_2O

$$P \times 2.29 = P_2O_5$$
$$P_2O_5 \times 0.44 = P$$

$$K \times 1.20 = K_2O$$
$$K_2O \times 0.83 = K$$

Fig. 13-7 Fertilizer conversion scales for phosphorus and potassium.

tilizer mixture and then what materials are to be used in making the mixture.

For example, make 1 ton of a 6—12—12 fertilizer using the following ingredients.

Ammonium nitrate
  33 percent nitrogen
Superphosphate
  20 percent available phosphoric acid

Muriate of potash
  60 percent water-soluble potash

The problem is to find out how much of each of these materials is needed. This may be done by use of the equation

$$X = \frac{A \cdot B}{C}$$

in which X equals pounds of carrier required, A equals pounds of mixed fertilizer required and, with nitrogen, B equals the percentage of nitrogen desired in the mixture, and C equals the percentage of nitrogen in the carrier (ammonium nitrate). By substituting these values in the above equation, the result is easily determined.

$$X = \frac{2000 \times 6}{33}$$

$X = 364$ pounds, the amount
    of ammonium nitrate required

Similarly, the required amounts of superphosphate and muriate of potash may be determined.

$$X = \frac{2000 \times 12}{20}$$

$X = 1200$ pounds, the weight
    of superphosphate required

$$X = \frac{2000 \times 12}{60}$$

$X = 400$ pounds, the weight of muriate of
    potash required

The total amount of materials used in this fertilizer mixture ($364 + 1200 + 400$) equals 1964 pounds. It is necessary to add 36 pounds of filler or physical conditioner to make a ton of the required mixture.

Liquid Fertilizer

A liquid fertilizer is simply a solution containing one or more water-soluble forms of nutrients. Materials similar to those used in the manufacture of liquid fertilizers have been added to soils for many years by dissolving them in irrigation water and as components of conventional dry fertilizers.

Advantages of liquid fertilizers over the dry fertilizers include (1) the saving of labor in handling where pumps and pipes can be used, (2) their convenience as foliar sprays, and (3) the convenience of adding pesticides. Disadvantages include (1) the increased fixation of the phosphorus, especially in mixed rather than banded application, (2) corrosion of metal containers and equipment, and (3) the need for special equipment for storage and application. A liquid-fertilizer attachment on a four-row planter is shown in Fig. 13-8.

A popular method used to manufacture liquid fertilizer consists of the neutralization of phosphoric acid with ammonia. Fertilizers with various ratios of nitrogen and phosphorus can be manufactured by varying the degree of neutralization. Potassium can be added to the fertilizer by the addition of KCl. The amount of potassium that can be added, without causing "salting out," is limited and makes the development of high-potash liquid fertilizers impractical.

The time of application and the placement of liquid fertilizer follow closely the principles that apply to the dry forms.

General Nature of Fertilizer Laws

In general, the nature of the laws controlling fertilizer sales in the various states is similar. They all require periodical registration of brands or analyses offered for sale and accurate labeling of the bags or packages. Most of the states require that the following information be printed on each fertilizer bag, or on an attached tag.

1. Name, brand, or trademark.
2. Analysis (guarantee) or chemical composition.

EVALUATION OF FERTILIZER NEEDS

Fig. 13-8 Four-row planter with liquid-fertilizer attachment. (Courtesy of John Deere, Moline, Ill.)

3. Net weight of fertilizer.

4. Name and address of manufacturer.

Fertilizers offered for sale may be sampled by inspectors any time during the year and at any point in the particular state. The samples are sent to the control laboratory and are analyzed to determine whether the goods are up to the guarantee. The results are checked against the guaranteed analysis; by this means the purchaser is protected from loss through the activities of unreliable companies. The inspection and analysis may be in the hands of the state department of agriculture, of the state agricultural experiment station, or of a state chemist. Once a year, in most states, the findings of the chemists are published in bulletin form, and copies of this bulletin may be obtained on request by any interested person.

Evaluation of Fertilizer Needs

The natural forests and grasslands that existed before the landscape was modified by agricultural pursuits is evidence that essentially all of the soils of the world are capable of supporting plants, provided the climate is favorable. Several important factors account for this situation. There is great diversity in terms of plant needs and tolerances to toxic elements. Nutrient recycling under natural conditions results in repeated use of the nutrients for plant growth. In the natural ecosystem there may also be abundant vegetation, which is the product of slow growth over many years. When people establish agriculture, the need for additional nutrients is inevitable if yields are to be maintained above a meager annual yield of 8 to 10 bushels of grain per acre. Cultivated crops also have different nutrient requirements than natural plants, and many of the nutrients are removed from the land by the harvesting of crops. Therefore, a need is created for evaluating the kind and amount of fertilizer to use, since unnecessary fertilizer is costly and the wrong fertilizer may be harmful.

Deficiency Symptoms and Tissue Tests for Evaluating Fertilizer Needs

Plants, like animals, exhibit peculiar symptoms that are associated with particular nutrient deficiencies. Color photographs of some of these are found in Chapter 12. When the sap of a plant showing a deficiency symptom is tested, a low or deficient amount of the deficient nutrient is usually found. In this way, deficiency symptoms and rapid tissue tests can be used to diagnose plant growth problems. By the time deficiency symptoms appear on a crop, however, the yield potential may be greatly reduced. In addition, tissue tests do not provide information on the quantity of fertilizer to apply. For these reasons, soil tests provide the soundest basis for evaluating the fertilizer needs of most field and vegetable crops.

Soil Tests for Fertilizer Recommendations

Soil testing is based on the concept that a crop's response (growth) to fertilizer will be related to the amount of *available* nutrients in the soil. For example, the exchangeable potassium is generally a good indication of the available potassium. The more exchangeable potassium there is in a soil, the less plants will respond to potassium fertilizer (Fig. 13-9).

Considerable experimental work is needed to be able to make fertilizer recom-

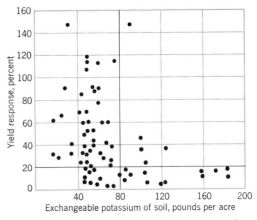

Fig. 13-9 Relation of alfalfa yield to amount of exchangeable potassium of several New York soils. Forty-four out of 57 soils gave a yield increase of 20 percent or more when the exchangeable potassium was less than 80 pounds per acre. (Data of Chandler, et al., 1945.)

mendations for specific situations. Fertilizer recommendations must take into account the different nutrient needs of crops, types of soil, and climatic conditions. In the United States the state agricultural experiment stations have taken the responsibility of working on the crops, soil conditions, and climatic variations that are locally important. The experimental work to develop fertilizer recommendations never ends, because new crop varieties are developed, soil-testing techniques are continually improved, and crop-yield expectations are generally increasing. Periodically, the latest research results are integrated with previous knowledge and are used to construct reference tables for making fertilizer recommendations.

The fertilizer recommendations in Table 13-4 take into account five important factors. These are soil test levels (in this case, pounds of exchangeable potassium per acre), type of crop, yield expectation, type of soil texture, and the climate (since the

Table 13-4
Potash—Potassium Recommendations for Field Crops on Loams, Clay Loams and Clays in Michigan

Available Soil Potassium—Pounds of Potassium per Acre				Pounds per Acre Annually Recommended	
				K₂O	Potassium
0–59	0–59	0–59	0–119	300	249
60–119	60–119	60–119	120–169	200	166
120–159	120–159	120–169	170–209	150	125
160–209	160–199	170–209	210–239	100	83
210+	200–239	210–249	240–279	60	50
	240+	250–269	280–299	30	25
		270+	300+	0	0
Barley 40–69 bu	Alfalfa 3–4T seeding	Alfalfa 5–6T seeding	Alfalfa 7+T topdressing		
Buckwheat	Alfalfa 3–4T topdressing	Alfalfa 5–6T topdressing	Potatoes 300–500 cwt		
Corn 60–89 bu	Birdsfoot Trefoil	Corn 150+bu			
Cover crops	Corn 90–119 bu	Corn silage 20–30T			
Field beans 15–29 bu (9–18 cwt)	Corn silage 10–14T	Potatoes 150–299 cwt			
Grass pasture (unimproved) 30–50 bu (19–30 cwt)	Field beans 30–50 bu	Sugar beets 24–28T			
Kidney beans	Corn silage 15–19T				
Grasses 30–50 bu (19–30 cwt)	Sugar beets 18–23T				
Timothy, Orchard, Brome	Wheat 65+bu				
Oats 80–120 bu					
Kidney beans 15–29 bu (9–18 cwt)					
Soybeans 40+bu					
Sorghum					
Millet					
Oats					
Sudangrass					
Wheat 50–79 bu					
Rye 40–65 bu					
Soybeans 20–40 bu					
Wheat 25–39 bu					

table is applicable for Michigan). For example, the potassium recommendation for sugar beets expected to yield 24 to 28 tons per acre growing on soils with loam or finer texture that have 190 pounds of exchangeable potassium per acre furrow slice is 100 pounds of potassium fertilizer expressed as K_2O or 83 pounds expressed as potassium.

One of the weakest steps in getting good fertilizer recommendations is obtaining a representative sample. A single soil sample of 1 pint in size, when used to represent 5 to 10 acres, means that 1 pint of soil is used to characterize 5000 to 10,000 tons of soil. Persons interested in testing soil should consult a testing lab for proper sampling procedures and care of samples after they have been collected.

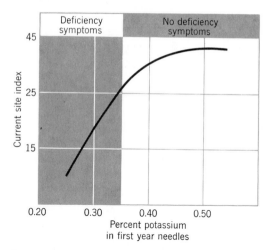

Fig. 13-10 Relationship between potassium content of needles and plant height (site index) and deficiency symptoms of Red Pine (*Pinus resinosa*). (From Stone, et al., 1958.)

Foliar Analysis for Evaluating Fertilizer Needs of Tree Crops

There are several conditions that limit the effectiveness of soil tests for making fertilizer recommendations for tree crops. The roots of trees are distributed through such a large volume of soil that sampling the upper soil layer or plow layer is of limited usefulness. Furthermore, trees conserve nutrients by the movement of nutrients from the leaves into the wood late in the year before the leaves fall. Deep roots of trees in unfrozen soil in cold winters may accumulate nutrients all winter long. Since the trees produce over many years, it is possible to evaluate the nutritional status by foliar analysis and find success by adding fertilizers. A good relationship has been found between potassium content of pine needles, plant height (site index), and potassium deficiency symptoms, as shown in Fig. 13-10 for sandy soils low in available potassium. Deficiency symptoms usually occurred when the potassium content of

the needles was less than 0.35 percent.

As with soil tests, foliar analysis must be correlated with yield response before good recommendations can be made. In many cases this research has been carried out and foliar analysis has become an important tool for horticulturists and foresters. Researchers have also determined the proper part of the plant to select for testing. Persons interested in use of foliar analysis should obtain instructions for sampling from a testing laboratory.

Fertilizer Nutrient Interactions

Frequently, the plant response to a fertilizer nutrient is affected by the use of some other fertilizer nutrient or the level of some other soil nutrient. If nitrogen fertilizer is applied and plant growth is stimulated, this increases the demand for all other plant nutrients. It is not uncommon to find that the response to one nutrient is related to the level of some other nutrient in the soil.

An interesting nitrogen and sulfur interaction exists for some crops in the rural areas of the southeastern United States. Several factors are involved that affect the supply of sulfur in the soil and the ability of plants to increase their growth when nitrogen fertilizer is used. In the rural areas only small amounts of sulfur are added to soils in precipitation (near cities rain washes out sulfur from industrial gases). In recent years ordinary superphosphate, which contains about half calcium sulfate, has been more and more often replaced by triple superphosphate, which does not contain sulfur. Consequently, the effectiveness of nitrogen fertilizer is limited in many cases by an insufficient supply of soil sulfur.

An interesting phosphorus-zinc interaction is shown in Fig. 13-11. Previous use of phosphorus fertilizer was associated with zinc deficiencies because the phosphorus reduced the plants' ability to utilize soil zinc. Interactions represent a complicating factor in making fertilizer recommendations and must be considered.

Use and Application of Fertilizers

For the farmer, fertilizer use is a matter of profit. As a result, little or no fertilizer is used on many of the world's most infertile soils, although considerable amounts of fertilizer are used on many of the world's best agricultural soils. Regardless of the situation, once a fertilizer recommendation has been made and fertilizer is to be used, the problem of proper use arises. This section considers several items that are important in efficient use of fertilizers.

Efficiency in the Use of Fertilizers

Crops grown on most soils, particularly in the humid regions, respond favorably to the use of well-chosen fertilizers. Fertilizers should not be expected to make up for every shortcoming of the soil and crop, such as poor seed, unadapted varieties of crops, unfavorable weather conditions, poor tillage practices, weeds, poor drainage, bad physical condition of the soil, low organic content, or insufficient lime. All

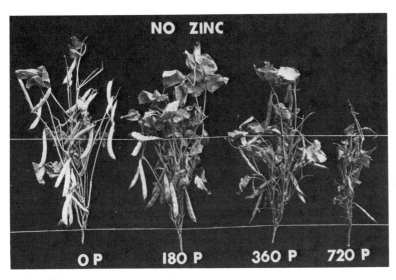

Fig. 13-11 Samples of white beans grown on soil that had received 0 to 720 pounds of phosphorus per acre in fertilizer during the previous 3 years. The poorer growth and delayed maturity of plants on which phosphorus had been used was because of the zinc deficiency caused by the use of phosphorus fertilizer.

these factors are important, because they all affect the efficiency of any fertilizer for any crop grown on any given soil. In other words, the proper use of fertilizer is only one, although a very important one, of the many phases of the scientific management of soil.

The most profitable return from fertilizer is nearly always obtained from soils that are in the best physical condition for plant growth. The most profitable results from fertilizers cannot be expected when they are used on soils that are too compact or too loose, too dry or too wet.

Application and Placement of Fertilizers

Another important factor in the efficiency of fertilizers is the method of applying them. The most efficient placement is direct application to the foliage. This avoids the problems of fixation, leaching, and denitrification, which are encountered in soil application. One pound of manganese applied as a spray was as effective as 10 pounds applied to the soil for vegetable production on an organic soil at the Michigan Agricultural Experiment Station. Foliar application is well suited to micronutrient fertilization, where rapid fixation in the soil occurs and, particularly, where it can be combined with another operation (as spraying to control disease).

Limitations of foliar application are related to the amount that can be applied in a single application or the possible burning effect on the leaves, and their application must wait until crops are established. Perhaps the most intensive use of foliar application has been developed for the long-growing period crop of pineapples in Hawaii. Here, about three fourths of the nitrogen and one half of the phosphorus and potassium are sometimes applied this way.

Fig. 13-12 This soybean plant was grown by the split-root technique. The seed was germinated in quartz sand, and the tiny seedling was transferred to this container filled with quartz sand. A root was pulled through the hole in the side of the can and immersed in a complete nutrient solution in the beaker. The growth of the plant shows that virtually all of the nutrients that a plant needs can be absorbed by a very small portion of the total root system. The quartz sand was kept moist, and abundant root growth occurred in it.

Most of the fertilizer used in the world is applied to the soil directly by machinery or airplane or as a component of irrigation water. The efficiency of applied fertilizer is dependent, among other factors, on its placement in relation to the seeds or roots and on its need. The proper placement and time of application minimizes losses by leaching or dentitrification and conversion to unavailable forms by fixation reactions. The entire root zone need not be fertilized, because a "few" roots are capable of absorbing the nutrients the plant needs (see Fig. 13-12). It is commonly recommended, for a row crop like corn, that a complete starter fertilizer be placed in a band about 2 inches to the side and 2 inches below the seed (Fig.

Fig. 13-13 Fertilizer placed 2 inches to the side and 2 inches deeper than the seed will be in the pathway of early developing corn roots. Diffusion of nutrients into the soil surrounding the fertilizer granules has caused root proliferation in this location.

13-13) at planting time. It will be in the pathway of young roots and quickly utilized, yet it will not cause injury or delay in germination. Band placement will minimize soil fertilizer contact and keep phosphorus and potassium fixation to a minimum. Furthermore, root proliferation in the fertilizer band increases the amount of root surface for absorption (Fig. 13-13). Another application, usually only nitrogen, is applied after emergence as a part of a normal cultivation operation. Nitrogen applied at this time reduces overstimulation of vegetative growth and reduces the possibility of loss before it is absorbed by the plant. Obviously, with hay, pasture crops, and orchards, this method of fertilizer application cannot be followed.

Shade and ornamental trees are commonly fertilized by boring holes about 2 feet apart and 12 inches deep within the drip line of the tree and adding some fertilizer to each hole. Ferric citrate capsules implanted in the trunks of iron chlorotic pin oak trees will supply the trees with iron for about 3 years.

Other factors are to be considered in many instances. As the overall fertility level of the soil is increased, the benefit of localized placement is reduced. Fertilizer purchased in bulk form is less expensive, and lower labor costs result in its use. This makes it possible for many farmers to apply more fertilizer in a less efficient manner and still obtain good results. When only nitrate nitrogen, which is mobile in the soil, is applied and is not fixed, almost any placement is satisfactory so long as the nitrogen permeates the soil and is not leached out of the rooting zone.

Salt Effect of Fertilizers on Germination

When fertilizers are properly placed an inch or so from the seed, germination is not affected, but early growth of the plant is stimulated. However, if seed and fertilizer are placed together, germination may be delayed or prevented. This is because fertilizers are basically soluble salts and, when they are dissolved in the soil solution, they increase the osmotic pressure of the soil solution. The effect on the osmotic pressure is related to the solubility of the fertilizer material. Potassium chloride and ammonium nitrate are very soluble and are very likely to prevent or delay germination of seed when placed in contact with seeds. In fact, dropping small amounts of these fertilizers in small areas on the lawn by accident or faulty spreading will kill the grass. Superphosphates have much less solubility by comparison and, when placed with seed, have little or no effect on seed germination if used at normal rates. The effect of placing some complete fertilizer with wheat seed on the germination of the wheat is shown in Fig. 13-14. Since the increased osmotic pressure of the soil solution produced by the fertilizer inhibits water absorption by seeds (and roots), one would expect the effects to be less damaging when the soil is kept moist, as is also shown in Fig. 13-14.

Time of Application

The nutrients in fertilizers are available or water soluble and, therefore, are likely to react with the soil and become unavailable or leach out of the root zone. Nitrate nitrogen is very soluble and mobile in soils and is subject to leaching. In fact, excessive use of nitrogen fertilizer can pollute the groundwater. Phosphorus, by contrast, is very immobile in soils. Phosphate ions react with other ions in the soil solution to form insoluble, unavailable compounds. Potassium is intermediate in that it is adsorbed to the "exchange," which limits its mobility, but the potassium remains available to plants. For these reasons fertilizers are most effective if applied near the time when plant needs are greatest. This is not always practical, since soil conditions, labor supply, and other factors affect the timing of fertilizer applications.

Effect of Fertilizers on Soil pH

Normally, fertilizers do not produce significant changes in soil pH. However, there are some cases where high rates of nitrogen fertilizer may adversely increase soil acidity and, in some cases, where proper selection

Fig. 13-14 Placing a complete fertilizer with wheat seed was more damaging on germination in dry soil than in moist soil.

Moist soil   Dry soil

of fertilizers and their use can help to bring about desirable changes in soil pH. A summary of the effects of fertilizers on soil pH follows.

1. The common potassium fertilizers, such as the muriate and sulfate of potash, have no permanent effect on soil acidity.

2. Superphosphates in general will have no permanent effect on soil reaction. Basic slag, bone meal, and rock phosphate have a tendency to neutralize soil acidity.

3. Fertilizers containing nitrogen in the form of ammonia or in other forms subject to nitrification (being changed to nitrate) will produce acidity unless sufficient liming material is present in the fertilizer to neutralize the acid formed. Some experimental fields that have received applications of sulfate of ammonia fairly regularly for many years without being treated with lime have become too acid to grow clover (see Fig. 13-15). This effect is more pronounced on soils such as sands and sandy loams.

4. Nitrogenous fertilizers in which the nitrogen is in the nitrate form and is combined with bases such as sodium or calcium will result, upon being utilized by plants, in decreased soil acidity (Fig. 13-15). Some of the fertilizers in this group are nitrate of soda and calcium nitrate. Calcium cyanamide should be placed in this group, although the nitrogen is not in the nitrate form, but the fertilizer carries a high content of lime. The acidity or basicity of several nitrogen fertilizers is given in Table 13-5.

In general, the systematic use of medium to large amounts of fertilizer at suitable times in the rotation will not appreciably affect soil acidity. Where it is desirable to increase soil acidity (to lower pH), ammonium sulfate can be used as a source of nitrogen.

Effect of Fertilizers on Plant Growth

One of the most important factors affecting the growth of plants is the weather (temperature and quantity and distribution of rainfall), and the response that any particular crop will make to the application of fertilizer is largely governed by the weather, particularly the moisture supply. In seasons when it is necessary to delay planting certain crops because of unfavorable weather conditions, the application of fertilizers may speed up the growth processes of the plant and thus offset somewhat the unfavorable effects of the season. But,

Fig. 13-15 Manganese concentration in the midribs of corn leaves at tasseling time and soil pH. Differences in soil pH and, consequently, in manganese in plants was caused by applying a total of 1500 pounds of nitrogen over a 5-year period on the Spinks loamy sand. The use of ammonium sulfate resulted in toxic concentration (over 400 to 500 parts per million) of manganese in plants.

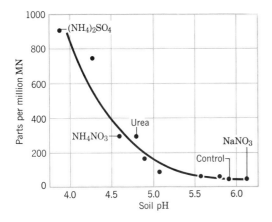

Table 13-5
Equivalent Acidity or Basicity of Several Nitrogen Fertilizers

Fertilizer Material	Nitrogen, %	Equivalent Acidity or Basicity, lb CaCO$_3$	
		Per Unit of Nitrogen	Per 100 lb of Fertilizer
Ammonium nitrate	33	37	59
Sulfate of ammonia	20	107	110
Ammo Phos	11	107	59
Anhydrous ammonia	82	36	148
Calcium nitrate	15	27[a]	20[a]
Crude nitrogen solution	44	24	53
Nitrate of soda	16	36[a]	29[a]
Urea	45	36	84
Calcium cyanamide	22	57[a]	63[a]

[a] Basicity.

looking at the fertilizer-weather relationships from another viewpoint, it is frequently observed that fertilizers in general stimulate early crop growth and, if dry weather prevails about midseason, the fertilizer may result in decreased yields. This is brought about because the soil moisture is more rapidly exhausted by the fertilized crop through increased growth and greater leaf development. On the other hand, fertilized crops may be more drought resistant because of more deeply penetrating root systems (Fig. 13-16).

The early growth of a crop should not be taken as a measure of the effect of a fertilizer on yield. At times fertilizers may stimulate early crop growth but, as the season advances, this difference disappears, and at harvest no increase is found. Fertilizers may also have little effect on the rate of growth of certain crops, but at harvest a decided increase in yield is noted.

Some of the most striking effects of fertilizers occur on tree crops. About $\frac{1}{2}$ pound of nitrogen applied around the base of Christmas trees in June may change the foliage color from yellow green to a dark green by cutting time. The value of the tree may be increased by an amount equal to 10 or more times the cost of the fertilizer.

In the forest nursery great savings occur if trees can be grown to plantable size a year sooner than usual. The pine plants shown in Fig. 13-17 and labeled 2-1 were grown as seedlings 2 years and as transplants in their present bed for 1 year. Proper use of fertilizer sometimes enables trees like these to be planted after growing for 3 years. Where plants can be grown to plantable size in 3 instead of 4 years, a saving of about $540 per acre may result.

Effect of Fertilizers on Quality and Composition of Crops

Some of the more general effects of nitrogen, phosphorus, and potassium on the quality of crops have already been discus-

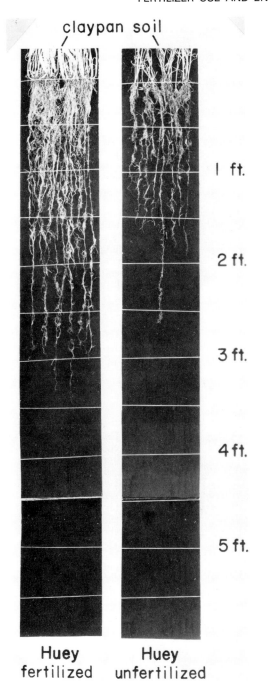

Fig. 13-16 Fertilizer increased the vigor and growth of tops and roots of wheat on the Huey soil in Illinois. Greater rooting depth of fertilized plants increased the supply of available nutrients and water. (Courtesy Dr. Joe Fehrenbacker, University of Illinois.)

sed (Chapter 12) (see also Fig. 13-18). The influence of fertilizers on the composition of the mineral matter of plants is exceedingly complex. Their influence is affected by the variety of crop, climate, water supply, and other environmental conditions. The influence that the presence of one element exerts on the absorptive powers of the plant for other mineral elements of the soil or fertilizer also complicates this problem.

Although it is possible to increase to a limited extent the content of mineral constituents of some crops by fertilization, one should not be misled by exaggerated claims for mineralizing human and animal foods. For a more detailed discussion of this subject, see pages 361–362.

Fertilizer Use and Environmental Quality

Algae are the most abundant plants growing in surface waters, and their growth is commonly limited by a lack of nitrogen and/or phosphorus. Consequently, any activity that enriches surface waters with nitrogen or phosphorus might contribute to excessive algal growth or an algal bloom. Furthermore, nitrogen as nitrate can be harmful in surface or groundwaters used for human or animal consumption. This section will consider some important factors related to the pollution of waters resulting from use of nitrogen and phosphorus fertilizers.

Natural Nitrate Content of Soil Percolate

Nitrogen fertilizers are mostly water soluble and the nitrate form of nitrogen is mobile in soils. During rains, nitrogen fertilizer is readily carried into the soil. For this

Fig. 13-17 High investment cost per acre makes it desirable to use fertilizer to reduce the number of years required to grow trees to plantable size. These plants were grown as seedlings two years and have been at their present location 1 year (2–1).

reason, nitrogen fertilizers are not likely to be carried into surface waters by runoff water. The nitrate from nitrogen fertilizer, however, might migrate downward through the soil with percolating water and be carried to the underground water reservoir or groundwater. In fact it is natural for some nitrate ions to move to the groundwater in natural ecosystems. This is because nitrate is a common form of available nitrogen in soils for plants and the nitrate may be carried to the water table in wet seasons

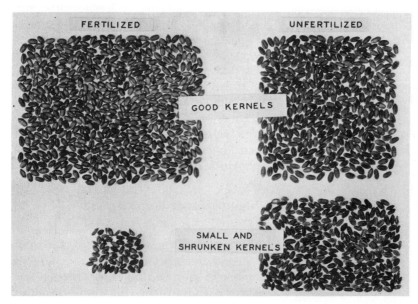

Fig. 13-18 Properly chosen fertilizer improves the quality of many crops. *(Left)* Twenty grams (549 kernels) of wheat grown on fertilized Brookston silt loam contained 53 shrunken grains. *(Right)* Twenty grams (617 kernels) of wheat from unfertilized soil contained 268 shrunken grains.

when plants are relatively inactive. A look at some experimental data should be helpful.

In Chapter 4 (see Table 4-2) we discussed some lysimeter data collected at Coshocton, Ohio, which showed nutrient losses in soil percolate. One lysimeter in 1948 grew a grass crop and no fertilizer or manure was applied. This situation can be considered comparable to a natural grassland. The loss of nitrate nitrogen per acre per year was 4.55 pounds and the amount of soil percolate was 6.2 inches. Since there are 4.4 acre inches of water in a million pounds, the parts per million of nitrate nitrogen in the soil percolate can be calculated from the following equation.

Parts per million nitrogen

$$= \frac{\text{pounds of nitrogen leached per acre} \times 4.4}{\text{inches of soil percolate}}$$

$$(4)$$

Substituting in the above equation and solving shows that the "natural" soil percolate had 3.3 parts per million of nitrate nitrogen. This value is very reasonable, since 2 to 3 parts per million of nitrogen are considered adequate for algal growth and this is also the nitrogen content commonly found in springs and shallow wells. The United States Health Service considers 10 parts per million of nitrate nitrogen the maximum allowable in water used to make baby formula.

There has always been a natural amount of nitrate in groundwater resulting from the natural reactions of the nitrogen cycle. Since ancient times the manner in which human and other animal wastes were disposed created varying degrees of nitrate pollution of water supplies. Today, however, the problem is different because of the magnitude of nitrogen used as fertilizer and the great concentrations of cattle that occur on large dairy and beef farms.

Factors Affecting Nitrate Pollution of Groundwater

The factors that affect the amount of nitrogen in drainage water or soil percolate can be represented in equation form as follows.

nitrogen in soil percolate

= amount of nitrogen available as nitrate

− amount of nitrogen immobilized   (5)

Factors affecting available nitrate	Factors affecting nitrogen immobilized
1. Organic matter mineralization	1. Kind of crop grown
2. Fixation from atmosphere	2. Yield or growth of crop
3. Nitrogen added in precipitation	3. Amount and distribution of rainfall
4. Nitrogen added as manure	
5. Nitrogen added in fertilizer	

A modest amount of nitrogen fertilizer could have no effect, increase or decrease the amount of nitrogen in the percolate, depending on the particular situation. If, however, an excessive amount of nitrogen fertilizer was applied and the soil's immobilization capacity was greatly exceeded, nitrate pollution of the groundwater would occur. The continued use of nitrogen at rates of 150 pounds or more per acre per year caused an accumulation of nitrate nitrogen in the Marshall silt loam in Missouri when the land was used continuously for corn production (see Fig. 13-19). The 0 or 100 pound rate produced similar amounts of nitrate nitrogen in the soil profile. Thus, we see that farmers

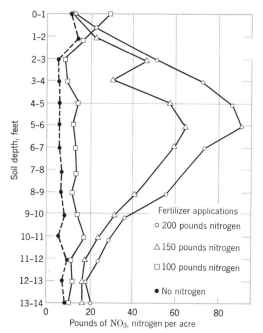

Fig. 13-19 Nitrate nitrogen in the Marshall silt loam after seven annual applications of nitrogen fertilizers in a continuous corn program. (From Linville, 1968.)

should avoid using too large amounts of nitrogen fertilizer, particularly when plant growth is inactive. The danger of nitrate pollution is also greatest on sandy soils with high percolation capacity. In these cases it is important to use modest amounts of nitrogen fertilizer and take care to apply fertilizer only when crops are actively growing.

Although excessive use of nitrogen fertilizer can cause nitrate pollution of groundwater and many persons credit fertilizer use for extensive pollution, the evidence at the present time suggests that use of fertilizers contribute little to groundwater pollution. In fact, studies in Missouri and Colorado show that nitrate pollution of groundwater is more related to numbers of livestock. Nitrate nitrogen is produced during the decomposition of manure in the

same ways that organic matter decomposition in soils produces nitrate. Where animals congregate in large numbers, as in feedlots, growing plants are absent and the nitrate produced from manure decomposition is not immobilized. Under these conditions rainwater can leach the nitrate out of the feedlot and transport the nitrates to the groundwater table. The same situation applies to corrals.

Phosphate Pollution of Surface Waters

Phosphate from fertilizer, in contrast to nitrate, reacts with soil constituents to form insoluble compounds that are immobile in soils. For this reason there is little possibility that groundwaters would become polluted from the use of phosphorus fertilizers. Erosion of soil particles can, however, carry phosphate adsorbed on soil particles into surface waters. This type of contamination occurs naturally as well as a result from erosion of agricultural land. In fact, considerable concern exists where unprotected sloping land lies exposed for long periods of time where highways or subdivisions are constructed. The real problem here is not so much the use of phosphorus fertilizer but, instead, erosion.

Erosion is a selective process. The finer particles are removed and these finer particles are richer in plant nutrients. Research at the University of Wisconsin showed that the fine, eroded soil material had 3.4 times more phosphorus than the soil itself. Where fertilizers promote a more vigorous plant cover and reduce soil erosion, fertilizers reduce the danger of enriching surface waters with phosphate.

Manure as a Fertilizer

After centuries during which animal manures were highly valued for their fertilizing quality, manure on many American farms has become a liability. Today, the nutrients in the manures can be purchased in fertilizers cheaper than manure can be stored and handled. Furthermore, farms have become more specialized and bigger so that most of the manure is produced on fewer larger farms instead of a large number of small farms. A farm with 10,000 head of cattle produces about 260 tons of manure a day. The magnitude of the manure-disposal problem is dramatized by the fact that the domestic animals in the United States produce body waste equivalent to 1.9 billion people. Emphasis has changed from application of manure on soils to increase yields to the use of the soil for animal waste disposal. This section considers the production, composition, storage, and handling of animal manures.

Components of Manure

Farm manure consists of two components, the solid and the liquid. The solid excrement, on the average, contains one half or more of the nitrogen, about one third of the potassium, and nearly all the phosphorus that is excreted by the animal. The nitrogen in the feces exists largely in two forms: first, the residual proteins that have resisted decomposition in the digestive processes; and, second, the proteins that have been synthesized in the cells of bacteria. Over one half of the nitrogen may be present as synthesized protein, and this form is readily broken down when added to soils, so that the nitrogen is available to plants. The solid excrement also contains large quantities of lignin. In other words, a large share of the organic matter in the feces is humified; a compound is formed similar to the humus that is formed in soils. As much as 50 percent of the organic matter in the solid excre-

ment may be in the humified state, and the nitrogen contained therein is only slowly available to plants when added to soils.

The liquid fraction or urine contains the plant nutrients that have been digested and utilized in the animal body and are later excreted. All the plant nutrients in this fraction are soluble and either are directly available to plants or readily become so. The liquid portion of manure differs from the solid not only in regard to the availability of nutrients but also in its low content of phosphorus and in its high content of potassium and nitrogen. The distribution of plant nutrients between liquid and solid portions of manure is shown in Fig. 13-20.

In general, the more digestible the food consumed by an animal, the greater is the proportion of its plant nutrients that appears in the urine. Furthermore, as a rule, the higher the food is in nitrogen, the greater the digestibility of the nitrogen and the larger the amounts of nitrogen in the urine.

When voided, the nitrogen in urine exists largely as urea and hipuric and uric acids. These compounds are not volatile at ordinary temperatures, but manure contains organisms that are capable of rapidly breaking these compounds down with the

Fig. 13-20 Distribution of plant nutrients between liquid and solid portions of a ton of average farm manure.

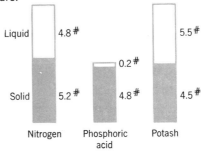

Figures represent pounds per ton

formation of ammonia, which combines with water and carbon dioxide to make ammonium carbonate. This compound is unstable; even in solution it tends to decompose, losing ammonia, especially at higher temperatures $[(NH_4)_2CO_3 \rightarrow 2NH_3 + H_2O + CO_2]$. This compound may lose all its ammonia on drying. The unstable nature of the nitrogen in urine presents a major problem in the handling of manure.

Quantity and Composition of Excrements

Many factors influence the quantity of manure produced and its composition, such as (1) the kind and age of the animal, (2) the kind and the amount of feed consumed, (3) the condition of the animal, and (4) the milk produced or the work performed by the animal. Wide variations are often found in the manure of animals even of a given class. Animals of different ages and doing different kinds of work require different amounts and proportions of nutrients to maintain them. A young animal, for example, that is building muscle and bone needs considerable phosphorus, nitrogen, calcium, and other elements, and the manure produced by such animals will contain much less of these elements. Since the composition of manure is so variable, data such as those presented in Table 13-6 can only be approximate.

The urine makes up only 20 percent of the total weight of the excrement of horses but 40 percent of that from hogs. These represent the two extremes. Since the urine makes up only 20 to 40 percent of the total weight of manure from any animal, and yet contains approximately two thirds of the total potash and somewhat less than one

Table 13-6
Quantity and Composition of Fresh Manure Excreted by Various Kinds of Farm Animals

Animal	Excrement	Pounds per Ton	Water, %	Nitrogen, lb	P_2O_5, lb	K_2O, lb	Tons Excreted[a] per year
Horse	Liquid	400	—	5.4	Trace	6.0	—
	Solid	1600	—	8.8	4.8	6.4	—
	Total	2000	78	14.2	4.8	12.4	9.0
Cow	Liquid	600	—	4.8	Trace	8.1	—
	Solid	1400	—	4.9	2.8	1.4	—
	Total	2000	86	9.7	2.8	9.5	13.5
Swine	Liquid	800	—	4.0	0.8	3.6	—
	Solid	1200	—	3.6	6.0	4.8	—
	Total	2000	87	7.6	6.8	8.4	15.3
Sheep	Liquid	660	—	9.9	0.3	13.8	—
	Solid	1340	—	10.7	6.7	6.0	—
	Total	2000	68	20.6	7.0	19.8	6.3
Poultry	Total	2000	55	20.0	16.0	8.0	4.3

Compiled from Van Slyke, 1932.
[a] Clear manure without bedding; tons excreted by 1000 lb of live weight of various animals.

half of the nitrogen, it is evident that, pound for pound, the urine is more concentrated and hence more valuable than the solid portion.

Losses of Liquid Manure

The loss of urine from manure is serious from the standpoint of plant-nutrient content. This loss occurs mostly from failure to use ample bedding, leakage through stable floors, seepage into earth floors, or drainage from manure heaps.

On the basis of the total plant food in manure, about 50 percent of the value is in the urine. It can be readily understood that large quantities of plant nutrients are lost and that the portions so lost, as already pointed out, are the most readily available forms of plant nutrients in manure.

Losses by Volatilization

Losses incurred by volatilization fall principally on nitrogen and organic matter. Large quantities of ammonia are produced in manure from urea and other nitrogenous compounds, and in the earlier stages of manure decomposition the ammonia is combined largely with carbonic acid as ammonium carbonate and bicarbonate. These ammonium compounds are rather unstable, and gaseous ammonia may be readily liberated (Fig. 13-21). The tendency to lose ammonia nitrogen increases with the increase in concentration of ammonium carbonate and with the increase in temperature.

At ordinary temperatures, little or no loss of ammonia from manure occurs at pH 7.0 or below. High temperatures, produced by aerobic decomposition in a loose manure pile, are conducive to very rapid loss of ammonia.

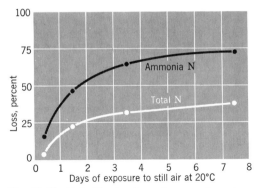

Fig. 13-21 Volatilization loss of ammonia and total nitrogen from fermented cow manure exposed to drying. (Based on data of Heck, 1931.)

Freezing also tends to increase the loss of ammonia by increasing the concentration of the solution through the crystallization of the water. This loss may be considerable when manure is spread and becomes frozen.

Air movement greatly affects the loss of ammonia. Wind movements hasten evaporation of water, which decreases the capacity of the water to hold ammonia. Thus, manure that is permitted to dry out may lose appreciable quantities of ammonia. This fact emphasizes the importance of ammonia loss due to air circulation in loosely piled manure heaps and in overfermented manure that is forked frequently. Losses may also be considerable if manure is spread and permitted to dry before plowing under.

It has been pointed out that when manure decomposes, it suffers important losses in organic matter. These losses occur mainly in the carbohydrate constituents. One of the important end products of carbohydrate decomposition is carbon dioxide, most of which is lost from manure by volatilization. Shrinkage that accompanies the partial decomposition of manure is evidence of organic matter loss.

Little or no loss of phosphoric acid or potash from manure occurs through volatilization.

Application of Manure

Prompt spreading of manure is generally considered most effective, although when in good storage it is likely to lose less value than if spread on the field without being plowed under or worked into the soil immediately. Losses of applied manure may occur in three ways: (1) volatilization of ammonia nitrogen as a result of drying or freezing; (2) surface runoff water carrying soluble portions of all three nutrients; or (3) leaching of nutrients.

Much experimental work with commercial fertilizers indicates that their effectiveness is decreased if they are applied a considerable time before the seeding of the crop. This effect is attributed to leaching losses and to the fixation in less soluble forms of the plant nutrients by the soil. Fresh and properly stored manures contain large amounts of soluble nutrients, and the same principles apply. Manure applied to corn land in the spring is likely to give greater returns than the same amount of manure applied in the fall. This effect will probably apply only to the crop of the first year and not to the later crops. If manure has already lost its readily soluble nutrients, the time of application is less important.

Little loss of nutrients from manure may be expected when applied on level, medium- to fine-textured soils. These soils are safe for fall and winter spreading of manure, although they fix considerable quantities of the nutrients in less available forms. It is not advisable to place manure on sandy soils or on hilly fields much ahead of plowing time because of leaching and erosion losses. It must be remembered, however, that manure applied on sloping fields decreases soil erosion losses, thereby counterbalancing to some extent the loss of nutrients from the manure. Illinois law, however, restricts application of manure (and fertilizer) on slopes over 5 percent when soils are frozen to protect water quality in streams and lakes.

The rate at which manure should be spread will depend on the amount produced on the farm. As much of the cultivated land should be covered as possible. It is usually better to cover an entire field with a light application than to give only a part of the field a heavy coating. In other words, the crop response from 50 tons over 10 acres proves larger than from 50 tons on 5 acres. It is wise to gauge the rate of application so as to extend the acreage covered for each crop.

Residual Effects of Manure

An application of manure usually shows a favorable influence on crop yields for several years. These beneficial effects are distributed over a longer time than those of chemical fertilizers. Striking results in showing the long-continued effects have been obtained (Fig. 13-22) by making heavy applications of manure for several years in succession and then discontinuing the application.

Organic versus Inorganic Sources of Fertilizer Nutrients

There have been extravagant claims made for the merits of organic materials as sources of fertilizer nutrients. Some people believe that the addition of nutrients to soils in organic materials produces better food than additions of nutrients in inorganic forms. Some people also believe in using

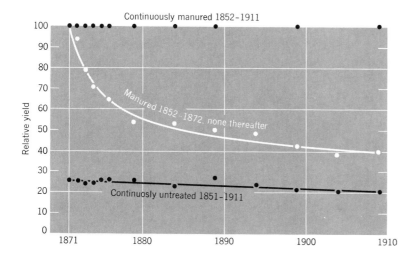

Continuously manured 1852–1911

Manured 1852–1872, none thereafter

Continuosly untreated 1851–1911

Relative yield

1871 1880 1890 1900 1910

Fig. 13-22 Residual effects of heavy applications of manure (14 tons annually) on yield of barley grown continuously on the soil of the Rothamsted Experiment Station in England. (From Salter, 1939.)

only natural inorganic sources such as rock phosphate and granite meal, for example. The nutrients in these sources are mainly insoluble and unavailable to plants. The potassium in KCl fertilizer was found to be eight times as effective as the same amount of potassium in granite meal for production of coastal Bermuda grass. Regardless of whether phosphorus and potassium (and other nutrients) are from organic or inorganic sources, plants absorb the same ions, $H_2PO_4^-$ and K^+. As far as I know there are no generally accepted scientific experiments that support the superiority of *either* organic or inorganic nutrient sources.

References

Azevedo, J., and P. R. Stout, "Farm Animal Manures: An Overview of Their Role in the Agricultural Environment," *Manual 44*, Cal. Agr. Exp. Sta., Davis, Calif., 1974.

Bear, F. E., "The Effect of Increasing Fertilizer Concentration on Exchangeable Cation Status of Soils," *Soil Sci. Soc. Am. Proc., 16*:327–330, 1952.

Brown, Lester R., "Human Food Production as a Process in the Biosphere," *The Biosphere*, Sci. Am., San Francisco, 1970, pp. 93–103.

Chandler, R. F., Jr., M. Peech, and C. W. Chang, "The Release of Exchangeable and Nonexchangeable Potassium from Different Soils upon Cropping," *Soil Sci. Soc. Am. Proc., 10*:141–146, 1945.

Committee on Elemental Guarantees of Soil Sci. Soc. Am.," N. P. K. Simpler Terms for Fertilizer," *Crops and Soils, 14* (6), 1962.

Heck, A. Floyd, "Conservation and Availability of the Nitrogen in Farm Manure," *Soil Sci., 31*:467–479, 1931.

Lamer, Mirko, *The World Fertilizer Economy*, Stanford Press, Stanford, Calif., 1957.

Linville, Kenneth W., *Residual Nitrate in Missouri Soils*, Master of Science Thesis, University of Missouri, 1968.

Massey, H. F., and M. L. Jackson, "Selective Erosion of Soil Fertility Constituents," *Soil Sci. Soc. Am. Proc., 16*:353–356, 1952.

Neely, Dan, "Pin Oak Chlorosis-Trunk Implantations Correct Iron Deficiency," *Jour. Forestry, 71*:340–342, 1973.

President's Science Advisory Committee, *The World Food Problem*, Vol. 2, The White House, Washington, D.C., May 1967.

Salter, Robert M., and C. J. Schollenberger, "Farm Manure," *Ohio Agr. Exp. Sta. Bull.*, *605*, 1939.

Stoeckeler, J. H., and H. F. Arneman, "Fertilizers in Forestry," *Adv. Agronomy*, *12*:127–195, 1960.

Stone, E. L., G. Taylor, and J. DeMent, "Soil and Species Adaptation: Red Pine Plantations in New York," *First Nor. Am. For. Soils Conf. Proc.*, Michigan State University, East Lansing, Mich., 1958, pp. 181–184.

Tisdale, Samuel L., and Werner L. Nelson, *Soil Fertility and Fertilizers*, 3rd ed., Macmillan, New York, 1975.

Van Slyke, L. L., *Fertilizers and Crop Production*, Orange Judd, New York, 1932.

Warncke, D. D., D. R., Christenson, and R. E. Lucas, "Fertilizer Recommendations for Vegetable and Field Crops in Michigan." *Ext. Bull.*, *E.*-550, Michigan State University, East Lansing, Mich., June 1976.

Wittwer, S. H., and G. F. G. Teubner, "Foliar Absorption of Mineral Nutrients," *An. Rev. Pl. Phy.*, *10*:13–32, 1959.

Wolcott, A. R., "The Acidifying Effects of Nitrogen Carriers," *Agr. Ammonia News*, Agr. Amm. Institute, Memphis, July–August, 1964.

14

SOILS, PLANT COMPOSITION, AND ANIMAL HEALTH

Considerable interest exists today in health foods and living in more natural ways. One important concern is whether the use of fertilizers affects food quality and health. "Do fertile soils produce better or more nutritious food than infertile soils?" "If a plant grows well and is healthy, will it be a satisfactory source of food for man or animals?" This chapter provides some insight into the role of soils and fertilizers in plant composition and some relationships between plant composition and animal health.

Factors Affecting Plant Composition

Plant composition is affected by soil, environmental conditions, and genetic factors, as summarized in Fig. 14-1. The genetic factor is probably the most important. The genetic factor not only controls the uptake and mineral composition, but also controls the amounts and kinds of proteins, vitamins, carbohydrates, and the like. It is interesting, however, that the vitamin C content of tomatoes is determined by the amount of sunlight striking the tomato.

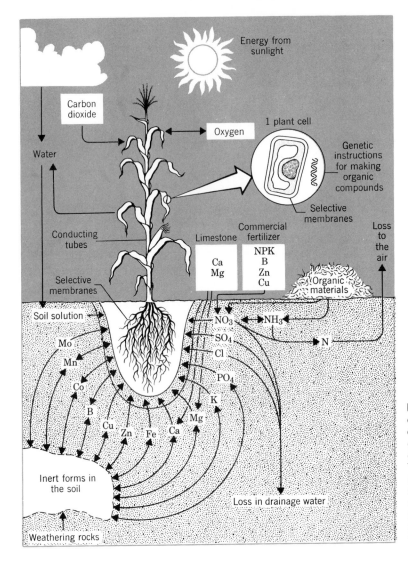

Fig. 14-1 The composition of plants is controlled by soil, genetic, and environmental factors. The plant processes operate according to a set of instructions determined by the genetic inheritance of the plant. (From Allaway, 1945.)

This discussion will emphasize the soil and genetic factors affecting the mineral composition of plants.

Plant Species and Plant Composition

Perhaps one of the most notable differences in plant composition exists between the grasses and legumes (Table 14-1). Legumes contain a higher percentage of nitrogen and calcium and a lower percentage of potassium than grasses grown under comparable conditions.

Tobacco is known for its high potassium content, which commonly exceeds 4 percent in the leaves. Certain varieties of cere-

Table 14-1

Average Composition of Legumes and Grasses

Plant	Percentage Composition			
	Nitrogen	Potassium	Calcium	Magnesium
Grasses	0.99	1.54	0.33	0.21
Legumes	2.38	1.13	1.47	0.38

Data of Snider, 1946.

als that have a high resistance to disease and insects have a silica encrusting layer in the epidermis and may contain up to 15 percent silica.

The higher base content of the foliage of hardwoods, as compared to spruce, was used to explain the higher pH of soils developed under the hardwood forest in Chapter 10. These examples support the fact that the composition of plants is influenced by genetic factors.

Level of an Individual Nutrient

An increase in the amount of a nutrient in the soil that is available or readily soluble may or may not cause an increase in the percentage of the nutrient in the plant. It depends on the extent to which the total growth of the plant is increased. Several possibilities exist. If the nutrient is in short supply and the growth of the plant is limited by it, addition of the nutrient would probably result in a great increase in the amount absorbed. A correspondingly large increase in plant growth could occur so that the percentage composition of the plant remained the same, as shown on the far left side of Fig. 14-2. An increase in growth and nutrient concentration or percentage composition can occur simultaneously in the zone of poverty adjustment (Fig. 14-2). A

point is finally reached with increasing nutrient supply where the maximum yield is obtained and further increases in the nutrient supply result in decreased yields. Plants may, however, continue to take in nutrients, and a sharp increase in nutrient concentration in the plant results. This is the luxury range shown in Fig. 14-2.

This principle has been used to effectively increase the protein content of crops. Nitrogen fertilization was used to increase the quality of wheat grain for milling purposes by increasing its protein content in Kansas. Fifty pounds of nitrogen per acre profitably increased yields. An additional 50 pounds of nitrogen per acre did not increase the grain yield, but it increased the protein content and quality enough to make it a profitable practice.

Cation Content of Plants a Constant

Potassium, calcium, and magnesium all play some of the same roles in the plant, for instance, their role in the buffer system of plant cells. In this regard, they can substitute for each other. This view is consistent with the common observation that the milliequivalents of these three cations in plant tissue tends to be a constant. Further support of the cation constancy in plants is that

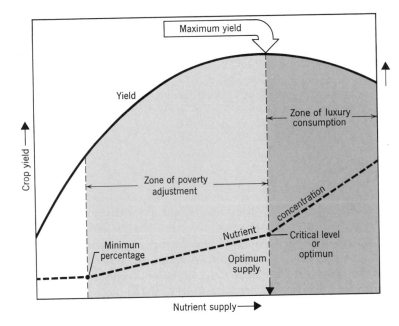

Fig. 14-2 Schematic graph of the manner in which nutrient concentration and crop yield varies with the supply of nutrients. (From Brown, 1970.)

a large application of potassium sometimes produces magnesium deficiency. Note in Fig. 14-3 that increasing the amount of potassium in the nutrient solution caused large and comparable decreases in the amount of calcium and magnesium in alfalfa plants; the total amounts of potassium, calcium, and magnesium in the tissue remained the same.

Corn grown on the highly calcareous Harpster soils of northcentral Iowa commonly has a severe potassium deficiency symptom so that during the summer these areas of soil can readily be located by observing the growth of the corn. The potassium-deficient plants were found to contain as much as 10 times more calcium and magnesium than potassium. Contrast this with the fact that grasses tend to contain as much or more potassium as the sum of calcium and magnesium (Table 14-1).

Fig. 14-3 The calcium, magnesium, and potassium content of alfalfa plants as a function of the potassium concentration in the nutrient solution. (Adapted from Wallace, 1948.)

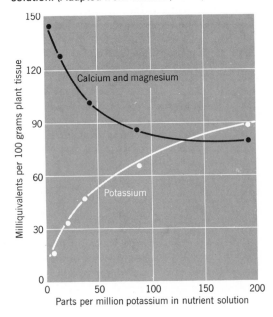

Plant Composition Changes with Age

Nutrient accumulation occurs at a faster rate than plant weight when the plant is young whereas the reverse is true when the plant approaches maturity. This causes a declining concentration of nutrients in plants with increasing age. It is a common practice for dairy farmers to cut forages when they are relatively immature in order to provide more nutritious feed.

The accumulation of dry matter in the various parts of the corn plant and the accumulation of nitrogen, phosphorus, and potassium through the season are given in Fig. 14-4. In this instance, corn plants had absorbed 40 percent of their nitrogen, 30 percent of their phosphorus, and over 50 percent of their potassium by the time they had made 20 percent of their growth.

Composition Varies with Plant Part

The pattern of organ development for corn is also shown in Fig. 14-5. The grain accounted for nearly one half of the weight of mature plants and was largely produced during the last one third of the growing season. During this time the plant accumulated over 40 percent of its dry matter and only 25 percent of its nitrogen. Thus, the production of the grain occurred at the expense or loss of some nitrogen from the earlier formed organs—stems, leaves, and roots (Fig. 14-5). The translocation of many nutrients and food is a continuing process in plant growth and different organs have different priority for the materials.

Fruit or seed production has the highest priority, and it is natural for nutrients accumulated in the vegetative parts to be translocated and used later for seed production. Nutrients in excess of fruit growth remain in other organs of the plant. This causes the seeds of plants to have a similar composition when grown under widely different conditions, whereas the composition of the vegetative parts may vary greatly.

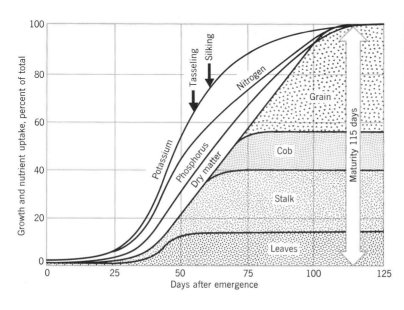

Fig. 14-4 Uptake of nutrients in relation to the accumulation of dry matter by corn plants. (From Hanway, 1960.)

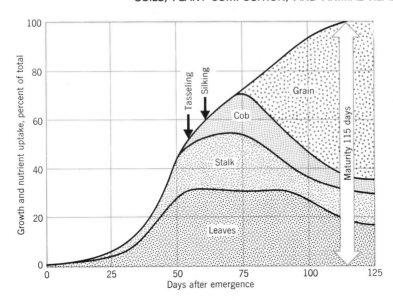

Fig. 14-5 Part of the nitrogen, as well as phosphorus, potassium, and some other nutrients, are translocated from vegetative tissue to the grain during the later part of the growing season. An amount of nitrogen equal to about half the nitrogen in the corn grain was absorbed during the time the grain was produced. A similar amount of nitrogen was translocated from the other organs to the grain during the same period. (From Hanway, 1960.)

Soils, Fertilizers, and Animal Health

We have noted that plants are commonly deficient in mineral elements and exhibit deficiency (disease) symptoms. It should be no surprise that plants may contain inadequate amounts of certain mineral elements for optimum animal health. One of the most striking cases is the phosphorus deficiency of cattle grazing on Oxisols low in available phosphorus in the tropics. I can remember seeing people with goiter and being given iodine tablets in grade school because I lived in the goiter belt in the central United States. In some cases plants contain toxic amounts of elements that create animal diseases. We must be cautious, however, in making a sweeping generalization about soil fertility and *human* health. There is a great variation in plant composition due to genetic differences and *human* diets commonly consist of foods from many locations with different soils and environments. Some major soil-plant-animal health relationships will be discussed in the following sections.

Plant and Animal Requirements Compared

Animals require over 20 different mineral elements, most of them are the same as those required for plants. Boron, however, is needed by plants and not by animals. The oldest and most common dietary supplement used by people is salt (sodium chloride) to supply sodium that is not essential for most plants. Animals, however, need in addition to the elements needed by plants chromium, iodine, and selenium. These elements are in soils and absorbed by plants even though they are not required. Thus plants can be "perfectly" healthy and unable to satisfy the dietary needs of animals.

Soil–Plant–Animal Health Relationships

Locations of mineral nutritional diseases in animals are shown in Fig. 14-6. The large iodine deficient or goiter belt in the northern states should be noted. Phosphorus deficiencies are widespread throughout the

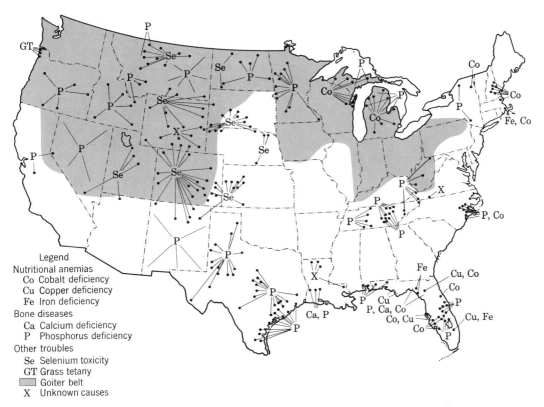

Fig. 14-6 Occurrence of mineral nutritional diseases in animals. The dots show the approximate locations of the observed deficiency. The lines not terminating in dots indicate a generalized area where specific locations have not been reported. The goiter region is also a generalized area.

states, and selenium toxicity is common in the western states. Almost every state is represented by either a deficiency or toxicity of one or more nutrients.

Phosphorus. Low soil phosphorus commonly limits plant growth and is perhaps the most critical mineral element deficiency for grazing livestock. Weak bones and phosphorus deficiency of cattle grazing on grass is quite common. Some of the complex relationships between phosphorus levels in soils and phosphorus deficiency in animals are shown in Fig. 14-7. As the amount of phosphorus fertilizer

increased (or available soil phosphorus increased), the phosphorus content of the oat grain and alfalfa hay were markedly increased. Since cattle require about 0.3 percent phosphorus in their diet for optimum growth, the oat grain contained sufficient phosphorus without fertilization. The oat straw remained phosphorus deficient for cattle, even with fertilization. Cattle or livestock with diets mainly of grasses usually need a phosphorus supplement to ensure adequate dietary phosphorus. The use of phosphorus fertilizer changed alfalfa hay from an inadequate source of phosphorus to an adequate source. By

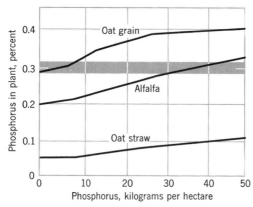

Fig. 14-7 Effect of rate of phosphorus fertilizer apllication on concentration of phosphorus in oats and alfalfa grown on Clarion loam. About 0.3 percent of phosphorus (dashed line) is required by dairy cattle for normal growth. (From Allaway, 1975.)

contrast, the racehorse industry has centered on the high phosphate soils of Kentucky and Florida.

Phosphorus is not a serious problem in human nutrition. Humans eat large amounts of cereal grain and meat, which are good phosphorus sources. For humans the use of phosphorus fertilizer is more important for increasing food supply than increasing the phosphorus content of plants.

Calcium.   Plants and animals need relatively large amounts of calcium. Calcium is important for formation of bones, teeth, and egg shells. Rarely or only occasionally are plants calcium deficient because some other soil factor usually becomes limiting before low soil calcium limits growth. Adequate calcium supply in animal diets appears importantly related to differences in plant species. Red clover grown on acid soils of the northeastern United States contains more calcium than grass grown on calcareous soils of the arid and semiarid

regions of the western United States. Snap beans and peas usually contain three to five times more calcium than corn or tomatoes. A diet adequate in calcium for humans depends on food selection. Liming acid soils enables people to grow more legume crops that normally have a higher calcium content. Lime increases the soil calcium supply, however, and tends to increase the calcium content of all crops.

Livestock diets commonly consist of large amounts of low calcium grasses and grains, and adding calcium to the diet is commonplace. Cows deficient in calcium produce less milk than cows on adequate calcium diets, but the concentration of calcium in the milk is similar. This means that the milk of cows remains similar in calcium regardless of its level in the cow's diet. Remedying a calcium deficiency in cattle diets results in more milk, not better milk.

Magnesium.   The similar behavior of calcium and magnesium in soils means that acid soils low in calcium are usually low in magnesium. Magnesium deficiency is fairly common for plants growing on acid, sandy soils. High levels of available soil potassium inhibit magnesium uptake by plants and sometimes causes magnesium deficiencies. As with calcium, the accumulation of magnesium in plants is strongly affected by species, leguminous plants being high in magnesium and grasses, tomatoes, corn, and other nonleguminous crops being low in magnesium regardless of soil magnesium level.

Grass tetany or grass staggers is the major animal disease caused by low magnesium in the forage of grazing cattle and sheep. Pregnant and lactating animals are the most susceptible. Animals become nervous, stagger, and fall. Injection of magnesium soon after the symptoms appear will result in recovery in a few minutes. Untreated cases severe enough to cause falling result in death if untreated. Magnesium is commonly added to diets of cattle grazing on green pastures.

High potassium fertilization has been associated with incidence of grass tetany resulting from the effect of high soil potassium on reduced uptake of magnesium (cation-constancy effect). Magnesium fertilizers can be used to increase plant magnesium and prevent tetany. Grass tetany tends to be more prevalent in the early spring and late fall when weather is cool because more of the forage is made up of grasses at that time.

Cobalt. Cobalt is required by bacteria that fix nitrogen in the nodules of legumes and is the only known need of cobalt by plants. Microorganisms in the ruminants incorporate cobalt into vitamin B_{12}. Nonruminants, including humans, cannot synthesize vitamin B_{12} and must obtain the vitamin in their diet.

Cattle and sheep that *are not* fed legumes almost always need supplementary cobalt. The relationship of soil cobalt and health of ruminants is one of the most striking examples of a soil–plant relationship to animal health. People get their vitamin B_{12} from animal products such as milk, cheese, meat, and eggs. Only persons on a strictly vegetation diet are likely to have a diet deficient in vitamin B_{12}.

Cobalt deficiency was observed by the earliest settlers in parts of New England, but the cause of the disease was not recognized. In New Hampshire, it was called "Chocorua's Curse" and "Albany Ail" from local place names. In southern Massachusetts, it was known as "neck's ail," not because the disease affected the neck of the animal, but because it was most common on

necks of land that projected into the bays:

It is now known that cobalt deficiency in New England has its origin in the geologic history of the soils. The low-cobalt soils of New Hampshire, for example, are formed in glacial deposits derived from granite of the White Mountains, which contained very little cobalt (see Fig. 14-8).

Moreover, the Spodosols (Podzol soils) in the area, formed from sandy glacial drift, tend to lose their cobalt by leaching. Cobalt moves downward at a more rapid rate than iron, especially in Spodosols of poor drainage.

Cobalt deficiency among grazing animals also is a problem on wet sandy soils of the lower Coastal Plain in the southeastern United States. The low-cobalt soils there are Humaquods—poorly drained sandy

Fig. 14-8 Low cobalt areas (shaded) in New England. The arrows indicate direction of glacial ice movement. (From Kubota, 1965.)

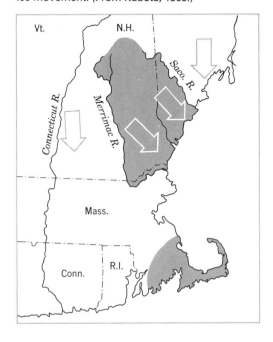

soils with organic pans. The original sandy deposits had little cobalt to begin with, and that was leached from the upper horizons as soil development proceeded. Consequently, there is little cobalt available to forage plants growing in these soils.

Copper. Plant copper deficiencies are most common on organic soils and very sandy soils (in Florida—see Fig. 14-6). Ruminants are sensitive to the copper deficiency that results from a diet of low copper feed. Copper fertilization of pastures will increase plant copper and prevent copper deficiency in animals. In parts of Australia livestock production was impossible until copper fertilizers were used on the pastures. Care must be used in applying copper to soils to prevent toxic copper levels in both plants and animals. Monograstric animals, including people, are less sensitive to low copper levels in feed, and no link has been found between copper in the soil and human health.

Iodine. Iodine is an element needed by animals but not essential to plants. The relationship between iodine levels in soils and plants and the incidence of goiter has been one of the most striking soil–plant–animal health relationships. The goiter belt where soils are low in iodine is shown in Fig. 14-6.

Iodine is taken up by plants and is passed along the food chain into animals. Iodine is volatized over oceans, carried in the atmosphere and carried down to the land in precipitation. The centers of continents are usually areas low in iodine because they receive very little iodine in precipitation. Coastal areas receiving ocean spray and ocean-grown plants have abundant iodine. Iodine in plants can be increased by soil fertilization with iodine; however, the addi-

tion of iodine to salt has proved to be a simple and inexpensive solution to goiter prevention. The development of iodized salt is considered by some to be one of the greatest scientific contributions to human health.

Iron. Iron deficiencies are common in both plants and humans. Some nutritionists consider iron deficiency or anemia to be the most frequently observed dietary deficiency in people. Animals generally make efficient use of body iron. Iron deficiency in people is closely related to loss of blood and inefficient body use of dietary iron. A major source of iron for humans is meat. The result is little or no relationship between soil-iron, plant-iron, and animal health (see Fig. 14-9).

Molybdenum. Plants and animals need very low amounts of molybdenum. Plants deficient in molybdenum for good animal nutrition have been found growing on certain acid soils. The major nutritional problem is molybdenum toxicity, molybdenosis, which develops in grazing animals when forage has over 10 to 20 parts per million of molybdenum. The toxicity problem arises from the fact that forage plants have a wide range of tolerance for molybdenum, but animals do not. Legumes accumulate more molybdenum than common grasses, and legumes take up more molybdenum in wet soils than in dry soils. There is no effective method to reduce molybdenum uptake from soils by plants. The data in Fig. 14-9 show that large increases in molybdenum can occur in plants with increases of molybdenum in the growing medium.

Most of the problem areas of molybdenum toxicity are related to the geologic origin and the wetness of the soils. The

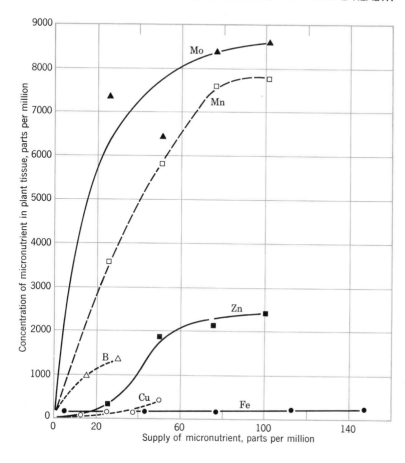

Fig. 14-9 The iron (and copper) content of tomato leaflets was little affected by a large increase in supply of nutrient in the growing medium. (From Beeson, 1955.)

common sources of molybdenum are granite, shales, and fine-grained sandstones.

Soils producing forages with high levels of molybdenum are generally confined to valleys of small mountain streams in the western United States. Only a very small part of any valley actually produces high molybdenum forages. These soils are wet or poorly drained, alkaline, and high in organic matter, and the alluvium from which they were formed was originally derived from granites or high-molybdenum shales. Molybdenosis in Nevada and California is associated with soils formed from the high-molybdenum-content granite of the Sierra Nevada.

A typical area where high-molybdenum forages grow and the effect of these forages on cattle are shown in Fig. 14-10. Molybdenum toxicity in cattle has also been found where dusts from molybdenum-processing industries or waste water from the tailings of uranium mines has contaminated pastures. Molybdenum toxicity has also occurred in cattle grazing pastures on organic soils in Florida. Molybdenum toxicity has been found in Hawaii on volcanic ash soils at high elevations where soil is wet most of the time.

Selenium. Selenium is not needed by plants but is needed in small amounts by

SOILS, PLANT COMPOSITION, AND ANIMAL HEALTH

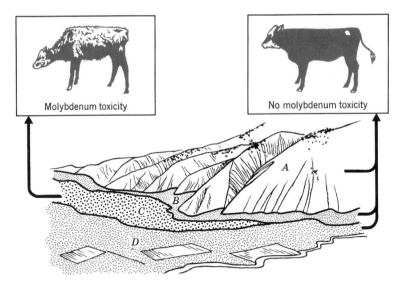

Molybdenum toxicity

No molybdenum toxicity

Fig. 14-10 Plants containing toxic levels of molybdenum are found only on wet soils formed from high-molybdenum parent materials, area C. Areas A and B have well-drained soils in which the molybdenum is not readily available to plants. Area D is wet but has soils formed from low molybdenum parent materials. (From Allaway, 1975.)

warm-blooded animals and probably people. Large areas of the United States have soil so low in selenium that the forages are deficient in this element for livestock. Selenium deficiency is a serious livestock problem. Acid soils formed from low-selenium parent materials are areas of likely selenium deficiency in livestock. These conditions are prevalent in the northwestern, northeastern, and southeastern states (Fig. 14-11). Much of the wheat in the United States is produced in areas where soils have adequate selenium, making bread a good source of selenium for people. People in the United States appear to have neither deficiencies or toxicities of selenium. Areas where forages and feed grains are likely to be deficient, adequate, or toxic for livestock are shown in Fig. 14-11.

In 1934 the mysterious livestock maladies on certain farms and ranches of the Plains and Rocky Mountain States were discovered to be due to plants with so much selenium that they were poisonous to ani-

mals grazing there. Affected animals had sore feet, lost some of their hair, and many died. Over the next 20 years scientists found that the high levels of selenium occurred only in soils derived from certain gelogical formations of high selenium content. Another important discovery was that a certain group of plants, called the selenium accumulators, had an extraordinary ability to extract selenium from the soil. These selenium accumulators were primarily shrubs or weeds native to semiarid and desert rangelands. They usually contained 50 parts per million or more of selenium, whereas range grasses and field crops growing nearby contained less than 5 parts per million of selenium.

An interesting story relating selenium toxicity to George Custer's fateful day at the Little Big Horn River in Montana on June 25, 1876 was written by E. V. Wilcox. He learned from an old horse packer that the horses used in the battle had been overwintered where soils were high in selenium and that the horses were allowed to forage

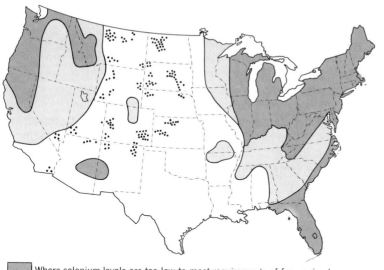

■ Where selenium levels are too low to meet requirements of farm animals

□ Where selenium is adequate to meet requirements of farm animals

▨ Where selenium is both adequate and inadequate in same locality

• Where selenium toxicity may be a problem

Fig. 14-11 Areas where forages and feed crops contain various levels of selenium. (From Allaway, 1975.)

native range plants because of a shortage of army hay. Enroute to the battle site a legume selenium accumulator, Astragalus bisulcatus, was in its succulent and palatable stage. The horses likely grazed on it enroute. The horse packer also related that the horses had peculiar symptoms similar to those now known to be selenium toxicity. Studies of the soils made in the 1930s in the area showed that Astragalus bisculatus had 140 times more selenium content than nonaccumulator plants growing alongside. Sitting Bull's horses came from an area outside the selenium toxicity belt. Wilcox suggests that one factor in the defeat was the selenium toxicity and poor performance of Custer's horses.

Zinc. Both plants and animals develop zinc deficiencies. Zinc deficiencies in crops have frequently been observed on fields where soils have been graded smooth so that irrigation water can be applied more uniformly. The deficiency occurs where the topsoil has been removed and a calcareous, low organic content soil is exposed. Research indicates that zinc fertilization of food and feed crops may be potentially useful in improving plants as sources of dietary zinc. No close relationship, however, has been found between available zinc in soils and the occurrence of zinc deficiency in people.

Soil Fertility Depletion and Nutritive Quality of Plants

We have noted that the cattle of the early settlers in New England suffered cobalt deficiency on soils of low but "natural" fertility. Shakespeare makes reference to goiter in mountainous regions now known to have soils low in iodine. It has been well

372

documented that animal and human health have been adversely affected by mineral nutrient deficiencies in food grown on natural soils or on soils little affected by humans. Some people are concerned that soil depletion results in changes in plant composition, causing a decline in human health. Some of the most dramatic cases in humans or animals of nutritional deficiencies associated with soils date back a long time and are due to naturally occurring deficiencies instead of to those of soil depletion.

Scientists at Michigan State University established an experiment on a farm where the land had been neglected; soils were very acid and depleted of nutrients from many years of cropping. Part of the land was used for crop production with no attempt made to restore the depleted nutrients or correct soil acidity by liming, and part of the land was limed and heavily fertilized. Two similar dairy herds were fed feed only from these two areas to test for differences in nutritional quality of the crops. At the end of 5 years, no differences were observed in the nutritional quality of the milk, health, or reproduction of the herds.

The average concentration of phosphorus in food crops grown on commercial farms is probably as high or higher than that in similar crops grown 50 years ago because phosphorus fertilizers have been extensively used. The concentration of iron and manganese in plants is controlled most often by factors that affect the ability of the plant to utilize these elements and seldom by the total amount of the element in the soil. The sodium, chlorine, and iodine required by people have been supplied by direct supplementation of diets, and changes in the levels of these elements in soil or in food crops would have no effect on human nutrition.

Of all the mineral elements required by people and animals, downward trends in concentration of zinc, magnesium, and possibly sulphur in food and feed crops would appear to be more likely than for any of the others. Even with these elements, any evidence of a decline must rest on circumstantial evidence, such as increasing reports of magnesium deficiency in cattle or the need for zinc or sulphur fertilizers to obtain optimum crop yields. There is no evidence that a decline, if there has really been a decline, in the concentration of zinc, magnesium, or sulphur in food crops has any effect on the nutritional status of people; this status is strongly dependent on food selection practices, dietary habits, and utilization of these elements in the diet.

In considering the effects of people's use of soils for agriculture, it is necessary to distinguish between depletion of the soil's supply of essential elements for plants and animals and deterioration of the soil due to the washing or blowing away of the surface soil to expose hard or rocky subsoil material. Depletion of the soil supply of essential nutrients has generally been recognized by agriculture research workers and has usually been corrected by proper use of fertilizers before there is any decline in the nutritional quality for people or animals of the crops produced. It is often much more difficult to correct or reclaim areas that have been damaged by excessive erosion or soil blowing. Some of the historical records of failure of settlements in certain parts of the world can be attributed to a failure of crop production from destruction of soil by erosion or by the "salting out" of irrigated lands. There are no records where failure of settlements can be attributed to crops of poor nutritional quality that result from depletion of the soil supply of essential elements.

Other Soil–Animal Health Concerns

There are several health concerns that are not closely related to the mineral nutrient composition of plants but are still related to soils. These include nitrate accumulation in plants, relation of soils to the vitamin content of plants, and potentially dangerous elements resulting from environmental pollution.

The Nitrate Problem

Nitrate by itself is not very toxic to animals when present in food and feed crops. Under certain conditions, however, nitrates in the digestive tracts of animals or in stored foods may be converted to nitrites. Nitrites are very toxic to animals; once they are absorbed into the blood they react with hemoglobin in a way that interferes with the transport of oxygen in the bloodstream.

High levels of nitrate in plants usually result from high levels of nitrate in soils, plus the impact of some environmental factor that interferes with the metabolism of nitrates into amino acids and protein. Nitrate accumulation in plants is favored by drought and cloudy weather. Extremely high levels of nitrate in soils, however, may cause high levels of nitrates in plants, even though environmental conditions are favorable for plant growth and the utilization of nitrates in the plant to form amino acids and proteins.

Different plants show different tendencies to accumulate nitrates. Leafy vegetables, cereals, and grasses cut at hay stage and some annual weeds are likely to contain high levels of nitrate. The pigweed is known as a nitrate accumulator. Perennial grasses and legumes are much less likely to be high in nitrate. The nitrate tends to remain high in leaves and low in seeds. Nitrates have not been a problem in the consumption of grains used as food or feed crops.

Forages high in nitrogen are especially dangerous to ruminants such as cattle and sheep. The rumen of these animals provides an environment conducive to reducing nitrate to the toxic nitrite. Losses of cattle and sheep because of nitrite poisoning have been a serious problem in livestock production for many years. The problem is particularly acute during seasons of cloudy weather or drought, which interrupts plant growth.

High levels of nitrate in soils may result from the excessive use of nitrogen fertilizers, excessive use of readily decomposible composts and manures, accumulation of nitrate from organic matter during fallow or drought periods, or breakdown of organic matter and green legume manure crops. However, plants containing high levels of nitrate often have been found growing on soils that have not received any fertilizer, manure, or compost. These plants are likely to be nitrate accumulators. A major step in an attempt to control nitrate accumulation in plants is to develop systems of soil management and the use of manures, fertilizers, and crop residues that will provide ample but no excessive supplies of available nitrates for plants.

Soil Fertility and Vitamins in Plants

Vitamins are organic compounds synthesized in plants; animals and people need at least 14 vitamins to remain healthy. About 30 years ago there was a growing concern about the effect of soils on the vitamin content of plants. Studies relating to carotene and vitamin C were numerous because of our great dependence on plants for these vitamins in our diet. It was found that yellow plants having an iron deficiency were low in carotene. When plants were green and grew normally, the carotene content was a function of the species and variety and not the soil. Since it is uneconomic to grow deficient plants low in carotene, there is no need for concern.

The vitamin C level in plants is dependent on the amount of sunlight striking the plant. Tomatoes heavily fertilized and shaded by luxuriant vegetative growth have less vitamin C than unfertilized plants with less foliage and more exposure to the sun. The other vitamins appeared to be controlled mainly by plant species or variety. Human vitamin deficiencies have been critical over the years, but no vitamin deficiencies have been known to be caused by a soil deficiency where the food is grown.

Potentially Toxic Elements in Plants from Environmental Pollution

Potential poisoning of humans by arsenic, cadmium, lead, and mercury are real concerns today. Arsenic has accumulated in soils from sprays used to control insects and weeds and to defoliate crops to facilitate harvest. Arsenic accumulates in soils and has injured crops, but has not created a hazard for humans or animals.

Cadmium poisoning has occurred in Japan from dumping of mine waste into rivers, where fish accumulated cadmium. Cadmium is an industrial waste and appears in sewage sludge. Using soils for sewage disposals represents a potential danger. No natural soils have harmful levels of cadmium.

Lead is discharged into the air from automobile exhausts and other sources and eventually reaches the soil. In the soil the lead is converted to forms unavailable to plants. The lead taken up tends to remain in the roots. Soils must become very polluted with lead before significant amounts move into the tops of plants and become a health hazard.

Mercury is discharged into the air and water from use in pesticides and industrial activities. Inorganic mercury is not highly toxic, but methyl mercury is. Under conditions of poor aeration, inorganic mercury is converted to methyl mercury. Plants do not take up mercury readily from soils, but soils should not be used to dispose of mercury because of the highly toxic nature of methyl mercury.

The soil is being used more and more for sewage disposal. The application of high levels of heavy metals to soils and their potential uptake by plants represents an important limitation in the use of soils for sewage disposal and industrial waste. Sewage known to be lacking in heavy metals, however, can be added to soils without any danger of heavy metal poisoning to humans or other animals.

References

Allaway, W. H., "The Effect of Soils and Fertilizers on Human and Animal Health," *USDA Agr. Information Bull.*, *378*, 1975.

Beeson, Kenneth C., "The Effect of Fertilizers on the Nutritional Quality of Crops," in *Nutrition of Plants, Animals, and Man*, Centennial Symposium Proc., Michigan State University, E. Lansing, Mich., 1955.

Brown, J. R., "Plant Analysis," *Missouri Agr. Exp. Sta. Bull.*, SB881, 1970.

Dexter, S. T. et al., "Nutritive Values of Crop and Cow's Milk as Affected by Soil Fertility," *Mich. Agr. Exp. Sta. Quart. Bul.*, *32*:352–359, 1950.

Grunes, D. L., P. R. Stout, and J. R. Brownell, "Grass Tetany of Ruminants," in *Advances in Agronomy*, Vol. 22, Academic, New York, 1970, pp. 331–374.

Hanway, John, "Growth and Nutrient Uptake by Corn," *Iowa State University Extension Pamphlet, 277,* Ames, Iowa, 1960.

Jackson, James E., and Glenn W. Burton, "An Evaluation of Granite Meal as a Source of Potassium for Coastal Bermudagrass," *Agron. Journ., 50*:307–308, 1958.

Kubota, Joe, "Soils and Animal Nutrition," *Soil Conservations,* November 1965, pp. 77–78.

Kubota, Joe, "Areas of Molybdenum Toxicity to Grazing Animals in the Western States," Jour. Range Management, *28*:252–256, 1975.

Kubota, Joe, "The Poisoned Cattle of Willow Creek," *Soil Conservation,* April 1975, pp. 18–21.

Smith, Floyd W., "Fertilizing Wheat for Profit," *Plant Food Review, 10*:4–6, 1964.

Snider, H. J., "Chemical Composition of Hay and Forage Crops as Affected by Various Soil Treatments," *Ill. Agr. Exp. Sta. Bull., 518,* 1946.

Stanford, George, and W. H. Pierre, "The Relation of Potassium Fixation to Ammonium Fixation," *Soil Sci. Soc. Am. Proc., 11*:155–160, 1946.

The Fertilizer Institute, "*Our Land and Its Care,*" *Washington, D.C.,* 1962.

Trelease, S. F., "Bad Earth," *Scientific Monthly, 54*:12–28, 1942.

Viets, Frank G., and Richard H. Hageman, "Factors Affecting the Accumulation of Nitrate in Soil, Water, and Plants," *USDA Agr. Handbook,* No. 413, Washington, D.C., 1971.

Wallace, A., et al., "Further Evidence Supporting Cation-Equivalent Constancy in Alfalfa," *Jour. Am. Soc. Agron., 40*:80–87, 1948.

Wilcox, E. V., "Selenium Versus General Custer," *Agr. Hist., 18*:105–107, 1944.

15

SOIL EROSION AND ITS CONTROL

In the development of a new country little attention is given to conservation. Usually the natural resources that the country affords are present in such quantity that they appear inexhaustible. The limitation of supply is in manpower and in the essentials of life that the country does not produce. Later, when population density has become great, the necessity for conserving resources becomes apparent and frequently acute. Often, as in the United States, the need for conserving soil is one of the last to be recognized. In fact, many people in our country do not yet realize the urgency for conserving soil resources. Our surplus production of a few crops, like cotton, wheat, and corn, tends to obscure the waste in soil productivity that is being sustained. The problem as a whole attracted no widespread attention until about 1933. Since that time, public interest in the work has developed rapidly.

Soil Erosion Defined

The basic definition of the word "erosion" is to wear away. Since the earth was first

377

formed, there has been a continual wearing away of the surface. Many agents are responsible, but the discussion here will be limited to cultivated fields. It will be restricted to water and wind erosion.

Geological versus Soil Erosion

Erosion that takes place under natural conditions (i.e., when the land surface and native vegetative cover have not been disturbed by human activities) is called natural or *geological erosion*. On the other hand, when timberland is cleared or grassland is broken up, processes of erosion are accelerated, and we have unnatural or soil erosion. Whenever erosion is speeded up as a result of human activities so that it removes all or part of the topsoil, we call the process *soil erosion.*

Geological erosion is a relatively slow process under many conditions, and soil formation may keep pace with the removal of the surface soil. Soil erosion, on the contrary, is very rapid when environmental factors favor erosion. For example, at the Statesville North Carolina Experiment Station, land planted to cotton year after year suffered an annual loss of 31 tons of soil per acre. The same kind of land, however, under natural conditions lost only 0.002 ton of soil per acre annually by geological erosion. Nevertheless, over long periods of time geological erosion moves very large quantities of soil material to lower levels, as pointed out in the following section.

Types of Water Erosion

Erosion by water may be divided into four categories: splash, sheet, rill, and gully. Strictly speaking, sheet erosion refers to the quite uniform removal of soil from the surface of an area in thin layers. For sheet erosion alone to occur it is necessary that there be a smooth soil surface, which is seldom the case. Usually a soil surface that is designated "smooth" contains small depressions in which water will accumulate. Overflowing from these at the lowest point, the water cuts a tiny channel as it moves down the slope. Duplicated at innumerable points, this process presently creates a surface cut by a multitude of very shallow trenches that are called rills. None of these may grow to appreciable size or depth, so the surface soil is rather uniformly removed from the field. Accordingly, sheet erosion and rill erosion work hand in hand; the combined process is usually called sheet erosion, as distinguished from gully formation.

Although sheet erosion may pass unnoticed by the average observer, gullies attract immediate attention. They disfigure the landscape and give the impression of land neglect and soil destruction (Fig. 15-1). Not only do gullies result in soil loss but also, as previously mentioned, the eroded material is usually deposited over more fertile soil at the foot of the slope. Also, fields dissected by gullies offer many problems in farming operations.

Fig. 15-1 Severe gully erosion cuts up fields and makes land unfit for crop production.

Gullying proceeds by three processes: (1) waterfall erosion, (2) channel erosion, and (3) erosion caused by alternate freezing and thawing. Usually more than one process is active in a gully. Water falling over a soil bank undermines the edges of the bank, which then caves in, and the waterfall moves upstream. This process produces U-shaped gullies, particularly if the underlying soil material is soft and easily cut. Gullies that are V-shaped are produced by *channel erosion* through the cutting away of the soil by water concentrated in a drainageway. This type of gully usually forms when the underlying soil horizons are of finer texture and more resistant to erosion than the surface horizons. Soil loosened from sides of gullies by alternate freezing and thawing sloughs off and is then carried away by heavy rains. Gullies are sometimes classified as small if they are less than 3 feet deep, medium if they are 3 to 15 feet deep, and large if they are over 15 feet in depth.

A type of erosion that received little attention until recent years is the splashing or scattering of the smaller soil particles by the impact of raindrops. At first thought this action seems trivial but, when consideration is given to the large number of raindrops that strike a square inch of soil surface during a 1-hour rain and the force with which they strike, it is seen that the net effect in loosening and moving soil particles may be considerable.

Geological Erosion a Natural Process

Geological erosion is a natural process that tends to bring the earth's surface to a uniform level. Whenever one part of the surface of the globe is elevated above the surrounding portions, erosion immediately begins the work of leveling off the high land. The leveling process may result in a very rough topography in the early stages by the cutting of gullies or of canyons in a mountainous region, but the ultimate result is a comparatively level surface. Evidence of this geological process is seen in peneplanes, mesas, valley fills, alluvial plains and deltas, extensive deposits of wind-laid material, and numerous other geological formations. Some concept of the extent of erosional activity may be gained from the fact that the Appalachian Mountains are about one half of their original height. The loss of surface soil is generally balanced by the gradual incorporation of less-weathered material into the lower part of the profile (see Fig. 15-2). The net result is the maintenance of soil fertility by the gradual incorporation of more nutrient-rich material within the root zone and the delay or inhibition in the evolution of highly weathered and infertile soils when viewed on a geologic time scale.

Much of the eroded material carried by rivers is derived by deep cutting of streams into relatively fresh, unweathered, and

Fig. 15-2 A landscape in the foothills of the Rocky Mountains where the rate of erosion nearly balances the formation of soil, resulting in thin soils.

nutrient-rich rocks and sediments. Deposition of these materials on alluvial plains has created large areas of fertile soil along the major rivers of the world (Fig. 15-3). The overflowing and deposition of silt along the Nile is a classic example. Most of the Chinese live on alluvial soil that is the by-product of erosion.

The pervasiveness of erosion gives rise to the concept of soil as "rock en route to the sea." En route to the sea, however, soil particles may participate in multiple cycles of erosion, deposition, and soil profile evolution. Each cycle results in a progressive depletion of weatherable minerals and increasing soil infertility. This can be observed along a traverse from the Appalachian Mountains to the Piedmont, to the Upper Coastal Plain and eventually to the Lower Coastal Plain near the Atlantic Ocean. It is obvious that there are many interesting aspects of water erosion. The discussion that follows will focus on erosion, which has accelerated as a result of the removal of vegetation in agriculture, forestry, and urbanization.

Reasons for Employing Water Erosion Control Practices

A relatively short trip through virtually any section of our country reveals evidences of soil mismanagement. In the midst of a productive agricultural area many farms will be observed that have been depleted in organic matter content, as evidenced by a lighter color and lower productivity than surrounding soil. In areas of undulating to

Fig. 15-3 Alluvial soils formed by the deposition of eroded material during floods are some of the most productive soils in the world.

rolling topography, slopes may be seen that have been denuded of the surface or plow soil, leaving the lighter-colored subsoil exposed. Again, many instances are seen of sandy soils that have been cleared of their forest cover, cropped for a few years until the virgin fertility was exhausted, and then abandoned to become covered with weeds and brush or to be blown about by the wind.

As pointed out in Chapter 16, many nations have inadequate food supplies, and a greatly increased food production will be needed to feed the expected world population in the future. This situation affords another reason for giving careful attention to soil conservation.

Damage Done to Agricultural Land by Soil Erosion

The damage done to farm lands in the United States since they were occupied by white humanity can only be estimated roughly. Of land used for crop production, it is estimated that around 50 million acres have been rendered useless for crop production, and a like amount is approaching that condition. An additional 100 million acres, although still being cultivated, have lost one half or more of the surface soil; on a similar acreage erosion is carrying on its insidious work of destruction. H. H. Bennett, former chief of the Soil Conservation Service, has expressed the opinion that erosion control "is the first and most essential step in the direction of correct land utilization on about 75 percent of the present and potential cultivated area of the nation."

In considering the damage done by erosion one should keep in mind the fact that a large share of the soil lost by this process is the surface or plow soil. It is this soil layer that contains the highest percentage, in an available condition, of many of the essential plant-food elements. Furthermore, studies of the soil eroded from fields in many parts of the country have shown these to be made up largely of the finer soil particles (clay, silt, very fine sand, and humus). These particles contain a higher percentage of several of the plant nutrients than do the coarser particles. A study in Wisconsin showed that compared to the original soil the eroded material contained 2.1 times more organic matter, 2.7 times more nitrogen, 3.4 times more available phosphorus, and 19.3 times more exchangeable potassium.

Erosion Damage Not Confined to Soil Loss

Erosion by water opens the way for at least five types of loss or damage.

1. The loss of the water causing the erosion. It might have been useful in crop production had it entered the soil instead of running off over the surface.

2. The soil carried away by erosion frequently ceases to be of value in crop production; furthermore, the remaining soil, denuded of the surface or plow soil, is much decreased in productivity.

3. The soil carried away frequently causes much damage. Especially during gully formation, a layer of infertile subsoil may be deposited over an area of productive soil, thus greatly reducing the crop-producing power.

4. Another damage resulting from gully formation is the cutting up of fields into irregular pieces. As these gullies get too deep to cross with farm implements, great inconvenience and loss of efficiency in cultivating the land and planting and harvesting crops result.

5. The soil removed through erosion may be deposited in streams, harbors, and reservoirs, thus increasing floods, impeding navigation, and reducing water-storage capacity.

6. Soil particles transport adsorbed nutrients and pesticides into water courses.

Rapid Sedimentation of Reservoirs

The amount of sediment carried by various streams is enormous, as is shown by the estimates made by the United States Geological Survey given in Table 15-1. When water storage reservoirs are built on rivers carrying large amounts of sediment, the reservoirs quickly fill and lose their water-storage capacity. An extreme example of the loss of storage capacity as a result of erosion is furnished by the Washington Mills Reservoir at Fries, Virginia. In the course of 33.5 years, 83 percent of the storage capacity had been lost.

A study of the Lake Decatur, municipal water-supply reservoir of Decatur, Illinois, showed that it had lost 1.0 percent of its storage capacity annually between the date of construction (1922) and 1936. Between 1936 and 1946 the annual rate of sedimentation had increased to 1.2 percent of the capacity. Also, in a land-utilization-project area near Pierre, South Dakota, the annual rate of silting of stock ponds was found to vary between 1.10 and 5.56 percent.

The Roosevelt Dam, which supplies water for electric power and for irrigation of the great Salt River Valley in Arizona, lost 7 percent of its storage capacity through sedimentation during the first 24 years of its existence. Likewise, the Elephant Butte Dam on the Rio Grande River in New Mexico decreased in storage capacity about 17 percent during the period between 1915 and 1947. Such losses of storage capacity are serious matters, especially when the supply of water is scarcely adequate to meet demands.

Other Environmental Consequences of Water Erosion

Sediment or soil is the major pollutant in the country exceeding by 500 to 700 times the amount of sewage discharged into waters. Over half of the population obtains its municipal water from surface water that is almost universally filtered to remove sediment. Harbors and streams are dredged to maintain navigable waterways. Sediment in streams has a negative effect on fish. Sediment buries fish eggs, reduces the penetration of sunlight, which in turn reduces plant growth (food), and reduces recreational quality and beauty. Sediment particles also serve as carriers of adsorbed pes-

Table 15-1
Tons of Sediment Carried by Several Rivers

River	Tons Per Year	River	Tons Per Year
Hudson	240,000	Savannah	1,000,000
Susquehanna	240,000	Tennessee	11,000,000
Roanoke	3,000,000	Missouri	176,000,000
Alabama	3,039,000		

ticide residues and phosphorus from the land to surface waters.

Predicting Water Erosion Losses on Agricultural Land

Since 1930 many controlled studies on field plots and small water sheds have been conducted to study the factors affecting erosion. This data forms the basis for the soil-loss prediction equation developed by Wischmeier and Smith. The soil-loss equation is:

$$A = R\ K\ L\ S\ C\ P$$

where A is the computed soil loss per unit area (tons per acre).

R, the rainfall factor, is the number of erosion-index units in a normal year's rain. The erosion index is a measure of the erosive force of specific rainfall.

K, the soil-erodibility factor, is the erosion rate per unit of erosion index for a specific soil in cultivated continuous fallow, on a 9-percent slope 72.6 feet long.

L, the slope-length factor, is the ratio of soil loss from the field slope length to that from a 72.6-foot length on the same soil type and gradient.

S, the slope-gradient factor, is the ratio of soil loss from the field gradient to that from a 9-percent slope.

C, the cropping-management factor, is the ratio of soil loss from a field with specified cropping and management to that from the fallow condition on which the factor K is evaluated.

P, the erosion-control practice factor, is the ratio of soil loss with contouring, stripcropping, or terracing to that with straight-row farming, up-and-down slope.

The six factors in the equation are the significant factors that influence soil loss by rainfall and each will be briefly discussed.

The Rainfall Factor (R)

The rainfall factor is a measure of the erosive force of specific rainfall. The erosive force or available energy is related to both quantity and intensity of rainfall. A 2-inch rain falling at 20 miles per hour would have 6 million foot-pounds of kinetic energy. The tremendous erosive power of such a rain is apparent when you consider that it is sufficient to raise a 7-inch acre furrow slice of soil 3 feet (see Fig. 15-4). The four most intense storms in a 10-year period at Clarinda, Iowa accounted for 40 percent of the erosion and only 3 percent of the runoff on plots in corn tilled up-and-down slope.

The rainfall or R factor is the sum of the kinetic energy times the maximum 30-minute intensity for each storm during the year. Rainfall factors have been computed for about 2000 locations in the states east of the Rocky Mountains and used to produce the iso-erodent map shown in Fig. 15-5. The values range from 50 in western North Dakota to 600 along the gulf coast.

The rainfall-erosion index measures only the erosivity of rainfall and associated runoff. Therefore, the equation does not predice soil loss that is due solely to thaw,

Fig. 15-4 Columns of soil capped by stones that absorbed the energy of raindrop impact and protected the underlying soil from erosion.

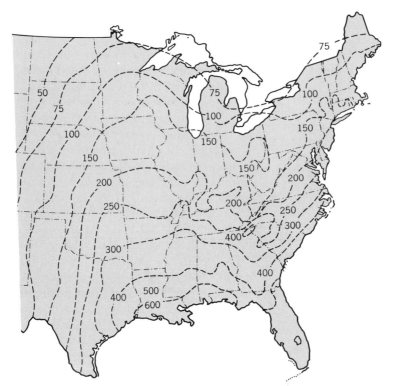

Fig. 15-5 Rainfall factors (R) for the eastern United States.

snowmelt, or wind. In areas where such losses are significant, they must be estimated separately and combined with those predicted by the equation for comparison with soil-loss tolerances.

The Soil-Erodibility Factor (K)

Soil factors that influence erodibility by water are (1) those that affect infiltration rate, permeability, and total water capacity, and (2) those that resist dispersion, splashing, abrasion, and transporting forces of the rainfall and runoff. The erodibility factor, K, has been determined experimentally for 23 major soils on which soil erosion studies were conducted since 1930. The soil loss from a plot 72.6 feet long on a 9-percent slope that is maintained in fallow with all tillage up and down the slope is determined and divided by the rainfall factor (for the storms producing the erosion) is the erodibility factor (K). Values of K determined for 23 major soils are listed in Table 15-2.

Slope Length (L) and Slope Gradient (S) Factors

Slope length is defined as the distance from the point of origin of overland flow to either a point where the slope decreases to the extent that deposition occurs or the point where runoff enters a well-defined channel. Runoff from the upper part of a slope contributes to the runoff produced on

Table 15-2
Computed K Values for Soils on Erosion-Research Stations

Soil	Source of Data	Computed K
Dunkirk silt loam	Geneva, N.Y.	0.69[a]
Keene silt loam	Zanesville, Ohio	0.48
Shelby loam	Bethany, Mo.	0.41
Lodi loam	Blacksburg, Va.	0.39
Fayette silt loam	LaCrosse, Wis.	0.38[a]
Cecil sandy clay loam	Watkinsville, Ga.	0.36
Marshall silt loam	Clarinda, Iowa	0.33
Ida silt loam	Castana, Iowa	0.33
Mansic clay loam	Hays, Kans.	0.32
Hagerstown silty clay loam	State College, Pa.	0.31[a]
Austin clay	Temple, Tex.	0.29
Mexico silt loam	McCredie, Mo.	0.28
Honeoye silt loam	Marcellus, N.Y.	0.28[a]
Cecil sandy loam	Clemson, S.C.	0.28[a]
Ontario loam	Geneva, N.Y.	0.27[a]
Cecil clay loam	Watkinsville, Ga.	0.26
Boswell fine sandy loam	Tyler, Tex.	0.25
Cecil sandy loam	Watkinsville, Ga.	0.23
Zaneis fine sandy loam	Guthrie, Okla.	0.22
Tifton loamy sand	Tifton, Ga.	0.10
Bath flaggy silt loam with surface stones 2 inches removed	Arnot, N.Y.	0.08[a]
Albia gravelly loam	Beemerville, N.J.	0.03

From Wischmeier and Smith, 1965.
[a] Evaluated from continuous fallow. All others were computed from row crop data.

the lower part of the slope. This increases the running over the lower part of the slope, creating more erosion on the lower part of the slope compared to the upper part. Studies have shown that erosion by water increases as the 0.5 power of slope length and is used as the basis for calculation of the slope length factor L. This results in about a 1.3 times greater soil loss per acre with a doubling of slope length.

As the gradient or percent of slope increases, the velocity of runoff water increases, which increases its erosive power. A doubling of velocity of runoff water increases the kinetic energy or erosive power four times and causes a 32-time increase in the amount of material of a given particle size that can be carried. Splash erosion, the splashing into the air of soil particles by raindrop impact, causes a net downslope movement of soil and also increases with slope gradient. Combined slope length and gradient factors (LS) for use in the soil loss prediction equation are given in Fig. 15-6.

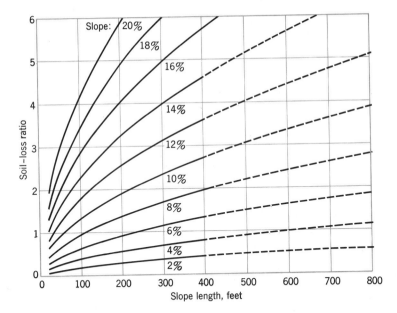

Fig. 15-6 Slope-effect chart for the topographic factor LS.

The Cropping-Management Factor (C)

We have observed the great erosive potential of rainfall on bare land on long and steep slopes. A vegetative cover, however, can absorb the kinetic energy of falling rain drops and defuse the rain's erosive potential. Furthermore, vegetation by itself retains a significant amount of the rain and slows the flow of runoff water. As a result, the presence or absence of a complete vegetative cover essentially determines whether erosion will be a problem or whether erosion will be zero, for all practical purposes (see Fig. 15-7).

The C factor measures the combined effect of all the interrelated cover and management variables including type of tillage, residue management, time of soil protection by vegetation, and so forth. Complicated tables have been devised for calculating the C factor. As an illustration, the C factor for a 4-year rotation of wheat-meadow-corn-corn in central Indiana with conventional tillage, average residue and other management, and average yields is 0.119.

Fig. 15-7 A continuous plant cover (e.g., bluegrass) is many times more effective for erosion control than cropping systems that leave the land bare most of the time. Continuous corn production resulted in the most erosion and the longest amount of time with soil unprotected by vegetation.

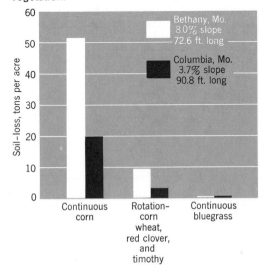

The Erosion Control Practice Factor (P)

In general, whenever sloping soil is to be cultivated and exposed to erosive rains, the protection offered by sod or close-growing crops in the systems needs to be supported by practices that will slow the runoff water and thus reduce the amount of soil carried. The most important of these practices for cropland are contour tillage, stripcropping on the contour, and terrace systems.

Limited field studies have shown that contouring alone is effective in controlling erosion during storms of low or moderate intensity but provides little protection against the occasional severe storm that causes breakovers of the contoured rows. Contouring alone appears to produce maximum average protection on slopes in the range of 3 to 7 percent. P values for contouring are given in Table 15-3. Stripcropping along with contouring provides more protection. In cases where both stripcropping and contour tillage are used, the P values listed in Table 15-3 are divided by 2.

Table 15-3
Practice Factor Values for Contouring

Land Slope, %	P value
1.1 to 2	0.60
2.1 to 7	0.50
7.1 to 12	0.60
12.1 to 18	0.80
18.1 to 24	0.90

From Wischmeier and Smith, 1965.

Terraces are an effective way to reduce slope length. To account for terracing, the slope length used to determine the LS factor should represent the distance between terraces.

Application of the Soil-Loss Equation

A Case Study

The procedure for computing the expected average annual soil loss from a given cropping system on a particular field is illustrated by the following example.

Assume that there is a field in Fountain County, Ind., on Russell silt loam, having an 8-percent slope about 200 feet long. The cropping system is a 4-year rotation of wheat-meadow-corn-corn (W-M-C-C), with tillage and rows on the contour and with corn residues disked for wheat seeding and turned under in the spring for second-year corn. Fertility and residue management on this farm are such that crop yields are rarely less than 85 bushels corn, 40 bushels wheat, or 4 tons alfalfa-brome hay, and the probability of meadow failure is slight.

The first step is to refer to the charts and tables discussed in the preceding section and to select the values of R, K, LS, C, and P that apply to the specific conditions on this particular field.

The value of the rainfall factor, R, is taken from Fig. 15-5. Fountain County, in west-central Indiana, lies between iso-erodents 175 and 200. By linear interpolation, R = 185. The value of the soil-erodibility factor, K, is taken from Table 15-2. Soil scientists in the north-central states consider Russell silt loam equal in erodibility to Fayette silt loam, for which Table 15-2 lists K = 0.38.

The slope-effect chart (Fig. 15-6) shows that, for an 8-percent slope, 200 feet long, LS = 1.41. For the productivity level and management practices assumed in this example, factor C for a W-M-C-C rotation in area 16 was shown to be 0.119.

Table 15-3 shows a practice-factor of 0.6 for contouring on 8-percent slope.

The next step is to substitute the selected numerical values for the symbols in the erosion equation and solve for A. In this example, $A = 185 \times 0.38 \times 1.41 \times 0.119 \times 0.6 = 7.1$ tons of soil loss per acre per year.

If planting had been up-and-down slope instead of on the contour, the factor P would have equaled 1.0, and the predicted soil loss for this field would have been $185 \times 0.38 \times 1.41 \times 0.119 \times 1.0 = 11.8$ tons per acre.

If contour farming had been combined with minimum tillage for all corn in the rotation, the value of the factor C would have been 0.075. The predicted average annual soil loss from the field would then have been $185 \times 0.38 \times 1.41 \times 0.075 \times 0.6 = 4.5$ tons per acre.

Permissible Soil Loss on Cultivated Land

Soils on level upland sites that have no erosion become infertile in humid regions because of intensive weathering and leaching. Soils on very steep slopes erode so fast that the soil remains thin and less fertile than if erosion was slower. Ideally, there is an optimum erosion rate for the maintenance of soil productivity. Unfortunately, this ideal rate is usually much less than the rate of erosion of sloping cultivated lands and the questions arises: "What is the maximum rate of erosion permissible that is consistent with long-time maintenance of soil productivity and economic use of land?"

The data in Fig. 15-7 show that a soil loss with continuous blue grass resulted in loss of about 0.03 tons annually or the loss of 1 inch of soil in about 5000 years (assuming 1 acre inch = 150 tons). By contrast, soil loss was 36 tons per acre annually for continuous corn, resulting in loss of 1 inch of soil in about 4 years. We cannot maintain all of the land in grass and produce the food we need. On the other hand, we obviously cannot tolerate the loss of 1 inch of soil every 3 to 4 years. Establishment of permissible soil losses for various soils is a matter of collective judgement based on soil properties and economic considerations. Maximum permissible soil losses for most soils in the United States have been judged to be between 1 and 5 tons per acre annually.

Loess is composed mainly of silt-sized particles that were transported and deposited as a result of wind action. The small size of the silt particle and lack of agents to stick the particles together makes loess very erosive. In northern China there is an extensive loess deposit that is locally over 250 feet thick. This loess is the major source of sediment for the Huang Ho or Yellow River and of the sediment deposited on the vast north China plain that supports hundreds of millions of people. The landscape is very distinctive with deeply entrenched streams and gullies. It has been estimated that the loess will be removed by erosion in 40,000 years, which represents a loss of about 15 tons per acre per year. The erosion rate is considered excessive by our standards but, in light of the need for food, the land must be cropped. Luckily, the loess has good physical properties and fertility compared to most other common soil parent materials.

The usefulness of the soil loss prediction equation rests in its ability to estimate soil losses and to aid in the formulation of managements systems consistent with the long-time maintenance of soil productivity.

Effect of Erosion on Crop Growth and Production Costs

Erosion results in loss of the soil with the greatest content of organic matter and nitrogen and, thus, erosion is particularly detrimental to nonleguminous grain crops. The reduced nitrogen-supplying power of the soil can be restored by the use of a nitrogen fertilizer; however, this increases the cost of production.

Exposure of argillic horizons because of erosion reduces infiltration and increases runoff, resulting in both more erosion and reduced available water for crops. Seed bed preparation is more difficult and stands are reduced with contant seeding rates. Summarization of data from many studies showed that an inch of soil loss reduced yields of wheat 5.3 percent, corn 6.3 percent, and grain sorghum 5.7 percent. Some effects of erosion of crop growth and production costs are given in Table 15-4. The increased cost of production resulting from erosion justifies a certain amount of investment to control erosion within the permissible limits.

Rainfall Erosion on Urban Lands

About half of the sediment in the streams and waters of the country originate on agricultural land; the other half originates on urban lands. We have observed that erosion rates can be very high on exposed land on steep slopes. In fact, exposure of land during highway and building construction and subdivision development commonly results in erosion rates many times greater than erosion that typically occurs on agricultural land. Urbanization may result in making half or more of the land surface impervious to water because of roads, buildings, and the like. Runoff is greatly increased. Erosion losses as large as 100,000 tons per square mile or about 1 inch per year have been reported during urbanization. Planning construction to leave land exposed for the shortest period of time is important to maintain quality in local waters (see Fig. 15-8).

Wind Erosion

Wind erosion is indirectly related to water conservation in that a lack of water leaves land barren and exposed to the wind. Wind

Table 15-4
Effects of Erosion on Corn Growth and Production Costs

	Degree of Erosion		
	Slight	Moderate	Severe
Depth of topsoil, inches	12	7	5
Decrease in height of plant during early stages of growth (from slight), percent	—	13	22
Stand at harvest, percent of planting rate	87	83	76
Corn yield, bushels per acre	112	96	87
Soybean yield, bushels per acre	43	29	16
Reduction in yield for all crops (from slight), percent	—	17	26
Increase in production cost (from slight), percent	—	20	56

From Beasley, 1974.

Fig. 15-8 Slurry truck spraying fertilizer and seed on freshly prepared seedbed along a recently completed road in Missouri. A thin layer of straw mulch will be secured with a thin spray of tar to protect the area until vegetation becomes established. (Photo USDA.)

erosion reaches its greatest extent in semiarid and arid regions. Nevertheless, much damage is caused to both crops and soils in humid areas by soil blowing, although the phenomenon is less spectacular and attracts comparatively little attention in these regions. In the United States more attention has been given to the control of wind erosion in the Great Plains than in other sections of the country. In North Dakota damage from this cause was reported as early as 1888, and in Oklahoma it was reported 4 years after breaking of the sod. Soil blowing was one of the hazards confronting the early settlers, and control of wind erosion was one of the first problems studied by agricultural experiment stations in the Plains states.

Three Types of Wind Erosion

Soil particles move in three ways during wind erosion. Fine soil particles, those from 0.1 to 0.5 millimeter in diameter, are rolled over the surface by direct wind pressure and then suddenly jump up almost vertically from a short distance to a foot or more.

Once in the air, the particles gain velocity and then descend in an almost straight line, not varying more than 6 to 12 degrees from the horizontal. The horizontal distance traveled by a particle is four to five times the height of its jump. Upon striking the surface, the particles may rebound into the air or knock other particles into the air and come to rest themselves. The major part of the soil carried by wind moves by this process, which is called *saltation*. It is interesting to note that around 93 percent of the total soil movement by wind takes place below a height of 1 foot, and probably 50 percent or more occurs between 0 and 2 inches.

Very fine dust particles are protected from wind action because they are too small to protrude above a minute viscous layer of air that clings to the soil surface. As a result, a soil composed entirely of extremely fine particles is very resistant to wind erosion. These dust particles are thrown into the air chiefly by the impact of particles moving in saltation but, once in the air, their movement is governed by wind action. They may be carried very high and over long distances by *suspension*.

Relatively large particles (between 0.5 and 1.0 millimeter in diameter) are too heavy to be lifted by wind action but are rolled or pushed along the surface by the impact of particles in saltation. This process is called *surface creep*. Between 50 and 75 percent of the soil is carried in saltation, 3 to 40 percent in suspension, and 5 to 25 percent in surface creep.

From these facts it is evident that wind erosion is due principally to the effect of wind on particles of a suitable size to move in saltation. Accordingly, wind erosion can be controlled (1) if the soil particles can be built up into clusters or granules of too large a size to move in saltation; (2) if the wind velocity near the soil surface can be reduced by ridging the land, by vegetable cover, or even by developing a cloddy surface; and (3) by providing strips of stubble or other vegetative cover sufficient to catch and hold the particles moving in saltation. Some management practices designed to provide these conditions are discussed in the following paragraphs.

Factors Affecting Wind Erosion and Its Control

The factors that affect wind erosion are contained in the wind erosion equation:

$$E = f(I, K, C, L, V)$$

where

    E = soil loss in tons per acre
    I = soil erodibility
    K = soil roughness
    C = climatic factor
    L = length of field
    V = quantity of vegetative cover

Soil erodibility is related primarily to texture and structure. As the clay content of soils increases, aggregation of the surface

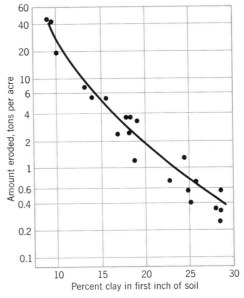

Fig. 15-9 Amount of wind erosion in relation to the amount of clay in the soil. (From Chepil et al., 1955.)

creates clods too large to be transported by the wind. A study in western Texas showed that soils with 10 percent clay eroded 30 to 40 times faster than soils with 25 percent clay (see Fig. 15-9). The effect of deep plowing of soils with argillic horizons on wind erosion was discussed in Chapter 3.

Rough surfaces reduce wind erosion by reducing wind velocity and trapping soil particles. Surface roughness can be increased with the tillage operations shown in Fig. 15-10. Tillage and rows are at right angles to the wind to be most effective.

Climate is a factor in wind erosion through wind frequency and velocity and wetness of soil during high-wind periods. Wind erosion and blowing dust are persistent features of deserts where plants are widely spaced and soil that is dry most of the time is exposed. Desert pavements are created as fine particles are blown away and

Fig. 15-10 Listing on the contour not only protects the listed field from wind action but also may collect much soil blowing from an adjoining field, as illustrated in this Oklahoma field. The velocity of the wind is indicated by the posture of the man in the foreground.

gravel accumulates (Fig. 15-11). Consequently, deserts are an important source of loess.

The Dust Bowl of the 1930s was associated with a period of unusually dry years. Both wheat yields and wind erosion were related to rainfall. The demise of the

Fig 15-11 Desert pavement created by wind erosion in the desert of the southwestern United States.

Dust Bowl coincided with years of increased rainfall.

Wind erosion increases from zero at the edge of a field to a maximum with increasing distance of soil exposed to the prevailing winds. As particles bounce and skip, they create an avalanche effect on soil movement. Windbreaks of trees and alternate strips of crops can be used to reduce field length. Finally, wind erosion is inversely related to degree of vegetative cover. Growing crops and the residues of previous crops when left on the surface are effective to control wind erosion.

The planting of crops in strips at right angles to the prevailing winds is valuable in reducing soil blowing. The width of the strips is determined by the nature of the soil, exposure to wind, and similar factors. On sandy soils the strips may be only 8 to 10 rods wide, but on fine-textured soils they may be from 15 to 20 rods in width.

Crop residues are an effective soil protection against wind erosion. Small-grain stubble reduces wind veolcity and also catches soil particles moving in saltation. Strips of stubble left at frequent intervals across a field being fallowed or fitted for a spring crop form effective barriers. The height of the stubble as well as wind velocity influence the width of strip needed.

Crops should be cultivated in such a way as to leave the residues on the surface; in other words, cultivate beneath them. This result can be accomplished by use of sweep-type cultivators. The stubble-mulch system of soil management has many possibilities and should be carefully investigated in areas where wind or water erosion is a serious hazard.

Trees for Windbreaks on Organic Soils

Organic soil areas of appreciable size are frequently protected from wind damage by planting windbrcaks of trees around them. In addition, rows of trees are often planted across the area at right angles to the prevailing wind. Willows have been found satisfactory for this purpose, especially since they make a very rapid growth in organic soil. The use of trees as windbreaks must be limited because of the large amount of soil they take out of crop production on account of their extensive root system. A number of shrubs also make good windbreaks. Spirea is frequently used.

Moist muck does not blow appreciably. Accordingly, use of an overhead irrigation system is helpful during windy periods before the crop cover is sufficient to prevent soil blowing. Rolling the muck with a very heavy roller induces moisture to rise more rapidly by capillarity and so dampens the surface layer. This practice is not effective unless the soil layers below the surface are quite moist.

Summary of Wind Erosion Control Principles

Wind erosion is a major problem of sandy soils when used for crop production in regions with ustic and aridic soil moisture regimes. Control is based on one or more of the following.

1. Trap moving soil particles with rough surface (tillage) and/or use of crop residues and strip cropping.
2. Deep-plow to increase clay content of surface soil.
3. Protect surface soil with complete vegetation cover.

References

Baver, L. D., "How Serious is Soil Erosion?", *Soil Sci. Soc. Am. Proc.*, *14*:1–5, 1950.

Beasley, R. P., "How Much Does Erosion Cost?", *Soil Survey Horizons, 15*:8–9, 1974.

Bennett, H. H., *Soil Conservation*, McGraw-Hill, New York, 1939.

Browing, G. M., R. A. Norton, A. G. McCall, and F. G. Bell, "Investigation in Erosion Control and the Reclamation of Eroded Land at the Missouri Valley Loess Conservation Experiment Station, Clarinda, Iowa," *USDA Tech. Bull., 959,* 1948.

Chepil, W. S., "Erosion of Soil By Wind," *Soil,* USDA Yearbook, Washington, D.C., 1957, pp. 308–314.

Chepil, W. S., N. P. Woodruff, and A. W. Zingg, "Field Study of Wind Erosion in Western Texas," *Kansas and Texas Agr. Expt. Sta., and USDA*, 1955.

Glymph, Louis M., and Herbert C. Storey, "Sediment—Its Consequences and Control," *Agr. and the Quality of Our Environment*, Am. Assoc. Adv. Sci., Pub. 85, Washington, D.C., 1967.

Lyles, Leon, "Possible Effects of Wind Erosion on Soil Productivity," *Jour. Soil and Water Con., 30*:279–283, 1975.

Robinson, A. R., "Sediment," *Jour. Soil and Water Con.*, *26*:61–62, 1971.

Massey, H. F., and M. L. Jackson, "Selective Erosion of Soil Fertility Constituents," *Soil Sci. Soc. Am. Proc.*, *16*:353–356, 1952.

Tuan, Yi-Fu, *China*, Aldine, Chicago, 1969.

Wischmeier, Walter H., and Dwight D. Smith, "Predicting Rainfall–Erosion Losses from Cropland East of the Rocky Mountains," *USDA Agr. Handbook, 282,* 1965.

Woodruff, N. P., and F. H. Siddoway, "A Wind Erosion Equation," *Soil Sci. Soc. Am. Proc.*, *29*:602–608, 1965.

16
SOIL RESOURCES AND POPULATION

As food, textiles, and housing materials are largely products of the soil, either directly or indirectly, an ample acreage of productive land is one of the most essential and most stabilizing resources a nation can have. A nation with soil resources so limited that it must depend mainly on imported food and fiber is always in a precarious economic position because not only war but also changes in trade conditions may curtail the supply of imports. Furthermore, manufacturers of many commodities must depend on agricultural workers to purchase a con-siderable portion of the manufactured products, even though the workers live in different countries. Unless the people who live on the land have reasonable incomes, their purchasing power is limited. It is extremely important, therefore, to the city producer that there be ample good land for farming purposes and that the land be maintained in a sufficiently productive state to afford a reasonable income to farmers and to provide an adequate food supply.

In considering the soil resources of the United States and other countries, the

395

adequacy of these resources to supply the needs of the people concerned will be discussed.

Acreages of Producing and Potential Cropland in the United States

An inventory of land resources in the United States was made by federal and state authorities. Within the United States, there are approximately 2.3 billion acres of land. Three fifths of this acreage is in farms or privately owned. About one third of the land is in federal ownership, and the state and local governments hold about 5 percent.

Land Use in the Continental United States

Approximately one third of the land in the continental United States is currently used for forest and woodland, one third for grassland and range, and one fourth is devoted mainly to crop land. Special and other uses account for 10 percent of the land use (Fig. 16-1).

From 1880 to 1950 the acreage of crop land more than doubled. This increase occurred primarily by conversion of available pasture and range land into crop land. Since 1920, the acreage of crop land has remained about the same and the acreage of pasture and range land has shown a small decrease. Since 1880, woodland and forest land have remained about the same. Important changes in land use have occurred since 1920, however, but some of these changes have, in large part, canceled each other. For instance, west of the Mississippi River the acreage of crop land has increased, and east of the Mississippi River it has decreased. For the country as a whole, the acreage has remained relatively stable since 1920. The slow gradual changes in land use from 1950 to 1980 are shown in Fig. 16-2. Note the modest decline in cropland from 1950 to 1970.

At present, the development of new farm land nearly equals the conversion of farm land to other uses. From 1910 to 1959 the increase in land devoted to cities, towns, parks, airports, and other uses rose from 53 to 113 million acres, or at a rate slightly over 1 million acres per year. Cities continue to absorb about one million acres a year at the present time.

Acreages of Land Suitable for Cultivation and Land Cultivated

The National Inventory of Soil and Water Conservation Needs taken in 1967 places land in one of eight classes. The range from class I, land suited for crops or most other agricultural uses, to class VIII, land suited only for recreation, wildlife, or water supply, having only limited agricultural use. The amount of land suitable for cultivation and the amount used for crops of the various classes are given in Table 16-1. The data in Table 16-1 show that only about 60 percent of the best land is currently being used for crops in the United States and that considerable other cultivatable land is available. For quite a few years to come, the supply of land for crops appears to be very adequate.

Distribution of Land of Different Quality Among the States

A study has been made with the purpose of dividing the land into five groups on the basis of the land's capacity to produce grains, grasses, and legumes without irriga-

ACREAGES OF PRODUCING AND POTENTIAL CROPLAND IN THE UNITED STATES

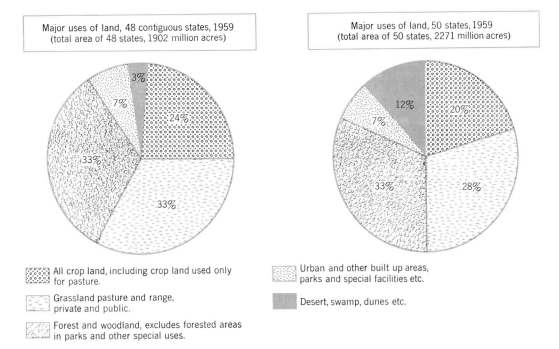

Major uses of land, 48 contiguous states, 1959
(total area of 48 states, 1902 million acres)

Major uses of land, 50 states, 1959
(total area of 50 states, 2271 million acres)

All crop land, including crop land used only for pasture.

Grassland pasture and range, private and public.

Forest and woodland, excludes forested areas in parks and other special uses.

Urban and other built up areas, parks and special facilities etc.

Desert, swamp, dunes etc.

Fig. 16-1 Major land uses in the United States.

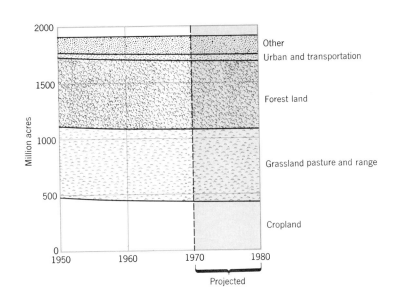

Cropland

Grassland pasture and range

Forest land

Urban and transportation

Other

Projected

Fig. 16-2 Major trends in land use in the United States (48 contiguous states), 1950 to 1980.

Table 16-1
Land Suited for Cultivation and Land Used for Crops in the United States in 1967

Land Class	Available Cultivatable Land, Million Acres	Land Used for Crops in 1967, Million Acres
Classes I, II and III (no limitations for farming)	630	360
Class IV (land suited for only limited farming)	180	49
Classes V to VIII (best suited for grazing, forests, recreation, wildlife, and water supply)	628	25

From "Fact Book of U.S. Agriculture," 1970. Data are for the entire United States, including Alaska and Hawaii.

tion, drainage, or fertilization. The group names used are excellent, good, fair, poor, and unfit for crop production. The distribution of the first three classes of land is also shown graphically in Fig. 16-3.

The high percentage of excellent and good land in the Corn Belt states and the states immediately adjoining them is worthy of consideration. The question may well be asked, "What conditions gave rise to the development of so much land of superior quality in that location?" It should be remembered, furthermore, that some of the land designated "fair" in this classification produces superior yields when fertilized. Likewise, the production on large acreages of land classified "fair" but that is irrigated is much greater than on high-quality land without irrigation. The extensive use of irrigation explains why California produces more agricultural products than any other state but has little excellent and good land, according to Fig. 16-3.

Acreages of Arable Land and Land Requirements

A knowledge of the acreage of farm land in different countries in relation to population leads to a better understanding of some of the problems confronting these nations. The interdependence of nations is also brought out by such a study. The more one learns about good farm land and the part it plays in the welfare of people, the more respect one has for the soil.

Meaning of Arable

A broad and variable interpretation of the term arable is possible on the basis of experience and viewpoint. Strictly speaking, arable land is land that is suitable for plowing or cultivation. No mention is made, however, of whether such land may be cultivated profitably at any set price level for agricultural products or whether the land will deteriorate rapidly under cultiva-

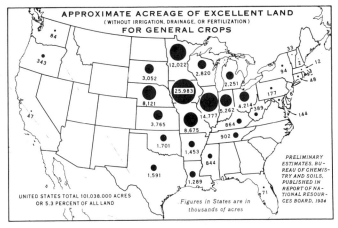

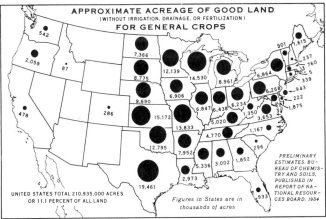

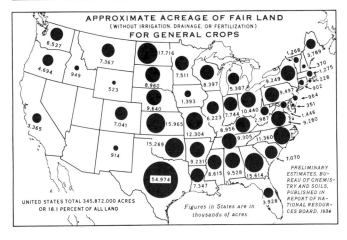

Fig. 16-3 The distribution of excellent, good, and fair land in the various states. (From *USDA Misc. Pub.*, 260.)

tion unless special tillage practices and cropping systems are used. Probably there is a tendency to apply the term arable from a purely physical viewpoint and to consider any cleared land arable that is not too stony, too rough topographically, or too badly in need of drainage to grow crops, or that is not in an area of very limited rainfall during the growing season. It is evident that land that one person or group of persons would consider arable might not be so classed by another group of persons. For this reason it is difficult to compare accurately the acreages of good land in different countries. In the following discussion the comparisons are based on the best data available, and arable land is considered the same as cultivatable land.

About One Tenth of the World's Land in Cultivation

In general, about one half of the land in the world is considered entirely unsuitable for cultivation. This portion is made up of areas permanently covered with ice and snow (11 percent), the tundra (4 percent), high mountains (16 percent), and deserts and semideserts (17 percent). Although appreciable areas of desert may be irrigated, this acreage is not large enough to change the general picture. By no means can all of the other one half of the world's land be easily placed under cultivation. Much of it is so rocky, so sandy, so hilly, so full of soluble salts, or so badly in need of drainage that it may never by cultivated. Estimates of the percentage of the world's land now in cultivation vary widely, some being greater and some less than 10 percent. Certainly not a large part of the land surface is being cultivated. How much this area can be practically increased depends on economic conditions and scientific developments, but any large increase in the immediate future appears doubtful.

Acres of Cultivatable Land per Person in the United States

We will consider the land data given in Table 16-1. Assuming a population of 215 million, there were about 1.7 acres of good cropland used per person and about 3 acres of good cropland available. Considering all the available cropland, there are about 6 acres per person. The data show that the United States yet has an abundance of cultivatable land.

Amount of Land Needed per Person

Harrison Brown estimated that in a hunting and food gathering situation there would need to be about 2 square miles of fertile land in the natural state to support one person, and that the world could support a population of 10 million people under these conditions. Shifting cultivators use about 50 acres per person or about one twenty-fifth as much land per person as food gatherers. One acre of cultivated land exists for each person in the world today. On Java, soils are fertile because of recent volcanism and land is used intensively. Eighty million people have 25 million acres of arable land or about $\frac{1}{3}$ acre per person.

A cultivated land-person ratio of $\frac{1}{3}$ is typical of the wet rice culture of southeast Asia and represents the most intensive land use on a large scale today. Technological inputs or use of cultural energy (use of fossil fuel), however, are low. The data in Fig. 16-4 indicate that the wet rice culture in southeast Asia returns 10 to 50 calories in the food produced for each calorie of cultural energy used in production. Agriculture in the United States is characterized by large inputs of land and energy (technology) and

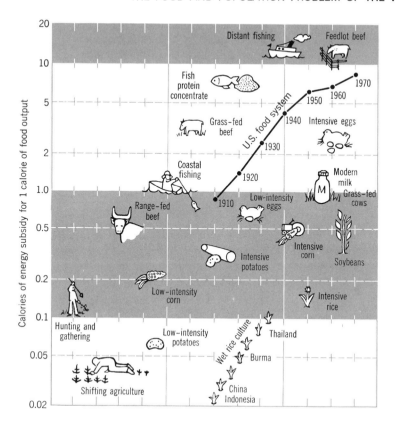

Fig. 16-4 The energy subsidy for various food crops, including the United States food system. Adapted from Carol and John Steinhart, *Energy Sources, Use and Role in Human Affairs*, Duxbury Press, North Scituate, Mass., 1974. Used by permission of authors.)

a low labor input. Intensive corn production in the United States produces 2 to 3 calories of food energy for each calorie used in production. Our food system, which converts a lot of grain into animal products, requires about 10 calories of energy for each food calorie (see Fig. 16-4).

Some interesting estimates of the amount of land needed per person have been made by de Wit. He estimated that an acre near the equator could produce calories equal to the dietary needs of about 40 people if nutrients and water are not limited. Assuming only plant calories for nutrition, no aminals except people, and no space used for urban or recreational use, de Wit estimated that the earth could support 1022

billion people. Taking into consideration urban and recreational needs, de Wit estimated that the earth could support only 146 billion people. The estimates are of particular interest because they point out that with current technology our food can be produced on a smaller piece of land that we need for living space. The estimates also help to redirect our thinking from "How much food can the earth produce?" to "How many people is there space for?"

The Food and Population Problem of the World

Political and social unrest are frequently related to population and food production problems. This provides a basis for concern about the possibilities for providing most of the world's population with an adequate diet.

Food Production Currently Increasing Faster than Population

World food production typically increases each year. A recent reversal occurred, however, when world food production in 1974 was less than in 1973. Even though world food production is increasing faster than world population, a problem exists. In the less-developed countries the increases in population are offsetting the gains made in food production, so that food per capita is remaining nearly constant (see Fig. 16-5). In the developed countries the food per capita has been increasing. The net effect has been to widen the food gap between the developed and less-developed nations. Although the number of persons receiving an adequate diet each year increases, the percentage of the world's population with an adequate diet decreases. When I (Foth) first became interested in the food-population problem after World War II, about one half of the world had adequate diets. Now only one third of the world's people have an adequate diet. The data in Fig. 16-5 show that population control as well as increased food production is needed to improve the world food-population situation.

Oceans as a Source of Fish

The oceans play a very important role in providing food; fish products are rich in protein, and they balance the calorie intake of starchy diets for many of the world's people. One factor commonly overlooked in food problems is that a person cannot eat enough low-protein food in the form of rice, wheat, or corn to satisfy the daily protein requirement. Stated another way, if a person ate all the corn he or she could, the corn would not contain sufficient protein to meet daily body needs. Consideration of increased food production from the oceans in the form of fish is, therefore, of more than ordinary importance because of the protein considerations.

A large increase in ocean-fish production can be expected with current technology. It is easy to be overoptimistic, however, because the oceans occupy three fourths of

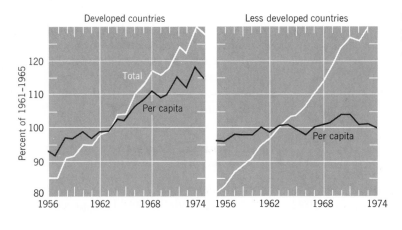

Fig. 16-5 World agricultural production for developed and less-developed countries, 1956 to 1974.

the earth's surface, and photosynthesis can occur at great depths. However, two factors are cited to point out the difficulties in achieving large and easy increases in world fish production. First, fish production is very inefficient in terms of the use of primary plant calories. The production of a pound of fish may require from 1000 to 100,000 pounds of primary plankton, depending on the number of times one organism is devoured by another before it is caught for food. Second, much of the ocean water is nutrient poor. Fish production is limited in large areas because of insufficient phosphorus or nitrogen for plankton production.

Potential for Increasing Arable Land

It has already been pointed out that only 60 percent of the "best" land in the United States is being used for crop production. What about the situation on a world basis?

The data clearly show that the world's supply of arable land could be doubled. Data in Table 16-2 indicate that 3.4 billion acres of land are now being cultivated and that 7.8 billion acres in the world could potentially be cultivated. However, it should be kept in mind that the best lands are being cultivated and that expanding the acreage of cultivated land will usually mean the use of poorer land and increased production costs.

On a world basis there is a very unequal distribution of the potentially arable land. Two continents are essentially "filled" with people in the sense that nearly all the arable land is now in use. These continents are Asia, with 83 percent of the arable land now in use, and Europe, with 88 percent of its arable land being used (see Table 16-2). Furthermore, the land used per person is low, less than 1 acre per person.

Large acreages of potentially arable land exist on the African continent in the Congo

Table 16-2
Present Cultivated Land on Each Continent Compared with Potentially Arable Land

Continent	Area, Billions of Acres			Acres of Cultivated Land per Person	Ratio of Cultivated to Potentially Arable Land, Percentage
	Total	Potentially Arable	Cultivated		
Africa	7.46	1.81	0.39	1.3	22
Asia	6.67	1.55	1.28	0.7	83
Australia and New Zealand	2.03	0.38	0.08	2.9	21
Europe	1.18	0.43	0.38	0.9	88
North America	5.21	1.15	0.59	2.3	51
South America	4.33	1.68	0.19	1.0	11
U.S.S.R.	5.52	0.88	0.56	2.4	64
Total	32.49	7.88	3.47	1.0	44

From "The World Food Problem," Vol. 2, The White House, May 1967.

Basin and in South America in the Amazon Basin. Here, great obstacles for expansion of cultivated land are found in the great infertility of the soils, lack of transportation, and so forth. The problem in Asia is that there is a very large population with little land per person and very little additional land that is potentially arable. Where living standards are good and considerable land is now being used per person, as in Australia, North America, and the Soviet Union, there is also considerable potential for future development of cultivated land. Thus, the uneven distribution of the cultivated and potentially cultivated land presents a difficult problem in terms of improving the food supply in regions of the world where food is in shortest supply.

Yields Can Be Increased

Crop yields appear to be a function of the overall economic development of a country. High acre yields generally depend on fertilizers, pesticides, and machinery, all of which are the products of an industrial society. Wheat yields in Britain were nearly constant or decreased slightly from 1100 to 1350 A.D. under the feudal system (Fig. 16-6). Conversion of land into private ownership and the use of grass to rejuvenate exhausted land caused a significant increase in wheat yields from 1350 to 1550. Very modest increases in yield occurred between the years 1550 and 1900 from improvements in farming evolved by trial and error. Note, however, that in a recent 60-year period the yields increased more than during the preceding 800 years (Fig. 16-6). There is no reason to doubt that, with the current level of knowledge, large increases can be made on a large portion of the cultivated land in the world today. It is very reasonable to again suggest what C. E. Kellogg pointed out in 1948, that "there is sufficient soil resources to feed the world's population, providing the economic, social, and political problems can be solved."

In the final sense, food production from the crop land is limited by the ultimate amount of carbon that is fixed in photosynthesis. Although the energy absorbed by the plant is used efficiently, as compared to the efficiency of man-made engines, only a slight percentage of the sun's energy is utilized when maximum yields are obtained. Some of the most important studies in agriculture today are those aimed at improving our knowledge of plant growth so that, ultimately, crop plants may be able to utilize more of the sun's energy through photosynthesis (Fig. 16-7).

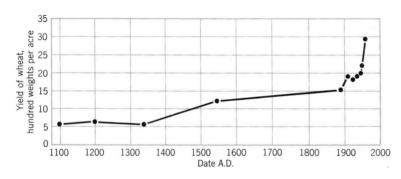

Fig. 16-6 Wheat yields in Britain 1100 to 1960. (From Jacks, 1962.)

Fig. 16-7 An experimental setup at Cornell University to study light intensity, carbon dioxide concentration of the air, temperature, and water utilization in relation to growth of corn. Such experiments provide knowledge that forms the basis for new practices that will enable us to increase the growth of plants.

Future Outlook Not Very Bright Unless Population Can Be Controlled

It has been mentioned that current soil resources and technology could result in sufficient food production to feed the people in the world today. The food-population problem is not bright, however. First, the current population of slightly over 4 billion is expected to be over 6 billion by the year 2000. Although the highly industrialized nations may have no problem in meeting their food needs for the remainder of the century, it appears that less and less food per person for millions of people will be the case because populations are increasing most rapidly in areas where food production increases are the most difficult to achieve. One of my objectives has been to bring to the attention of the reader the fact that soil resources and technology are sufficient for the twentieth century and that a major goal of a nation such as the United States should be to bring about those economic, social, and political changes that will result in a better life for an ever increasing portion of the world's people. The task is enormous and cannot be ignored if there is to be world peace.

References

Brown, Harrison, *The Challenge of Man's Future*, Viking, New York, 1954.

Cotner, Melvin L., and Louise N. Samuel, "Competition for Land Resources," in *Contours of Change*, USDA Yearbook, Washington, D.C., 1970, pp. 204–212.

deWit, C. T., "Photosynthesis: Its Relationship to Overpopulation," in *Harvesting the Sun*, Anthony San Pietro, Frances A. Greer, Thomas J. Army, editors, Academic, New York, 1967, pp. 315–332.

Jacks, G. V., "Man the Fertility Maker," *Jour. Soil and Water Cons., 17*:147–148, 1962.

Kellogg, C. E., "Modern Soil Science," *Am. Scientist, 38*:517–536, 1948.

Kellogg, Charles E., and Arnold C. Orvedal, "Potentially Arable Soils of the World and Critical Measures for Their Use," in *Advances in Agronomy*, 21, pp. 109–170, Academic, New York, 1969.

Office of Information, "Fact Book of US Agriculture," *Mesc. Pub. No. 1063,* USDA, Washington, D.C., 1970.

Steinhart, John S., and Carol E. Steinhart, "Energy Use in the U.S. Food System," *Science, 184*:307–316, 1974.

The White House, *The World Food Problem*, Vol. 2, U.S. Government Printing Office, Washington, D.C., 1967.

Wooten, Hugh H., Karl Gertel, and William C. Pendleton, "Major Uses of Land and Water in the United States," *Agr. Econ. Report No. 13*, USDA, Washington, D.C., 1962.

To Convert Column 1 into Column 2, Multiply by	Column 1	Column 2	To Convert Column 2 into Column 1, Multiply by
Length			
0.621	kilometer, km	mile, mi	1.609
1.094	meter, m	yard, yd	0.914
0.394	centimeter, cm	inch, in	2.54
Area			
0.386	kilometer^2, km^2	mile^2, mi^2	2.590
247.1	kilometer^2, km^2	acre, acre	0.00405
2.471	hectare, ha	acre, acre	0.405
Volume			
0.00973	meter^3, m^3	acre-inch	102.8
3.532	hectoliter, hl	cubic foot, ft^3	0.2832
2.838	hectoliter, hl	bushel, bu	0.352
0.0284	liter	bushel, bu	35.24
1.057	liter	quart (liquid), qt	0.946
Mass			
1.102	ton (metric)	ton (U.S.)	0.9072
2.205	quintal, q	hundredweight, cwt (short)	0.454
2.205	kilogram, kg	pound, lb	0.454
0.035	gram, g	ounce (avdp), oz	28.35
Pressure			
14.50	bar	lb/inch^2, psi	0.06895
0.9869	bar	atmosphere, atm	1.013
0.9678	kg (weight)/cm^2	atmosphere, atm	1.033
14.22	kg (weight)/cm^2	lb/inch^2, psi	0.07031
14.70	atmosphere, atm	lb/inch^2, psi	0.06805
Yield or Rate			
0.446	ton (metric)/hectare	ton (U.S.)/acre	2.240
0.892	kg/ha	lb/acre	1.12
0.892	quintal/hectare	hundredweight/acre	1.12

continued overleaf

To Convert Column 1 into Column 2, Multiply by	Column 1	Column 2	To Convert Column 2 into Column 1, Multiply by

Temperature

$\left(\frac{9}{5}°C\right)+32$	Celsius	Fahrenheit	$\frac{5}{9}(°F-32)$
	−17.8C	0F	
	0C	32F	
	20C	68F	
	100C	212F	

Water Measurement

8.108	hectare-meters, ha-m	acre-feet	0.1233
97.29	hectare-meters, ha-m	acre-inches	0.01028
0.08108	hectare-centimeters, ha-cm	acre-feet	12.33
0.973	hectare-centimeters, ha-cm	acre-inches	1.028
0.00973	meters^3, m^3	acre-inches	102.8
0.981	hectare-centimeters/hour, ha-cm/hour	feet^3/sec	1.0194
440.3	hectare-centimeters/hour, ha-cm/hour	U.S. gallons/min	0.00227
0.00981	meters^3/hour, m^3/hour	feet^3/sec	101.94
4.403	meters^3/hour, m^3/hour	U.S. gallons/min	0.227

Reproduced from *Soil Science Society of America Journal, 40,* November–December 1976, by permission of Soil Science Society of America.

GLOSSARY

A Horizon. The surface horizon of a mineral soil having maximum biological activity, or eluviation (removal of materials dissolved or suspended in water), or both.

ABC Soil. A soil with a complete profile, including an A, a B, and a C horizon.

AC Soil. A soil with an incomplete profile, including an A and a C horizon, but no B horizon. Commonly such soils are young, like those developing from alluvium or on steep, rocky slopes.

Acidity, Residual. Soil acidity that is neutralized by lime or other alkaline materials, but which cannot be replaced by an unbuffered salt solution.

Acidity, Salt-replaceable. The aluminum and hydrogen that can be replaced from an acid soil by an unbuffered salt solution such as KCl or NaCl.

Acidity, Total. The total acidity in a soil or clay. Usually it is estimated by a buffered salt determination of (cation-exchange capacity—exchangeable bases) = total acidity. It also is approximated by the sum of salt replaceable acidity + residual acidity.

Actinomycetes. A nontaxonomic term applied to a group of organisms with characteristics intermediate between the simple bacteria and the true fungi.

Aerobic. (1) Having molecular oxygen as a part of the environment. (2) Growing only in the presence of molecular oxygen, as aerobic organisms. (3) Occurring only in the presence of molecular oxygen (said of certain chemical or biochemical processes such as aerobic decomposition).

Agric Horizon. A mineral soil horizon in which clay, silt, and humus derived from an overlying cultivated and fertilized layer have accumulated. The wormholes and illuvial clay, silt, and humus occupy at least 5 percent of the horizon by volume. The illuvial clay and humus occur as horizontal lamellae or fibers, or as coatings on ped surfaces or in wormholes.

Air-dry. The state of dryness (of a soil) at equilibrium with the moisture content in the surrounding atmosphere.

Albic Horizon. A mineral soil horizon from which clay and free iron oxides have been removed or in which the oxides have been segregated to the extent that the color of the horizon is determined primarily by the color of the primary sand and silt particles instead of by coatings on these particles.

Adapted from *Glossary of Soil Science Terms,* published by Soil Science Society of America, Madis, 1975.

Albolls. Mollisols that have an albic horizon immediately below the mollic epipedon. These soils have an argillic or natric horizon and mottles, iron-manganese concretions, or both, within the albic, argillic, or natric horizon. (A suborder in the USDA soil taxonomy.)

Alfisols. Mineral soils that have umbric or ochric epipedons, argillic horizons, and hold water at less than 15-bars tension during at least 3 months when the soil is warm enough for plants to grow outdoors. Alfisols have a mean annual soil temperature of less than 8°C or a base saturation in the lower part of the argillic horizon of 35 percent or more when measured at pH 8.2. (An order in the USDA soil taxomony.)

Ammonification. The biochemical process whereby ammoniacal nitrogen is released from nitrogen-containing organic compounds.

Ammonium Fixation. Adsorption of ammonium ions (NH_4^+) by the mineral fraction of the soil in forms that cannot be replaced by a neutral potassium salt solution (e.g., $1N$ KCl).

Anaerobic. (1) The absence of molecular oxygen. (2) Growing in the absence of molecular oxygen (such as anaerobic bacteria). (3) Occurring in the absence of molecular oxygen (as a biochemical process).

Andepts. Inceptisols that have formed either in vitric pyroclastic materials, or have low bulk density and large amounts of amorphous materials, or both. Andepts are not saturated with water long enough to limit their use for most crops. (A suborder in the USDA soil taxonomy.)

Anion-exchange Capacity. The sum total of exchangeable anions that a soil can adsorb. Expressed as milliequivalents per 100 grams of soil (or of other adsorbing material such as clay).

Anthropic Epipedon. A surface layer of mineral soil that has the same requirements as the mollic epipedon with respect to color, thickness, organic carbon content, consistence, and base saturation, but that has more than 250 parts per million of P_2O_5 soluble in 1 percent citric acid, or is dry more than 10 months (cumulative) during the period when not irrigated. The anthropic epipedon forms under long continued cultivation and fertilization.

Antibiotic. A substance produced by one species of organism that, in low concentrations, will kill or inhibit growth of certain other organisms.

Aqualfs. Alfisols that are saturated with water for periods long enough to limit their use for most crops other than pasture or woodland unless they are artificially drained. Aqualfs have mottles, iron-manganese concretions, or gray colors immediately below the A1 or Ap horizons and gray colors in the argillic horizon. (A suborder in the USDA soil taxonomy.)

Aquents. Entisols that are saturated with water for periods long enough to limit their use for most crops other than pasture unless they are artifically drained. Aquents have low chromas or distinct mottles within 50 centimeters of the surface, or are saturated with water at all times. (A suborder in the USDA soil taxonomy.)

Aquepts. Inceptisols that are saturated with water for periods long enough to limit their use for most crops other than pasture or woodland unless they are artificially drained. Aquepts have either a histic or umbric epipedon and gray colors within 50 centimeters, or an ochric epipedon underlain by a cambic horizon with gray colors, or have sodium saturation of 15 percent or more. (A suborder in the USDA soil taxonomy.)

Aquic. A mostly reducing soil moisture regime nearly free of dissolved oxygen due to saturation by ground water or its capillary fringe and occurring at periods when the soil temperature at 50 centimeters is above 5°C.

Aquods. Spodosols that are saturated with water for periods long enough to limit their use for most crops other than pasture or woodland unless they are artificially drained. Aquods may have a histic epipedon, an albic horizon that is mottled or contains a duripan, or mottling or gray colors within or immediately below the spodic horizon. (A suborder in the USDA soil taxonomy.)

Aquolls. Mollisols that are saturated with water for periods long enough to limit their use for most crops other than pasture unless they are artificially drained. Aquolls may have a histic epipedon, a sodium saturation in the upper part of the mollic epipedon of more than 15 percent, that decreases with depth or mottles or gray colors within or immediately below the mollic epipedon. (A suborder in the USDA soil taxonomy.)

Aquox. Oxisols that have continuous plinthite near the surface, or that are saturated with water sometime during the year if not artificially drained. Aquox have either a histic epipedon, or mottles or colors indicative or poor drainage within the oxic horizon, or both. (A suborder in the USDA soil taxonomy.)

Aquults. Ultisols that are saturated with water for periods long enough to limit their use for most crops other than pasture or woodland unless they are artificially drained. Aquults have mottles, iron-manganese concretions, or gray colors immediately below the A1 or Ap horizons and gray colors in the argillic horizon. (A suborder in the USDA soil taxonomy.)

Arents. Entisols that contain recognizable fragments of pedogenic horizons that have been mixed by mechanical disturbance. Arents are not saturated with water for periods long enough to limit their use for most crops. (A suborder in the USDA soil taxonomy).

Argids. Aridisols that have an argillic or a natric horizon. (A suborder in the USDA soil taxonomy.)

Argillic Horizon. A mineral soil horizon that is characterized by the illuvial accumulation of layer-lattice silicate clays. The argillic horizon has a certain minimum thickness, depending on the thickness of the solum, a minimum quantity of clay in comparison with an overlying eluvial horizon depending on the clay content of the eluvial horizon, and usually has coatings of oriented clay on the surface of pores or peds or bridging sand grains.

Aridic. A soil moisture regime that has no moisture available for plants for more than half the cumulative time that the soil temperature at 50 centimeters is above 5°C, and has no period as long as 90 consecutive days when there is moisture for plants, while the soil temperature at 50 centimeters is continuously above 8°C.

Aridisols. Mineral soils that have an aridic moisture regime, an ochric epipedon, and other pedogenic horizons, but no oxic horizon. (An order in the USDA soil taxonomy.)

Autotrophic. Capable of utilizing carbon dioxide and/or carbonates as a sole or major source of carbon and of obtaining energy for carbon reduction and biosynthetic processes from radiant energy (photoautotroph) or oxidation of inorganic substances (chemoautotroph).

Available Nutrient. That quantity of a nutrient element or compound in the soil that can be readily absorbed and assimilated by growing plants.

Available Water. The portion of water in a soil that can be readily absorbed by plant roots. Considered by most workers to be that water held in the soil against a pressure of up to approximately 15 bars. See **field capacity** and **moisture tension.**

Azonal Soils (Obsolete). Soils without distinct genetic horizons. A soil order in 1949 system.

B Horizon. A soil horizon, usually beneath an A horizon, or surface soil, in which (1) clay, iron, or aluminum, with accessory organic matter, have accumulated by receiving suspended material from the A horizon above it or by clay development in place; (2) the soil has a blocky or prismatic structure; or (3) the soil has some combination of these features. In soils with distinct profiles, the B horizon is roughly equivalent to the general term "subsoil."

Bacteroid. An irregular form of cells of certain bacteria. Refers particularly to the swollen, vaculated cells of *Rhizobium* in nodules of leguminous plants.

Bar. A unit of pressure equal to 1 million dynes per square centimeter.

Base-saturation Percentage. The extent to which the adsorption complex of a soil is saturated with exchangeable cations other than hydrogen. It is expressed as a percentage of the total cation-exchange capacity.

BC Soil. A soil profile with B and C horizons but with little or no A horizon.

Bedding. Preparing a series of flat-topped, parallel ridges usually no wider than two crop rows, separated by shallow trenches usually less than the width between crop rows. The resultant structures are beds.

Biome. A large, easily recognized community unit formed by the interaction of regional climates with regional biota and substrates. In a given biome the life form of the climatic climax vegetation is uniform. Thus, the climax vegetation of the grassland biome is grass, although the dominant species of grass may vary in different parts of the biome.

Biosequence. A sequence of related soils that differ, one from the other, primarily because of differences in kinds and numbers of *soil organisms* as a soil-forming factor.

Bleicherde. The light-colored, leached A2 horizon of Podzol soils.

Bog Soil. A great soil group of the intrazonal order and hydromorphic suborder. Includes muck and peat. (1949 classification system).

Boralfs. Alfisols that have formed in cool places. Boralfs have frigid or cryic but not pergelic temperature regimes, and have udic moisture regimes. Boralfs are not saturated with water for periods long enough to limit their use for most crops. (A suborder in the USDA soil taxonomy.)

Borolls. Mollisols with a mean annual temperature of less than 8°C and are never dry for 60 consecutive days or more within the 3 months following the summer solstice. Borolls do not contain material that has more than 40 percent CaCO₃ equivalent unless they have a calcic horizon, and they are not saturated with water for periods long enough to limit their use for most crops. (A suborder in the USDA soil taxonomy.)

Brown Forest Soils. A great soil group of the intrazonal order and calcimorphic suborder, formed on calcium-rich parent materials under deciduous forest, and possessing a high base status but lacking a pronounced illuvial horizon.

Brown Podzolic Soils. A zonal great soil group similar to Podzols but lacking the distinct A2 horizon characteristic of the Podzol group.

Brown Soils. A great soil group of the temperate to cool arid regions, composed of soils with a brown surface and a light-colored transitional subsurface horizon over calcium carbonate accumulation.

Brunizem. Synonymous with **prairie soils.**

Buffer Compounds, Soil. The clay, organic matter, and compounds such as carbonates and phosphates that enable the soil to resist appreciable change in pH.

Bulk Density, Soil. The mass of dry soil per unit bulk volume. The bulk volume is determined before drying to constant weight at 105°C.

Buried soil. Soil covered by an alluvial, loessal, or other deposit, usually to a depth greater than the thickness of the solum.

C Horizon. A mineral horizon or layer, excluding bedrock, that is either like or unlike the material from which the solum is presumed to have formed, relatively little affected by pedogenic processes.

Calcareous Soil. Soil containing sufficient free calcium carbonate or calcium-magnesium carbonate to effervesce visibly when treated with cold 0.1N hydrochloric acid.

Calcic Horizon. A mineral soil horizon of secondary carbonate enrichment that is more than 15 centimeters thick, has a calcium carbonate equivalent of more than 15 percent, and has at least 5 percent more calcium carbonate equivalent than the underlying C horizon.

Caliche. A layer near the surface, more or less cemented by secondary carbonates of calcium or magnesium precipitated from the soil solution. It may occur as a soft thin soil horizon, as a hard thick bed just beneath the solum, or as a surface layer exposed by erosion (see **croute calcaire**).

Cambic Horizon. A mineral soil horizon that has a texture of loamy, very fine sand or finer, has soil structure instead of rock structure, contains some weatherable minerals, and is characterized by the alteration or removal of mineral material as indicated by mottling or gray colors, stronger chromas or redder hues than in underlying horizons, or the removal of carbonates. The cambic horizon lacks cementation or induration and has too few evidences of illuviation to meet the requirements of the argillic or spodic horizon.

Capillary Fringe. A zone just above the water table (zero gauge pressure) that remains almost saturated.

Carbon-nitrogen Ratio. The ratio of the weight of organic carbon to the weight of total nitrogen (mineral plus organic forms) in soil or organic material.

Catena. A sequence of soils of about the same age, derived from similar parent material, and occurring under similar climatic conditions, but having different characteristics because of variation in *relief* and in *drainage*. See **clinosequence** and **toposequence.**

Cation Exchange. The interchange between a cation in solution and another cation on the surface of any surface-active material such as clay colloid or organic colloid.

Cation-exchange Capacity (CEC). The sum total of exchangeable cations that a soil can adsorb. Expressed in milliequivalents per 100 grams or per gram of soil (or of other exchangers such as clay).

Chernozem. A zonal great soil group consisting of soils with a thick, nearly black or black, organic matter-rich A horizon high in exchangeable calcium, underlain by a lighter colored transitional horizon above a zone of calcium carbonate accumulation; occurs in a cool sub-humid climate under a vegetation of tall and midgrass prairie.

Chestnut Soil. A zonal great soil group consisting of soils with a moderately thick, dark-brown A horizon over a lighter colored horizon that is above a zone of calcium carbonate accumulation.

Chlorosis. A condition in plants resulting from the failure of chlorophyll to develop caused by a deficiency of an essential nutrient. Leaves of chlorotic plants range from light green through yellow to almost white.

Chroma. The relative purity, strength, or saturation of a color; directly related to the dominance of the determining wavelength of the light and inversely related to grayness; one of the three variables of color. See **Munsell color system, hue,** and **value, color.**

Chronosequence. A sequence of related soils that differ, one from the other, in certain properties primarily as a result of *time* as a soil-forming factor.

Clay. (1) A soil separate consisting of particles < 0.002 millimeter in equivalent diameter. (2) A textural class.

Clay Films. Coatings of clay on the surfaces of soil peds and mineral grains and in soil pores. (Also called clay skins, clay flows, illuviation cutans, argillans, or tonhäutchen.)

Claypan. A dense, compact layer in the subsoil having a much higher clay content than the overlying material, from which it is separated by a sharply defined boundary; formed by downward movement of clay or by synthesis of clay in place during soil formation. Claypans are usually hard when dry, and plastic and sticky when wet. Also, they usually impede the movement of water and air and the growth of plant roots.

Climosequence. A sequence of related soils that differ, one from the other, in certain properties primarily as a result of the effect of *climate* as a soil-forming factor.

Clinosequence. A group of related soils that differ, one from the other, in certain properties primarily as a result of the effect of the *degree of slope* on which they were formed. See **toposequence.**

Colloid. The term colloid is used in reference to matter, both organic and inorganic, having very small particle size and high specific surface.

Colluvium. A deposit of rock fragments and soil material accumulated at the base of steep slopes as a result of gravitational action.

Concretion. A local concentration of a chemical compound, such as calcium carbonate or iron oxide, in the form of a grain or nodule of varying size, shape, hardness, and color.

Consistency. (1) The resistance of a material to deformation or rupture. (2) The degree of cohesion or adhesion of the soil mass.

Consumptive Use. The water used by plants in transpiration and growth, plus water vapor loss from adjacent soil or snow, or from intercepted precipitation in any specified time. Usually expressed as equivalent depth of free water per unit of time.

Contour Tillage. Performing the tillage operations and planting on the contour within a given tolerance.

Cover Crop. A crop used to cover the soil surface; to decrease erosion and leaching, shade the ground, and offer protection to the ground from excessive freezing and heaving.

Cradle Knoll. A small knoll formed by earth that is raised and left by an uprooted tree. (A microrelief term.)

Creep. Slow mass movement of soil and soil material down relatively steep slopes primarily under the influence of gravity, but facilitated by saturation with water and by alternate freezing and thawing.

Crop Rotation. A planned sequence of crops growing in a regularly recurring succession on the same area of land, as contrasted to continuous culture of one crop or growing different crops in haphazard order.

Crotovina. A former animal burrow in one soil horizon that has been filled with organic matter or material from another horizon (also spelled "krotovina").

Croute Calcaire. Hardened caliche, often found in thick masses or beds overlain by only a few inches of earth. See **caliche.**

Cryic. A soil temperature regime that has mean annual soil temperatures of more than 0°C but less than 8°C, more than 5°C difference between mean summer and mean winter soil temperatures at 50 centimeters, and cold summer temperatures.

Crystal Structure. The orderly arrangement of atoms in a crystalline material.

Darcy's Law. A law describing the rate of flow of water through porous media.

Deflation. The removal of fine soil particles from soil by wind.

Deflocculate. (1) To separate the individual components of compound particles by chemical and/or physical means. (2) To cause the particles of the *disperse phase* of a colloidal system to become suspended in the *dispersion medium.*

Denitrification. The biochemical reduction of nitrate or nitrite to gaseous nitrogen either as molecular nitrogen or as an oxide of nitrogen.

Desert Pavement. The layer of gravel or stones left on the land surface in desert regions after the removal of the fine material by wind erosion.

Desert Soil. A zonal great soil group consisting of soils with a very thin, light-colored surface horizon, which may be vesicular and is ordinarily underlain by calcareous material; formed in arid regions under sparse shrub vegetation.

Desert Varnish. A glossy sheen or coating on stones and gravel in arid regions.

Diatoms. Algae having siliceous cell walls that persist as a skeleton after death. Any of the microscopic unicellular or colonial algae constituting the class Bacillariaceae. They occur abundantly in fresh and salt waters and their remains are widely distributed in soils.

Diatomaceous Earth. A geologic deposit of fine, grayish siliceous material composed chiefly or wholly of the remains of diatoms.

Diffuse Layer. A system, when referred to soils, that consists of a charged (negative) particle surface and an equal amount of counterions (positive) accumulated in the liquid near the particle surface.

Disperse. (1) To break up compound particles, such as aggregates, into the individual component particles. (2) To distribute or suspend fine particles, such as clay, in or throughout a dispersion medium, such as water.

Dryland Farming. The practice of crop production in low-rainfall areas without irrigation.

Duff Mull. A type of forest humus transitional between mull and mor; H and F layers as well as the A1 horizon.

Durinodes. Weakly cemented to indurated soil nodules cemented with SiO_2. Durinodes break down in concentrated KOH after treatment with HCl to remove carbonates, but do not break down on treatment with concentrated HCl alone.

Duripan. A mineral soil horizon that is cemented by silica, usually opal or microcrystalline forms of silica, to the point that air-dry fragments will not slake in water or HCl. A duripan may also have accessory cement such as iron oxide or calcium carbonate.

Dust Mulch. A loose, finely granular or powdery condition on the surface of the soil, usually produced by shallow cultivation.

EC. The electrical conductivity of an extract from saturated soil, normally expressed in units of millimhos per centimeter at 25°C.

Ecology. The science that deals with the interrelations of organisms and their environment.

Ecosystem. A community of organisms and the surroundings in which they live.

Ectotrophic Mycorrhiza. A mycorrhizal association in which the fungal hyphae form a compact mantle on the surface of the roots. Mycelial strands extend inward between cortical cells and outward from the mantle to the surrounding soil.

Edaphic. (1) Of or pertaining to the soil. (2) Resulting from or influenced by factors inherent in the soil or other substrate instead of climatic factors.

Electrokinetic (Zeta) Potential. (1) The difference in electrical potential between the immobile liquid layer attached to the surface of a charged particle and the bulk liquid phase. (2) The work done in bringing a unit charge from infinite (bulk solution) to the plane of shear in the diffuse double layer.

Eluvial Horizon. A soil horizon that has been formed by the process of eluviation.

Eluviation. The removal of soil material in suspension (or in solution) from a layer or layers of a soil. (Usually, the loss of material in *solution* is described by the term "leaching.")

Endotrophic. Nourished or receiving nourishment from within, as fungi or their hyphae receiving nourishment from plant roots in a mycorrhizal association.

Entisols. Mineral soils that have no distinct pedogenic horizons within 1 meter of the soil surface. (An order in the USDA soil taxonomy.)

Equivalent Weight of a Soil Colloid. The weight of clay or organic colloid that has a combining power equivalent to 1 gram atomic weight of hydrogen.

Erosion. (1) The wearing away of the land surface by running water, wind, ice, or other geological agents, including processes such as gravitational creep. (2) Detachment and movement of soil or rock by water, wind, ice, or gravity. The following terms are used to describe different types of water erosion.

accelerated erosion. Erosion much more rapid than normal, natural, geological erosion, primarily as a result of the influence of the activities of people or, in some cases, of animals.

geological erosion. The normal or natural erosion caused by geological processes acting over long geologic periods and resulting in the wearing away of mountains, the building up of flood plains, coastal plains, etc. Synonymous with *natural erosion.*

gully erosion. The erosion process whereby water accumulates in narrow channels and, over short periods, removes the soil from this narrow area to considerable depths, ranging from 1 or 2 feet to as much as 75 to 100 feet.

natural erosion. Wearing away of the earth's surface by water, ice, or other natural agents under natural environmental conditions of climate, vegetation, etc., undisturbed by humans. Synonymous with *geological erosion.*

normal erosion. The gradual erosion of land used by people that does not greatly exceed natural erosion. See *natural erosion.*

rill erosion. An erosion process in which numerous small channels of only several inches in depth are formed; occurs mainly on recently cultivated soils. See **rill.**

sheet erosion. The removal of a fairly uniform layer of soil from the land surface by runoff water.

splash erosion. The spattering of small soil particles caused by the impact of raindrops on very wet soils. The loosened and spattered particles may or may not be subsequently removed by surface runoff.

Eutrophic. Having concentrations of nutrients optimal, or nearly so, for plant or animal growth. (Said of nutrient or soil solutions and bodies of water.)

Evapotranspiration. The combined loss of water from a given area and during a specified period of time by evaporation from the soil surface and by transpiration from plants.

Exchange Capacity. The total ionic charge of the adsorption complex active in the adsorption of ions. See **anion-exchange capacity** and **cation-exchange capacity.**

Exchangeable-Cation Percentage. The extent to which the adsorption complex of a soil is occupied by a particular cation. It is expressed as follows.

$$ECP = \frac{\text{Exchangeable cation (meq/100 g soil)}}{\text{Cation-exchange capacity (meq/100 g soil)}} \times 100$$

Exchangeable Phosphate. The phosphate anion reversibly attached to the surface of the solid phase of the soil in such form that it may go into solution by anionic equilibrium reactions with isotopes of phos-

phorus or with other anions of the liquid phase without solution of the colloid phase to which it was attached.

Exchangeable Potassium. The potassium that is held by the adsorption complex of the soil and is easily exchanged with the cation of neutral nonpotassium salt solutions.

Exchangeable-Sodium Percentage. The percentage of the cation-exchange capacity of a soil occupied by sodium. It is expressed as follows.

$$ESP = \frac{\text{Exchangeable sodium (meq/100 g soil)}}{\text{Cation-exchange capacity (meq/100 g soil)}} \times 100$$

F Layer. A layer of partially decomposed litter with portions of plant structures still recognizable. Occurs below the L layer (O11 horizon) on the forest floor in forest soils. It is the fermentation layer or the O12 layer. See **L layer.**

Fallow. The practice of leaving land uncropped and weed-free for periods of time to accumulate and retain water and mineralized nutrient elements.

Family, Soil. In soil classification one of the categories.

Ferrods. Spodosols that have more than six times as much free iron (elemental) than organic carbon in the spodic horizon. Ferrods are rarely saturated with water or do not have characteristics associated with wetness. (A suborder in the USDA soil taxonomy.)

Fertility, Soil. The status of a soil with respect to its ability to supply the nutrients essential to plant growth.

Fertilizer. Any organic or inorganic material of natural or synthetic origin that is added to a soil to supply one or more elements essential to the growth of plants.

Fertilizer Grade. The guaranteed minimum analysis, in percent, of the major plant nutrient elements contained in a fertilizer material or in a mixed fertilizer. (Usually refers to the percentage of N-P_2O_5-K_2O, but proposals are pending to change the designation to the percentage of N-P-K)

Fibrists. Histosols that have a high content of undecomposed plant fibers and a bulk density less than about 0.1. Fibrists are saturated with water for periods long enough to limit their use for most crops unless they are artificially drained. (A suborder in the USDA soil taxonomy.)

Field Capacity (field moisture capacity). Obsolete in technical work). The percentage of water remaining in a soil 2 or 3 days after having been saturated and after free drainage has practically ceased. (The percentage may be expressed on the basis of weight or volume.)

Fifteen-atmosphere Percentage. See **moisture tension.**

Fifteen-bar Percentage. See **moisture tension.**

Film Water. A layer of water surrounding soil particles and varying in thickness from 1 or 2 to perhaps 100 or more molecular layers. Usually considered as that water remaining after drainage has occurred, because it is not distinguishable in saturated soils.

Fixation. The process of conversion of an element in the soil essential to plants from a readily available to a less available form.

Flood Plain. The land bordering a stream, built up of sediments from overflow of the stream and subject to inundation when the stream is at flood stage.

Fluvents. Entisols that form in recent loamy or clayey alluvial deposits, are usually stratified, and have an organic carbon content that decreases irregularly with depth. Fluvents are not saturated with water for periods long enough to limit their use for most crops. (A suborder in the USDA taxonomy.)

Foliar Diagnosis. An estimation of mineral nutrient deficiencies (excesses) of plants based on examination of the chemical composition of selected plant parts, and the color and growth characteristics of the foliage of the plants.

Folists. Histosols that have an accumulation of organic soil materials mainly as forest litter that is less than 1 meter deep to rock or to fragmental materials with interstices filled with organic materials. Folists are not saturated with water for periods long enough to limit their use if cropped. (A suborder in the USDA soil taxonomy.)

Forest Floor. All dead vegetable or organic matter, including litter and unincorporated humus, on the mineral soil surface under forest vegetation.

Fragipan. A natural subsurface horizon with high bulk density relative to the solum above, seemingly cemented when dry, but when moist showing a moderate to weak brittleness. The layer is low in organic matter, mottled, slowly or very slowly permeable to water, and usually shows occasional or frequent bleached cracks forming polygons. It may be found in profiles of either cultivated or virgin soils but not in calcareous material.

Friable. A consistency term pertaining to the ease of crumbling of soils. See **consistency.**

Frigid. A soil temperature regime that has mean annual soil temperatures of more than 0°C but less than 8°C, more than 5°C difference between mean summer and mean winter soil temperatures at 50 centimeters and warm summer temperatures. Isofrigid is the same except the summer and winter temperatures differ by less than 5°C.

G Horizon. (Obsolete as a major horizon; now used to modify a major horizon, as Bg). A layer of intense reduction, with ferrous iron and gray to brown mottlings.

Gilgai. The microrelief of soils produced by expansion and contraction with changes in moisture. Found in soils that contain large amounts of clay which swells and shrinks considerably with wetting and drying. Usually a succession of microbasins and microknolls in nearly level areas or of microvalleys and microridges parallel to the direction of the slope.

Glacial Drift. Rock debris that has been transported by glaciers and deposited, either directly from the ice or from the melt-water. The debris may or may not be heterogeneous.

Gypsic Horizon. A mineral soil horizon of secondary calcium sulfate enrichment that is more than 15 centimeters thick, has at least 5 percent more gypsum than the C horizon, and in which the product of the thickness in centimeters and the percent calcium sulfate is equal to or greater than 150 percent centimeters.

Glaciofluvial Deposits. Material moved by glaciers and subsequently sorted and deposited by streams flowing from the melting ice. The deposits are stratified and may occur in the form of outwash plains, deltas, kames, eskers, and kame terraces. See **glacial drift and till, (1).**

Gleyzation. A soil-forming process resulting in the development of gley soils.

Gley Soil. (Obsolete in the United States). Soil developed under conditions of poor drainage, resulting in reduction of iron and other elements and in gray colors and mottles.

Gravitational Water. Water that moves into, through, or out of the soil under the influence of gravity.

Gray-Brown Podzolic Soil. A zonal great soil group consisting of soils with a thin, moderately dark A1 horizon and with a grayish-brown A2 horizon underlain by a B horizon containing a high percentage of bases and an appreciable quantity of illuviated silicate clay; formed on relatively young land surfaces, mostly glacial deposits, from material relatively rich in calcium, under deciduous forests in humid temperate regions.

Great Soil Group. One of the categories in the system of soil classification that has been used in the United States for many years.

Green-Manure Crop. A crop grown for use as green manure; to be incorporated into soil when green.

Ground Water. The portion of the total precipitation that at any particular time is either passing through or standing in the soil and the underlying strata and is free to move under the influence of gravity.

Ground-Water Laterite Soil. A great soil group of the intrazonal order and hydromorphic suborder, consisting of soils characterized by hardpans or concretional horizons rich in iron and aluminum (and sometimes manganese) that have formed immediately above the water table.

Ground-Water Podzol Soil. A great soil group of the intrazonal order and hydromorphic suborder, consisting of soils with an organic mat on the surface over a very thin layer of acid humus material underlain by a whitish-gray leached layer, which may be as much as 2 or 3 feet in thickness, and is underlain by a brown, or very dark-brown, cemented hardpan layer; formed under various types of forest vegetation in cool to tropical, humid climates under conditions of poor drainage.

Gypsum Requirement. The quantity of gypsum or its equivalent required to reduce the exchangeable-sodium percentage of a given increment of soil to an acceptable level.

H Layer. A layer occurring in mor humus consisting of well-decomposed organic matter of unrecognizable origin. The O2 horizon.

Half-Bog Soil. A great soil group, of the intrazonal order and hydromorphic suborder, consisting of soil with dark-brown or black peaty material over grayish and rust mottled mineral soil; formed under conditions of poor drainage under forest, sedge, or grass vegetation in cool to tropical humid climates.

Halophytic Vegetation. Vegetation requiring or tolerating a saline environment.

Hardpan. A hardened soil layer, in the lower A or in the B horizon, caused by cementation of soil particles with organic matter or with materials such as silica, sesquioxides, or calcium carbonate. The hardness does not change appreciably with changes in moisture content, and pieces of hard layer do not slake in water.

Hemists. Histosols that have an intermediate degree of plant fiber decomposition and a bulk density between about 0.1 and 0.2. Hemists are saturated with water for periods long enough to limit their use for most crops unless they are artificially drained. (A suborder in the USDA soil taxonomy.)

Heterotrophic. An organism capable of deriving energy for life processes from the oxidation of organic compounds.

Histic Epipedon. A thin organic soil horizon that is saturated with water at some period of the year unless artificially drained and that is at or near the surface of a mineral soil. The histic epipedon has a maximum thickness, depending on the kind of materials in the horizon and the lower limit of organic carbon is the upper limit for the mollic epipedon.

Histosols. Organic soils that have organic soil materials in more than half of the upper 80 centimeters, or that are of any thickness if overlying rock or fragmental materials that have interstices filled with organic soil materials. (An order in the USDA soil taxonomy.)

Horizon. See **soil horizon.**

Hue. One of the three variables of color. It is caused by light of certain wavelengths and changes with the wavelength. See **Munsell color system, chroma,** and **value, color.**

Humic Acid. A mixture of dark-colored substances of indefinite composition extracted from soil with dilute alkali and precipitated by acidification.

Humic Gley Soil. Soil of the intrazonal order and hydromorphic suborder that includes Wisenboden and related soils, such as Half-Bog soils, which have a thin muck or peat O2 horizon and an A1 horizon. Developed in wet meadow and in forested swamps.

Humification. The processes involved in the decomposition of organic matter and leading to the formation of humus.

Humin. The fraction of the soil organic matter that is not dissolved upon extraction of the soil with dilute alkali.

Humods. Spodosols that have accumulated organic carbon and aluminum, but not iron, in the upper part of the spodic horizon. Humods are rarely saturated with water or do not have characteristics associated with wetness. (A suborder in the USDA soil taxonomy.)

Humox. Oxisols that are moist all or most of the time and that have a high organic carbon content within the upper meter. Humox have a mean annual soil temperature of less than 22°C and a base saturation within the oxic horizon of less than 35%, measured at pH 7. (A suborder in the USDA soil taxonomy.)

Humults. Ultisols that have a high content of organic carbon. Humults are not saturated with water for periods long enough to limit their use for most crops. (A suborder in the USDA soil taxonomy.)

Humus. (1) That more or less stable fraction of the soil organic matter remaining after the major portion of added plant and animal residues have decomposed. Usually it is dark colored. (2) Includes the F and H layers in undisturbed forest soils.

Hydraulic Conductivity. An expression of the readiness of water to flow through a soil in response to a given water potential gradient.

Hydrous Mica. A silicate clay with 2:1 lattice structure, but of indefinite chemical composition, since usually part of the silicon in the silica tetrahedral layer has been replaced by aluminum, and containing a considerable amount of potassium that serves as an additional bonding between the crystal units, resulting in particles larger than normal in montmorillonite and, consequently, in a lower cation-exchange capacity. Sometimes referred to as illite.

Hygroscopic Water. Water adsorbed by a dry soil from an atmosphere of high relative humidity, water remaining in the soil after "air-drying," or water held by the soil when it is in equilibrium with an atmosphere of a specified relative humidity at a specified temperature, usually 98 percent relative humidity at 25°C.

Hyperthermic. A soil temperature regime that has mean annual soil temperatures of 22°C or more and more than 5°C difference between mean summer and mean winter soil temperatures at 50 centimeters. Isohyperthermic is the same except the summer and winter temperatures differ by less than 5°C.

Igneous Rock. Rock formed from the cooling and solidification of magma, and that has not been changed appreciably since its formation.

Illite. A hydrous mica.

Illuvial Horizon. A soil layer or horizon in which material carried from an overlying layer has been precipitated from solution or deposited from suspension. The layer of accumulation. See **eluvial horizon.**

Illuviation. The process of deposition of soil material removed from one horizon to another in the soil; usually from an upper to a lower horizon in the soil profile. See **eluviation.**

Immature Soil. A soil with indistinct or only slightly developed horizons because of the relatively short time it has been subjected to the various soil-forming processes. A soil that has not reached equilibrium with its environment.

Immobilization. The conversion of an element from the inorganic to the organic form in microbial tissues or in plant tissues.

Inceptisols. Mineral soils that have one or more pedogenic horizons in which mineral materials other than carbonates or amorphous silica have been altered or removed but not accumulated to a significant degree. Under certain conditions, inceptisols may have an ochric, umbric, histic, plaggen, or mollic epipedon. Water is available to plants more than half of the year or more than 3 consecutive months during a warm season. (An order in the USDA soil taxonomy.)

Indicator Plants. Plants characteristic of specific soil or site conditions..

Infiltration. The downward entry of water into the soil.

Intergrade. A soil that possesses moderately well-developed distinguishing characteristics of two or more genetically related great soil groups.

Intrazonal Soils (Obsolete). (1) One of the three orders in soil classification in 1949 system. (2) A soil with more or less well-developed soil characteristics that reflect the dominating influence of some local factor of relief, parent material, or age, over the normal effect of climate and vegetation.

Iron-Pan. An indurated soil horizon in which iron oxide is the principal cementing agent.

Irrigation. The artificial application of water to the soil for the benefit of growing crops.

Irrigation Efficiency. The ratio of the water actually consumed by crops on an irrigated area to the amount of water diverted from the source onto the area.

Isomorphous Substitution. The replacement of one atom by another of similar size in a crystal lattice without disrupting or changing the crystal structure of the mineral.

Kame. An irregular ridge or hill of stratified glacial drift.

Kaolin. (1) An aluminosilicate mineral of the 1:1 crystal lattice group; that is, consisting of one silicon tetrahedral layer and one aluminum oxide-hydroxide octahedral layer. (2) The 1:1 group or family of aluminosilicates.

L Layer (Litter). The surface layer of the forest floor consisting of freshly fallen leaves, needles, twigs, stems, bark, and fruits. This layer may be very thin or absent during the growing season. The O1 horizon.

Landscape. All the natural features, such as fields, hills, forests, water, etc., that distinguish one part of the earth's surface from another part. Usually that portion of land or territory that the eye can comprehend in a single view, including all its natural characteristics.

Lateritic Soil. A suborder of zonal soils formed in warm, temperate, and tropical regions and including the following great soil groups: Yellow Podzolic, Red Podzolic, Yellowish-Brown Lateritic, and Lateritic.

Latosol. A suborder of zonal soils including soils formed under forested, tropical, humid conditions and characterized by low silica-sesquioxide ratios of the clay fractions, low base-exchange capacity, low activity of the clay, low content of most primary minerals, low content of soluble constituents, a high degree of aggregate stability, and usually having a red color.

Lattice. A three-dimensional grid of lines connecting the points *representing* the centers of atoms or ions in a crystal.

Leaching. The removal of materials in solution from the soil. See **eluviation.**

Leaching Requirement. The fraction of applied irrigation water that should be passed through the root zone to control soil salinity at a specified level.

Lime Requirement. The mass of agricultural lime, or the equivalent of other specified liming material, required to raise the pH of the soil to a specifiee value under field conditions.

Liquid Limit. The minimum percentage (by weight) of moisture at which a small sample of soil will barely flow under a standard treatment.

Lithic Contact. A boundary between soil and continuous, coherent, underlying material. The underlying material must be sufficiently coherent to make hand-digging with a spade impractical. If mineral, it must have a hardness of 3 or more (Mohs scale), and gravel-sized chunks that can be broken out do not disperse with 15 hours shaking in water or sodium hexametaphosphate solution.

Lithosequence. A group of related soils that differ, one from the other, in certain properties primarily as a result of differences in the *parent rock* as a soil-forming factor.

Lithosols. A great soil group of azonal soils characterized by an incomplete solum or no clearly expressed soil morphology and consisting of freshly and imperfectly weathered rock or rock fragments.

Loam. A soil textural class.

Loess. Material transported and deposited by wind and consisting of predominantly silt-sized particles.

Luxury Uptake. The absorption by plants of nutrients in excess of their need for growth. Luxury concentrations during early growth may be utilized in later growth.

Macronutrient. A chemical element necessary in relatively large amounts (usually > 500 parts per million in the plant) for the growth of plants. These elements consist of C, H, O, Ca, Mg, K, P, S, and N.

Made Land. Areas filled with earth, or with earth and trash mixed, usually by or under the control of humans. A miscellaneous land type.

Mass Flow (Nutrient). The movement of solutes associated with net movement of water.

Mature Soil. A soil with well-developed soil horizons produced by the natural processes of soil formation and essentially in equilibrium with its present environment.

Mesic. A soil temperature regime that has mean annual soil temperatures of 8°C or more but less than 15°C, and more than 5°C difference between mean summer and mean winter soil temperatures at 50 centimeters. Isomesic is the same except the summer and winter temperatures differ by less than 5°C.

Metamorphic Rock. Rock derived from preexisting rocks but that differ from them in physical, chemical, and mineralogical properties as a result of natural geological processes, principally heat and pressure, originating within the earth. The preexisting rocks may have been igneous, sedimentary, or another form of metamorphic rock.

Microclimate. (1) The climatic condition of a small area resulting from the modification of the general climatic conditions by local differences in elevation or exposure. (2) The sequence of atmospheric changes within a very small region.

Micronutrient. A chemical element necessary only in extremely small amounts (usually < 50 parts per million in the plant) for the growth of plants. These elements consist of B, Cl, Cu, Fe, Mn, Mo, and Zn.

Microrelief. Small-scale, local differences in topography, including mounds, swales, or pits that are only a few feet in diameter and with elevation differences of up to 6 feet. See **cradle knoll** and **gilgai**.

Mineralization. The conversion of an element from an organic form to an inorganic state as a result of microbial decomposition.

Mineralogical Analysis. The estimation or determination of the kinds or amounts of minerals present in a rock or in a soil.

Mineral Soil. A soil consisting predominantly of, and having its properties determined predominantly by, mineral matter. Usually contains < 20 percent organic matter, but may contain an organic surface layer up to 30 centimeters thick.

Moisture Equivalent. The weight percentage of water retained by a previously saturated sample of soil 1 centimeter in thickness after it has been subjected to a centrifugal force of 1000 times gravity for 30 minutes.

Moisture Tension (or Pressure). The equivalent negative pressure in the soil water. It is equal to the equivalent pressure that must be applied to the soil water to bring it to hydraulic equilibrium, through a porous permeable wall or membrane, with a pool of water of the same composition. The pressures used and the corresponding percentages most commonly determined are:

Fifteen-atmosphere Percentage. The percentage of water contained in a soil that has been saturated, subjected to, and is in equilibrium with, an applied pressure of 15-atmosphere. (Pressure applied in a pressure membrane or ceramic pressure plate apparatus. Usually expressed as a weight percentage. Approximately the same as 15-bar percentage.)

Fifteen-bar Percentage. The percentage of water contained in a soil that has been saturated, subjected to, and is in equilibrium with, an applied pressure of 15 bars. (Pressure applied in a pressure membrane or ceramic pressure plate apparatus. Usually expressed as a weight percentage but may be expressed as a volume percentage. Approximately the same as the 15-atmosphere percentage.)

One-third-atmosphere Percentage. The percentage of water contained in a soil that has been saturated, subjected to, and is in equilibrium with, an applied pressure of ⅓ atmosphere. (Pressure applied in a ceramic plate apparatus. Usually expressed as a weight percentage, but may be expressed as a volume percentage. Approximately the same as ⅓ bar percentage. Also, for medium- to coarse-textured soils approximately numerically equal to moisture equivalent.)

One-third bar Percentage. The percentage of water contained in a soil that has been saturated, subjected to, and is in equilibrium with, an applied pressure of ⅓ bar. (Pressure applied in a ceramic plate apparatus. Usually expressed as a weight percentage. Approximately the same as ⅓ atmosphere percentage. Also, for medium- to coarse-textured soils approximately numerically equal to moisture equivalent.)

Sixty-centimeter Percentage. The percentage of water contained in a soil that has been saturated, subjected to, and is in equilibrium with, an applied pressure (or tension) equivalent to a column of water 60 centimeters high. (The pressure may be applied in a pressure plate apparatus or, as a tension, on a tension table. May be expressed on a weight or volume basis. Considered by many to approximate the "field-moisture capacity," especially in medium- to coarse-textured soils.)

Moisture Volume Percentage. The ratio of the volume of water in a soil to the total bulk volume of the soil.

Moisture Weight Percentage. The moisture content expressed as a percentage of the oven-dry weight of soil.

Mollic Epipedon. A surface horizon of mineral soil that is dark colored and relatively thick, contains at least 0.58 percent organic carbon, is not massive and hard or very hard when dry, has a base saturation of more than 50 percent when measured at pH 7, has less than 250 parts per million of P_2O_5 soluble in 1 percent citric acid, and is dominantly saturated with bivalent cations.

Mollisols. Mineral soils that have a mollic epipedon overlying mineral material with a base saturation of 50 percent or more when measured at pH 7. Mollisols may have an argillic, natric, albic, cambic, gypsic, calcic, or petrocalcic horizon, a histic epipedon, or a duripan, but not an oxic or spodic horizon. (An order in the USDA soil taxonomy.)

Montmorillonite. An aluminosilicate clay mineral with a 2:1 expanding crystal structure; that is, with two silicon tetrahedral layers enclosing an aluminum octahedral layer. Considerable expansion may be caused along the C axis by water moving between silica layers of contiguous units.

Montmorillonite-Saponite Group. Clay minerals with 2:1 crystal lattice structure; that is, two silicon tetrahedral layers enclosing an aluminum octahedral layer. Consists of montmorillonite, beidellite, nontronite, saponite, and others.

Mor. A type of forest humus in which the H layer is present and in which there is practically no mixing of surface organic matter with mineral soil; that is, the transition from the H layer to the A1 horizon is abrupt.

Mottled Zone. A layer that is marked with spots or blotches of different color or shades of color. The pattern of mottling and the size, abundance, and color contrast of the mottles may vary considerably and should be specified in soil description.

Mottling. Spots or blotches of different color or shades of color interspersed with the dominant color.

Muck. Highly decomposed organic material in which the original plant parts are not recognizable. Contains more mineral matter and is usually darker in color than peat. See **muck soil, peat,** and **peat soil.**

Muck Soil. (1) A soil containing between 20 and 50 percent of organic matter. (2) An organic soil in which the organic matter is well decomposed (US usage).

Mulch. (1) Any material such as straw, sawdust, leaves, plastic film, loose soil, etc., that is spread on the surface of the soil to protect the soil and plant roots from the effects of raindrops, soil crusting, freezing, evaporation, etc. (2) To apply mulch to the soil surface.

Mulch Farming. A system of farming in which the organic residues are not plowed into or otherwise mixed with the soil, but are left on the surface as a mulch.

Mull. A type of forest humus in which the F layer may or may not be present and in which there is no H layer. The A1 horizon consists of an intimate mixture of organic matter and mineral soil with gradual transition between the A1 and the horizon beneath.

Munsell Color System. A color designation system that specifies the relative degrees of the three simple variables of color: hue, value, and chroma. For example: 10YR 6/4 is a color (of soil) with a hue = 10YR, value = 6, and chroma = 4.

Mycorrhiza. Literally, "fungus root." The association, usually symbiotic, of specific fungi with the roots of higher plants.

Natric Horizon. A mineral soil horizon that satisfied the requirements of an argillic horizon, but that also has prismatic, columnar, or blocky structure and a subhorizon having more than 15 percent saturation with exchangeable sodium.

Nitrification. Biological oxidation of ammonium to nitrate and nitrite, or a biologically induced increase in the oxidation state of nitrogen.

Nitrogen Assimilation. The incorporation of nitrogen into organic cell substances by living organisms.

Nitrogen Cycle. The sequence of biochemical changes undergone by nitrogen wherein it is used by a living organism, liberated upon the death and decomposition of the organism, and converted to its original state of oxidation.

Nitrogen Fixation. Biological conversion of molecular dinitrogen (N_2) to organic combinations or to forms utilizable in biological processes.

Nutrient, Diffusion. The movement of nutrients in soil that results from a concentration gradient.

O Horizon. Organic horizon of mineral soils.

Ochrepts. Inceptisols formed in cold or temperate climates and that commonly have an ochric epipedon and a cambic horizon. They may have an umbric or mollic epipedon less than 25 centimeters thick or a fragipan or duripan under certain conditions. These soils are not dominated by amorphous materials and are not saturated with water for periods long enough to limit their use for most crops. (A suborder in the USDA soil taxonomy.)

Ochic Epipedon. A surface horizon of mineral soil that is too light in color, too high in chroma, too low in organic carbon, or too thin to be a plaggen, mollic, umbric, anthropic or histic epipedon, or that is both hard and massive when dry.

Order. The highest category in soil classification.

Organic Soil. A soil that contains a high percentage (> 15 or 20 percent) of organic matter throughout the solum.

Orthents. Entisols that have either textures of very fine sand or finer in the fine earth fraction, or textures of loamy fine sand or coarser and a coarse fragment content of 35 percent or more and that have an organic carbon content that decreases regularly with depth. Orthents are not saturated with water for periods long enough to limit their use for most crops. (A suborder in the USDA taxonomy.)

Orthids. Aridisols that have a cambic, calcic, petrocalcic, gypsic, or salic horizon or a duripan but that lack an argillic or natric horizon. (A suborder in the USDA taxonomy.)

Orthods. Spodosols that have less than six times as much free iron (elemental) than organic carbon in the spodic horizon, but the ratio of iron to carbon is 0.2 or more. Orthods are not saturated with water for periods long enough to limit their use for most crops. (A suborder in the USDA soil taxonomy.)

Orthox. Oxisols that are moist all or most of the time and that have a low to moderate content of organic carbon within the upper 1 meter or a mean annual soil temperature of 22°C or more. (A suborder in the USDA soil taxonomy.)

Ortstein. An indurated layer in the B horizon of Podzols in which the cementing material consists of illuviated sesquioxides (mostly iron) and organic matter.

Osmotic. A type of pressure exerted in living bodies as a result of unequal concentration of salts on both sides of a cell wall or membrane. Water will move from the area having the least salt concentration through the membrane into the area having the highest salt concentration and, therefore, exert additional pressure on one side of the membrane.

Oven-dry Soil. Soil that has been dried at 105°C until it reaches constant weight.

Oxic Horizon. A mineral soil horizon that is at least 30 centimeters thick and characterized by the virtual absence of weatherable primary minerals or 2:1 lattice clayss, the presence of 1:1 lattice clays and highly insoluble minerals such as quartz sand, the presence of hydrated oxides of iron and aluminum, the absence of water-dispersible clay, and the presence of low cation exchange capacity and small amounts of exchangeable bases.

Oxisols. Mineral soils that have an oxic horizon within 2 meters of the surface or plinthite as a continuous phase within 30 centimeters of the surface, and that do not have a spodic or argillic horizon above the oxic horizon. (An order in the USDA soil taxonomy.)

Paleosol. A soil formed on a landscape during the geologic past and subsequently buried by sedimentation.

Pans. Horizons or layers in soils that are strongly compacted, indurated, or very high in clay content. See **caliche, claypan, fragipan,** and **hardpan.**

Pan, Genetic. A natural subsurface soil layer of low or very low permeability, with a high concentration of small particles, and differing in certain physical and chemical properties from the soil immediately above or below the pan. See **claypan, fragipan,** and **hardpan,** all of which are genetic pans.

Pan, Pressure or Induced. A subsurface horizon or soil layer having a higher bulk density and a lower total porosity than the soil directly above or below it, as a result of pressure that has been applied by normal tillage operations or by other artificial means. Frequently referred to as plowpan, plowsole, or traffic pan.

Paralithic Contact. Similar to a lithic contact except that the mineral material below the contact has a hardness of less than 3 (Mohs scale), and gravel-sized chunks that can be broken out will partially disperse within 15 hours shaking in water or sodium hexametaphosphate solution.

Parent Material. The unconsolidated and more or less chemically weathered mineral or organic matter from which the solum of soils is developed by pedogenic processes.

Particle Density. The mass per unit volume of the soil particles. In technical work, usually expressed as grams per cubic centimeter.

Particle-size Analysis. Determination of the various amounts of the different separates in a soil sample, usually by sedimentation, sieving, micrometry, or combinations of these methods.

Parts per Million (ppm). Weight units of any given substance per 1 million equivalent weight units of oven-dry soil; or, in the case of soil solution or the solution, the weight units of solute per million weight units of solution.

Peat. Unconsolidated soil material consisting largely of undecomposed, or only slightly decomposed, organic matter accumulated under conditions of excessive moisture.

Peat Soil. An organic soil containing more than 50 percent organic matter. Used in the United States to refer to the stage of decomposition of the organic matter, "peat" referring to the slightly decomposed or undecomposed deposits and "muck" to the highly decomposed materials.

Ped. A unit of soil structure such as an aggregate, crumb, prism, block, or granule, formed by natural processes (in contrast with a clod, which is formed artificially).

Pedalfer (Obsolete). A subdivision of a soil order comprising a large group of soils in which sesquioxides increased relative to silica during soil formation.

Pedocal (Obsolete). A subdivision of a soil order comprising a large group of soils in which calcium accumulated during soil formation.

Pedon. A three-dimensional body of soil with lateral dimensions large enough to permit the study of horizon shapes and relations. Its area ranges from 1 to 10 square meters. Where horizons are intermittent or cyclic, and recur at linear intervals of 2 to 7 meters, the pedon includes one-half of the cycle. Where the cycle is less than 2 meters, or all horizons are continuous and of uniform thickness, the pedon has an area of approximately 1 square meter. If the horizons are cyclic, but recur at intervals greater than 7 meters, the pedon reverts to the 1 square meter size, and more than one soil will usually be represented in each cycle.

Percolation, Soil Water. The downward movement of

water through soil. Especially, the downward flow of water in saturated or nearly saturated soil at hydraulic gradients of the order of 1.0 or less.

Pergelic. A soil temperature regime that has mean annual soil temperatures of less than 0°C. Permafrost is present.

Permafrost. (1) Permanently frozen material underlying the solum. (2) A perennially frozen soil horizon.

Permafrost Table. The upper boundary of the permafrost [see permafrost, (1)], coincident with the lower limit of seasonal thaw.

Permeability, Soil. The ease with which gases, liquids, or plant roots penetrate or pass through a bulk mass of soil or a layer of soil. Since different soil horizons vary in permeability, the particular horizon under question should be designated.

Petrocalcic Horizon. A continuous, indurated calcic horizon that is cemented by calcium carbonate and, in some places, with magnesium carbonate. It cannot be penetrated with a spade or auger when dry, dry fragments do not slake in water, and it is impenetrable to roots.

Petrogypsic Horizon. A continuous, strongly cemented, massive, gypsic horizon that is cemented by calcium sulfate. It can be chipped with a spade when dry. Dry fragments do not slake in water and it is impenetrable to roots.

pH, Soil. The negative logarithm of the hydrogen-ion activity of a soil. The degree of acidity (or alkalinity) of a soil as determined by means of a glass, quinhydrone, or other suitable electrode or indicator at a specified moisture content or soil-water ratio, and expressed in terms of the pH scale.

pH-Dependent Charge. The portion of the cation or anion exchange capacity that varies with pH.

Phase, Soil. A subdivision of a soil type or other unit of classification having characteristics that affect the use and management of the soil but that do not vary sufficiently to differentiate it as a separate type. A variation in a property or characteristic such as degree of slope, degree of erosion, content of stones, etc.

Placic Horizon. A black to dark reddish mineral soil horizon that is usually thin but that may range from 1 to 25 millimeters in thickness. The placic horizon is commonly cemented with iron and is slowly permeable or impenetrable to water and roots.

Plaggen Epipedon. A synthetic surface horizon more than 50 centimeters thick that is formed by long-continued manuring and mixing.

Plaggepts. Inceptisols that have a plaggen epipedon. (A suborder in the USDA soil taxonomy.)

Planosol. A great soil group of the intrazonal order

and hydromorphic suborder consisting of soils with eluviated surface horizons underlain by B horizons more strongly eluviated, cemented, or compacted than associated normal soil.

Plastic Soil. A soil capable of being molded or deformed continuously and permanently, by relatively moderate pressure, into various shapes. See **consistency.**

Plinthite. A nonindurated mixture of iron and aluminum oxides, clay, quartz and other diluents that commonly occurs as red soil mottles usually arranged in platy, polygonal, or reticulate patterns. Plinthite changes irreversibly to ironstone hardpans or irregular aggregates on exposure to repeated wetting and drying.

Podzol. A great soil group of the zonal order consisting of soils formed in cool-temperate to temperate, humid climates, under coniferous or mixed coniferous and deciduous forest, and characterized particularly by a highly leached, whitish-gray (Podzol) A2 horizon.

Prairie Soils. A zonal great soil group consisting of soils formed under temperate to cool-temperate, humid regions under tall grass vegetation.

Primary Mineral. A mineral that has not been altered chemically since deposition and crystallization from molten lava. See **secondary mineral.**

Prismatic Soil Structure. A soil structure type with prismlike aggregates that have a vertical axis much longer than the horizontal axes. See **soil structure types.**

Productivity, Soil. The capacity of a soil, in its normal environment, for producing a specified plant or sequence of plants under a specified system of management. The "specified" limitations are necessary, since no soil can produce all crops with equal success nor can a single system of management produce the same effect on all soils. Productivity emphasizes the capacity of soil to produce crops and should be expressed in terms of yields.

Profile, Soil. A vertical section of the soil through all its horizons and extending into the parent material.

Psamments. Entisols that have textures of loamy fine sand or coarser in all parts, have less than 35 percent coarse fragments, and that are not saturated with water for periods long enough to limit their use for most crops. (A suborder in the USDA soil taxonomy.)

R Horizon. Underlying consolidated bedrock, such as granite, sandstone, or limestone.

Red Desert Soil. A zonal great soil group consisting of soils formed under warm-temperate to hot, dry regions under desert-type vegetation, mostly shrubs.

Red-Yellow Podzolic Soils. A combination of the zonal great soil groups, Red Podzolic and Yellow Podzolic, consisting of soils formed under warm-temperate to tropical, humid climates, under deciduous or coniferous forest vegetation and usually, except for a few members of the Yellow Podzolic Group, under conditions of good drainage.

Regolith. The unconsolidated mantle of weathered rock and soil material on the earth's surface; loose earth materials above solid rock. (Approximately equivalent to the term "soil" as used by many engineers.)

Regosol. Any soil of the azonal order without definite genetic horizons and developing from or on deep, unconsolidated, soft mineral deposits such as sands, loess, or glacial drift.

Regur. An intrazonal group of dark calcareous soils high in clay, which is mainly montmorillonitic, and formed mainly from rocks low in quartz; occurring extensively on the Deccan Plateau of India.

Rendolls. Mollisols that have no argillic or calcic horizon but that contain material with more than 40 percent $CaCO_3$ equivalent within or immediately below the mollic epipedon. Rendolls are not saturated with water for periods long enough to limit their use for most crops. (A suborder in the USDA soil taxonomy.)

Rendzina. A great soil group of the intrazonal order and calcimorphic suborder consisting of soils with brown or black friable surface horizons underlain by light gray to pale yellow calcareous material; developed from soft, highly calcareous parent material under grass vegetation or mixed grasses and forest in humid and semiarid climates.

Reticulate Mottling. A network of streaks of different color; most commonly found in the deeper profiles of Lateritic soils.

Rhizobia. Bacteria capable of living symbiotically in roots of legumes, from which they receive energy and often utilize molecular nitrogen. Collective common name for the genus *Rhizobium.*

Rhizosphere. The zone of soil where the microbial population is altered both quantitatively and qualitatively by the presence of plant roots.

Runoff. The portion of the precipitation on an area that is discharged from the area through stream channels. That which is lost without entering the soil is called *surface runoff,* and that which enters the soil before reaching the stream is called *ground water runoff* or *seepage flow* from ground water. (In soil science "runoff" usually refers to the water lost by surface flow; in geology and hydraulics "runoff" usually includes both surface and substrate flow.)

Salic Horizon. A mineral soil horizon of enrichment with secondary salts more soluble in cold water than gypsum. A salic horizon is 15 centimeters or more in thickness, contains at least 2 percent salt, and the product of the thickness in centimeters and percent salt by weight is 60 percent centimeters or more.

Saline Soil. A nonsodic soil containing sufficient soluble salt to impair its productivity. The electrical conductivity of the saturation extract is > 2 mmhos per centimeter at 25°C.

Saline-Sodic Soil. A soil containing a combination of soluble salts and exchangeable sodium sufficient to interfere with the growth of most crop plants. The electrical conductivity and sodium-adsorption ratio of the saturation extract are > 2 millimoles per centimeter at 25°C. and > 15, respectively. The pH is usually 8.5 or less in the saturated soil paste. (Formerly called saline-alkali soil.)

Salinization. The process whereby soluble salts accumulate in soil.

Salt Balance. The relation between the quantity of dissolved salts carried to an area in irrigation water and the quantity of dissolved salts removed by drainage water.

Sand. (1) A soil particle between 0.05 and 2.0 millimeters in diameter. (2) Any one of five soil separates: very coarse sand, coarse sand, medium sand, fine sand, very fine sand. See **soil separates.** (3) A soil textural class.

Saprists. Histosols that have a high content of plant materials so decomposed that original plant structures cannot be determined and a bulk density of about 0.2 or more. Saprists are saturated with water for periods long enough to limit their use for most crops unless they are artificially drained. (A suborder in the USDA soil taxonomy.)

Saturation Extract. An increment of solution obtained from a saturated soil paste.

Secondary Mineral. A mineral resulting from the decomposition of a primary mineral or from the reprecipitation of the products of decomposition of a primary mineral. See **primary mineral.**

Self-Mulching Soil. A soil in which the surface layer becomes so well aggregated that it does not crust and seal under the impact of rain but, instead, serves as a surface mulch upon drying.

Sierozem. A zonal great soil group consisting of soils with pale grayish A horizons grading into calcareous material at a depth of 1 foot or less, and formed in temperate to cool, arid climates under a vegetation of desert plants, short grass, and scattered brush.

Silica-Alumina Ratio. The molecules of silicon dioxide (SiO_2) per molecule of aluminum oxide (Al_2O_3) in clay minerals or in soils.

Silica-Sesquioxide Ratio. The molecules of silicon dioxide (SiO_2) per molecule of aluminum oxide (Al_2O_3) plus ferric oxide (Fe_2O_3) in clay minerals or in soils.

Silt. (1) A soil separate consisting of particles between 0.05 and 0.002 millimeters in equivalent diameter. See **soil separates.** (2) A soil textural class.

Site Index. (1) A quantitative evaluation of the productivity of a soil for forest growth under the existing or specified environment. (2) The height in feet of the dominant forest vegetation taken at or calculated to an index age, usually 50 or 100 years.

Sllickensides. Polished and grooved surfaces produced by one mass sliding past another. Slickensides are common in Vertisols.

Slick Spots. Small areas in a field that are slick when wet because of a high content of alkali or exchangeable sodium.

Sodic Soil. (1) A soil containing sufficient exchangeable-sodium to interfere with the growth of most crop plants. (2) A soil in which the sodium-adsorption ratio of the saturation extract is 15 or more.

Sodication. The process whereby the exchangeable sodium content of a soil is increased.

Sodium-Adsorption Ratio (SAR). A relation between soluble sodium and soluble divalent cations that can be used to predict the exchangeable-sodium percentage of soil equilibrated with a given solution. It is defined as follows.

$$SAR = \frac{\text{sodium, mmoles/liter}}{(\text{calcium} + \text{magnesium})^{1/2} (\text{mmoles/liter})^{1/2}}$$

Soil. (1) The unconsolidated mineral material on the immediate surface of the earth that serves as a natural medium for the growth of land plants. (2) The unconsolidated mineral matter on the surface of the earth that has been subjected to and influenced by genetic and environmental factors of: *parent material, climate* (including moisture and temperature effects), *macro-* and *microorganisms,* and *topography,* all acting over a period of *time* and producing a product—soil—that differs from the material from which it is derived in many physical, chemical, biological, and morphological properties and characteristics.

Soil Association. (1) A group of defined and named taxonomic soil units occurring together in an individual and characteristic pattern over a geographic region, comparable to plant associations in many ways. (Sometimes called "natural land type.") (2) A mapping unit used on general soil maps, in which two or more defined taxonomic units occurring together in a characteristic pattern are combined because the scale of the map or the purpose for which it is being made does not require delineation of the individual soils.

Soil Genesis. (1) The mode or origin of the soil with special reference to the processes or soil-forming factors responsible for the development of the solum, or true soil, from the unconsolidated parent material. (2) A division of soil science concerned with soil genesis (1).

Soil Geography. A specialization of soil science concerned with the areal distributions of soil types.

Soil Horizon. A layer of soil or soil material approximately parallel to the land surface and differing from adjacent genetically related layers in physical, chemical, and biological properties or characteristics such as color, structure, texture, consistency, kinds and numbers of organisms present, degree of acidity or alkalinity, etc.

Soil Management Groups. Groups of taxonomic soil units with similar adaptations or management requirements for one or more specific purposes, such as adapted crops or crop rotations, drainage practices, fertilization, forestry, and highway engineering.

Soil Map. A map showing the distribution of soil types or other soil mapping units in relation to the prominent physical and cultural features of the earth's surface.

Soil Moisture Regimes. See **aquic, aridic, torric, udic, ustic, xeric.**

Soil Monolith. A vertical section of a soil profile removed from the soil and mounted for display or study.

Soil Morphology. (1) The physical constitution, particularly the structural properties, of a soil profile as exhibited by the kinds, thickness, and arrangement of the horizons in the profile, and by the texture, structure, consistency, and porosity or each horizon. (2) The structural characteristics of the soil or any of its parts.

Soil Science. The science dealing with soils as a natural resource on the surface of the earth, including soil formation, classification, and mapping, and the physical, chemical, biological, and fertility properties of soils *per se;* and these properties in relation to their management for crop production.

Soil Separates. Mineral particles, < 2.0 millimeters in equivalent diameter, ranging between specified size limits. The names and size limits of separates recognized in the United States are: *very coarse sa id,* 2.0 to 1.0 millimeters; *coarse sand,* 1.0 to 0.5 millimeters; *medium sand,* 0.5 to 0.25 millimeters; *fine sand,* 0.25 to 0.10 millimeters; *very fine sand,* 0.10 to 0.05 millimeters; *silt,* 0.05 to 0.002 millimeters; and *clay,* < 0.002 millimeters.

Soil Series. The basic unit of soil classification being a subdivision of a family and consisting of soils that are essentially alike in all major profile characteristics except the texture of the A horizon.

Soil Solution. The aqueous liquid phase of the soil and its solutes.

Soil Structure. The combination or arrangement of primary soil particles into secondary particles, units, or peds. These secondary units may be, but usually are not, arranged in the profile in such a manner as to give a distinctive characteristic pattern. The secondary units are characterized and classified on the basis of size, shape, and degree of distinctness into classes, types, and grades, respectively. See **soil structure classes, soil structure grades,** and **soil structure types.** Also see Table 1.

Soil Survey. The systematic examination, description, classification, and mapping of soils in an area. Soil surveys are classified according to the kind and intensity of field examination.

Soil Temperature Regimes. See **pergelic, cryic, frigid, mesic, thermic, hyperthermic.**

Soil Texture. The relative proportions of the various soil separates in a soil.

Soil Type. (1) The lowest unit in the natural system of soil classification; a subdivision of a soil series and consisting of or describing soils that are alike in all characteristics including the texture of the A horizon. (2) In Europe, roughly equivalent to a great soil group.

Soil Water Potential. The amount of work that must be done per unit quantity of pure water in order to transport reversibly and isothermally an infinitesimal quantity of water from a pool of pure water, at a specified elevation and at atmospheric pressure, to the soil water (at the point under consideration). The total potential (of soil water) consists of the following:

Osmotic Potential. The amount of work that must be done per unit quantity of pure water in order to transport reversibly and isothermally an infinitesimal quantity of water from a pool of pure water, at a specified elevation and at atmospheric pressure, to a pool of water identical in composition to the soil water (at the point under consideration), but in all other respects being identical to the reference pool.

Gravitational Potential. The amount of work that must be done per unit quantity of pure water in order to transport reversibly and isot hermally an infinitesimal quantity of water, identical in composition to the soil water, from a pool at a specified elevation and at atmospheric pressure, to a similar pool at the elevation of the point under consideration.

Capillary Potential. The amount of work that must be done per unit quantity of pure water in order to transport reversibly and isothermally an infinitesimal quantity of water, identical in composition to the soil water, from a pool at the elevation and the external gas pressure of the point under consideration.

Gas Pressure Potential. This potential component is to be considered only when *external* gas pressure differs from atmospheric pressure as, for example, in a pressure membrane apparatus. A specific term and definition is not given.

Solodized Soil. A soil that has been subjected to the processes responsible for the development of a Soloth and having at least some of the characteristics of a Soloth.

Solonchak. A great soil group of the intrazonal order and halomorphic suborder, consisting of soils with gray, thin, salty crust on the surface, and with fine granular mulch immediately below being underlain with grayish, friable, salty soil; formed under subhumid to arid, hot or cool climate, under conditions of poor drainage, and under a sparse growth of halophytic grasses, shrubs, and some trees.

Solonetz A great soil group of the intrazonal order and halomorphic suborder, consisting of soils with a very thin, friable, surface soil underlain by a dark, hard columnar layer usually highly alkaline; formed under subhumid to arid, hot to cool climates, under better drainage than Solonchaks, and nder a native vegetation of halophytic plants.

Solum (plural: Sola). The upper and most weathered part of the soil profile; the A and B horizons.

Sombric Horizon. A subsurface mineral horizon that is darker in color than the overlying horizon but lacks the properties of a spodic horizon. Common in cool, moist soils of high altitude in tropical regions.

Spodic Horizon. A mineral soil horizon that is characterized by the illuvial accumulation of amorphous materials composed of aluminum and organic carbon with or without iron. The spodic horizon has a certain minimum thickness and a minimum quantity of extractable carbon plus aluminum in relation to its content of clay.

Spodosols. Mineral soils that have a spodic horizon or a placic horizon that overlies a fragipan. (An order in the USDA soil taxonomy.)

Structural Charge. The negative charge on a clay mineral that is caused by isomorphous substitution within the layer. (Expressed as equivalents per formula weight of clay or milliequivalents per 100 grams or per gram of clay.)

Subsoiling. Breaking of compact subsoils, without inverting them, with a special knifelike instrument (chisel) that is pulled through the soil at depths usually of 12 to 24 inches and at spacings usually of 2 to 5 feet.

Sulfur Cycle. The sequence of transformations whereby sulfur is oxidized or reduced through both organic or inorganic products.

Surface-charge Density. The excess of negative or positive charge per unit of surface area of soil or soil mineral.

Talus. Fragments of rock and other soil material accumulated by gravity at the foot of cliffs or steep slopes.

Tensiometer. A device for measuring the negative pressure (or tension) of water in soil *in situ;* a porous, permeable ceramic cup connected through a tube to a manometer or vacuum gauge.

Thermic. A soil temperature regime that has mean annual soil temperatures of 15°C or more but less than 22°C, and more than 5°C difference between mean summer and mean winter soil temperatures at 50 centimeters. Isothermic is the same except the summer and winter temperatures differ by less than 5°C.

Thermosequence. A sequence of related soils that differ, one from the other, primarily as a result of *temperature* as a soil-formation factor.

Till. (1) Unstratified glacial drift deposited directly by the ice and consisting of clay, sand, gravel, and boulders intermingled in any proportion. (2) To plow and prepare for seeding; to seed or cultivate the soil.

Tillage. The mechanical manipulation of soil for any purpose; but in agriculture it is usually restricted to the modifying of soil conditions for crop production.

Tilth. The physical condition of soil as related to its ease of tillage, fitness as a seedbed, and its impedance to seedling emergence and root penetration.

Toposaic. A photomap on which topographic or terrain-form lines are shown, as on standard topographic quadrangles.

Toposequence. A sequence of related soils that differ, one from the other, primarily because of *topography* as a soil-formation factor. See **clinosequence.**

Topsoil. (1) The layer of soil moved in cultivation. See **Surface Soil.** (2) The A horizon. (3) The A1 horizon. (4) Presumably fertile soil material used to topdress roadbanks, gardens, and lawns.

Torrerts. Vertisols of arid regions that have wide, deep cracks that remain open throughout the year in most years. (A suborder in the USDA soil taxonomy.)

Torric. A soil moisture regime defined like aridic moisture regime but used in a different category of the soil taxonomy.

Torrox. Oxisols that have a torric soil moisture regime. (A suborder in the USDA soil taxonomy.)

Tropepts. Inceptisols that have a mean annual soil temperature of 8°C or more, and less than 5°C difference between mean summer and mean winter temperatures at a depth of 50 centimeters below the surface. Tropepts may have an ochric epipedon and a cambic horizon, or an umbric epipedon, or a mollic epipedon under certain conditions but no plaggen epipedon, and are not saturated with water for periods long enough to limit their use for most crops. (A suborder in the USDA soil taxonomy.)

Truncated. Having lost all or part of the upper soil horizon or horizons.

Tuff. Volcanic ash usually more or less stratified and in various states of consolidation.

Tundra. A level or undulating treeless plain characteristic of arctic regions.

Tundra Soils. (1) Soils characteristic of tundra regions. (2) A zonal great soil group consisting of soils with dark-brown peaty layers over grayish horizons mottled with rust and having continually frozen substrata; formed under frigid, humid climates, with poor drainage, and native vegetation of lichens, moss, flowering plants, and shrubs.

Udalfs. Alfisols that have a udic soil moisture regime and mesic or warmer soil temperature regimes. Udalfs generally have brownish colors throughout and are not saturated with water for periods long enough to limit their use for most crops. (A suborder in the USDA soil taxonomy.)

Uderts. Vertisols of relatively humid regions that have wide, deep cracks that usually remain open continuously for less than 2 months or intermittently for periods that total less than 3 months. (A suborder in the USDA soil taxonomy.)

Udic. A soil moisture regime that is neither dry for as long as 90 cumulative days nor for as long as 60 consecutive days in the 90 days following the summer solstice at periods when the soil temperature at 50 centimeters is above 5°.

Udolls. Mollisols that have a udic soil moisture regime with mean annual soil temperatures of 8° or more. Udolls have no calcic or gypsic horizon, and are not saturated with water for periods long enough to limit their use for most crops. (A suborder in the USDA soil taxonomy.)

Udults. Utisols that have low or moderate amounts of organic carbon, reddish or yellowish argillic horizons, and a udic soil moisture regime. Udults are not saturated with water for periods long enough to limit their use for most crops. (A suborder in the USDA soil taxonomy.)

Ultisols. Mineral soils that have an argillic horizon with a base saturation of less than 35 percent when measured at pH 8.2. Ultisols have a mean annual soil temperature of 8° or higher. (An order in the USDA soil taxonomy.)

Umbrepts. Inceptisols formed in cold or temperate climates that commonly have an umbric epipedon, but they may have a mollic or an anthropic epipedon 25 centimeters or more thick under certain conditions. These soils are not dominated by amorphous materials and are not saturated with water for periods long enough to limit their use for most crops. (A suborder in the USDA soil taxonomy.)

Umbric Epipedon. A surface layer of mineral soil that has the same requirements as the mollic epipedon with respect to color, thickness, organic carbon content, consistence, structure, and P_2O_5 content, but that has a base saturation of less than 50 percent when measured at pH 7.

Ustalfs. Alfisols that have an ustic soil moisture regime and mesic or warmer soil temperature regimes. Ustalfs are brownish or reddish throughout and are not saturated with water for periods long enough to limit their use for most crops. (A suborder in the USDA soil taxonomy.)

Usterts. Vertisols of temperate or tropical regions that have wide, deep cracks that usually remain open for periods that total more than 3 months but do not remain open continuously throughout the year, and have either a mean annual soil temperature of 22° or more or a mean summer and mean winter soil temperature at 50 centimeters that differ by less than 5° or have cracks that open and close more than once during the year. (A suborder in the USDA soil taxonomy.)

Ustic. A soil moisture regime that is intermediate between the aridic and udic regimes and common in temperate subhumid or semiarid regions, or in tropical and subtropical regions with a monsoon climate. A limited amount of moisture is available for plants but occurs at times when the soil temperature is optimum for plant growth.

Ustolls. Mollisols that have an ustic soil moisture regime and mesic or warmer soil temperature regimes. Ustolls may have a calcic, petrocalcic, or gypsic horizon, and are not saturated with water for periods long enough to limit their use for most crops. (A suborder in the USDA soil taxonomy.)

Ustox. Oxisols that have an ustic moisture regime and either hyperthermic or isohyperthermic soil temperature regimes or have less than 20 kilogram organic carbon in the surface cubic meter. (A suborder in the USDA soil taxonomy.)

Ustults. Ultisols that have low or moderate amounts of organic carbon, are brownish or reddish throughout, and have an ustic soil moisture regime. (A suborder in the USDA soil taxonomy.)

Value, Color. The relative lightness or intensity of color and approximately a function of the square root of the total amount of light. One of the three variables of color. See **Munsell color system, hue,** and **chroma.**

Varve. A distinct band representing the annual deposit in sedimentary materials regardless of origin and usually consisting of two layers, one a thick, light-colored layer of silt and fine sand and the other a thin, dark-colored layer of clay.

Vertisols. Mineral soils that have 30 percent or more clay, deep wide cracks when dry, and either gilgai microrelief, intersecting slickensides, or wedgeshaped structural aggregates tilted at an angle from the horizontal. (An order in the USDA soil taxonomy.)

Vesicular Structure. Soil structure characterized by round or egg-shaped cavities or vesicles.

Water-stable Aggregate. A soil aggregate that is stable to the action of water such as falling drops, or agitation as in wet-sieving analysis.

Water Table. The upper surface of ground water or that level below which the soil is saturated with water; locus of points in soil water at which the hydraulic pressure is equal to atmospheric pressure.

Water Table, Perched. The water table of a saturated layer of soil that is separated from an underlying saturated layer by an unsaturated layer.

Weathering. All physical and chemical changes produced in rocks, at or near the earth's surface, by atmospheric agents.

Wilting Point. Same as "permanent wilting percentage" as defined in standard plant physiology texts. See **fifteen-atmosphere percentage** and **fifteen-bar percentage.**

Xeralfs. Alfisols that have a xeric soil moisture regime. Xeralfs are brownish or reddish throughout. (A suborder in the USDA soil taxonomy.)

Xererts. Vertisols of Mediterranean climates that have wide, deep cracks that open and close once each year and usually remain open continuously for more than 2 months. Xererts have a mean annual soil temperature of less than 22°. (A suborder in the USDA soil taxonomy.)

Xeric. A soil moisture regime common to Mediterranean climates that have moist cool winters and warm dry summers. A limited amount of moisture is present but does not occur at optimum periods for plant growth. Irrigation or summerfallow is commonly necessary for crop production.

Xerolls. Mollisols that have a xeric soil moisture regime. Xerolls may have a calcic, petrocalcic, or gypsic horizon, or a duripan. (A suborder in the USDA soil taxonomy.)

Xerophytes. Plants that grow in or on extremely dry soils or soil materials.

Xerults. Ultisols that have low or moderate amounts of organic carbon, are brownish or reddish throughout, and have a xeric soil moisture regime. (A suborder in the USDA soil taxonomy.)

Zonal Soil (Obsolete). Anyone of the great groups of soils having well-developed soil characteristics that reflect the influence of the active factors of soil genesis—climate and living organisms, chiefly vegetation (1949 system).

INDEX